Passive Micro-Optical Alignment Methods

OPTICAL ENGINEERING

Founding Editor

Brian J. Thompson

University of Rochester
Rochester, New York

1. Electron and Ion Microscopy and Microanalysis: Principles and Applications, *Lawrence E. Murr*
2. Acousto-Optic Signal Processing: Theory and Implementation, *edited by Norman J. Berg and John N. Lee*
3. Electro-Optic and Acousto-Optic Scanning and Deflection, *Milton Gottlieb, Clive L. M. Ireland, and John Martin Ley*
4. Single-Mode Fiber Optics: Principles and Applications, *Luc B. Jeunhomme*
5. Pulse Code Formats for Fiber Optical Data Communication: Basic Principles and Applications, *David J. Morris*
6. Optical Materials: An Introduction to Selection and Application, *Solomon Musikant*
7. Infrared Methods for Gaseous Measurements: Theory and Practice, *edited by Joda Wormhoudt*
8. Laser Beam Scanning: Opto-Mechanical Devices, Systems, and Data Storage Optics, *edited by Gerald F. Marshall*
9. Opto-Mechanical Systems Design, *Paul R. Yoder, Jr.*
10. Optical Fiber Splices and Connectors: Theory and Methods, *Calvin M. Miller with Stephen C. Mettler and Ian A. White*
11. Laser Spectroscopy and Its Applications, *edited by Leon J. Radziemski, Richard W. Solarz, and Jeffrey A. Paisner*
12. Infrared Optoelectronics: Devices and Applications, *William Nunley and J. Scott Bechtel*
13. Integrated Optical Circuits and Components: Design and Applications, *edited by Lynn D. Hutcheson*
14. Handbook of Molecular Lasers, *edited by Peter K. Cheo*
15. Handbook of Optical Fibers and Cables, *Hiroshi Murata*
16. Acousto-Optics, *Adrian Korpel*
17. Procedures in Applied Optics, *John Strong*
18. Handbook of Solid-State Lasers, *edited by Peter K. Cheo*
19. Optical Computing: Digital and Symbolic, *edited by Raymond Arrathoon*

Passive Micro-Optical Alignment Methods

Robert A. Boudreau

Inplane Photonics
South Plainfield, New Jersey, U.S.A.

Sharon M. Boudreau

Department of Chemistry
Corning Painted Post High School
Corning, New York, U.S.A.

CRC Press
Taylor & Francis Group
Boca Raton   London   New York

CRC Press is an imprint of the
Taylor & Francis Group, an **informa** business

A TAYLOR & FRANCIS BOOK

Credits: The following figures are reprinted with permission from IEEE. Copyright IEEE 2005.

Figures 1.1, 1.2, 1.4, 1.6, 1.7, 1.8, 1.13
Figures 4.28, 4.29
Figures 5.1, 5.2, 5.3, 5.5, 5.6, 5.7, 5.8, 5.9, 5.11, 5.12, 5.13, 5.16-5.23, 5.29, 5.30, 5.31a, 5.31b
Figures 6.1-6.13
Figures 8.1-8.8, 8.9b, 8.10-8.12, 8.17-8.18, 8.21-8.25, 8.29, 8.33, 8.35-8.39, 8.41, 8.43
Figures 9.1, 9.3, 9.4, 9.6, 9.8, 9.9, 9.13, 9.29-9.31, 9.32, 9.33-9.41
Figures 10.1, 10.7, 10.8, 10.10-10.14, 10.16-10.18, 10.21, 10.22, 10.24, 10.26

CRC Press
Taylor & Francis Group
6000 Broken Sound Parkway NW, Suite 300
Boca Raton, FL 33487-2742

First issued in paperback 2019

© 2005 by Taylor & Francis Group, LLC
CRC Press is an imprint of Taylor & Francis Group, an Informa business

No claim to original U.S. Government works

ISBN-13: 978-0-8247-0706-4 (hbk)
ISBN-13: 978-0-367-39260-4 (pbk)
Library of Congress Card Number 2005001132

Library of Congress Cataloging-in-Publication Data

Passive micro-optical alignment methods / edited by Robert A. Boudreau & Sharon M. Boudreau.
     p. cm. -- (Optical engineering ; 98)
  Includes bibliographical references.
  ISBN 0-8247-0706-0
  1. Optical tooling. 2. Optical instruments. I. Boudreau, Robert A. II. Boudreau, Sharon M. III. Optical engineering (Marcel Dekker, Inc.) ; v. 98.

TJ153.P3125 2005
621.36--dc22
                                         2005001132

Visit the Taylor & Francis Web site at
http://www.taylorandfrancis.com

and the CRC Press Web site at
http://www.crcpress.com

Preface

The most expensive step in the manufacture of micro-optical components and fiber optics is the optical alignment of the parts during manufacture. It is very difficult to align microscopic semiconductor lasers, lenses, and other optical elements to the necessary micrometer positional tolerances. Frequently, the alignment occurs near the end of a product's manufacturing process, risking catastrophic loss from failure. Recently, with the advent of sophisticated miniaturization, micromachining, and advanced optoelectronics, a need has arisen for a fresh new approach toward low-cost alignments. This book describes a resulting major new manufacturing approach, known as passive alignment, aimed at addressing this issue. The approach is just now beginning to pass from the laboratory phase to the manufacturing phase. Various methods from around the world are presented in this book, but the competition between methods is only just beginning, and new techniques and materials are rapidly emerging. The objective of this book is to provide the reader with a vision of what future micro-optic packaging may be like and what may be possible to stimulate further development.

The first chapter provides a scope for the topic looking at current activities and outlines the requirements for optical alignments, the emergence of passive alignment, and its comparison with the currently used but costly method of active alignment. Passive alignment is defined as any method that permits the successful positioning and bonding of the micro-optics without having to power the devices involved. Active alignment, in contrast, requires powering the devices, which is often awkward to do, followed by finding the optical signals and peaking the signals by manipulating the position of the components while they are operating. For many systems this may not even be possible.

The remainder of the book is divided into three sections: Mechanical Passive Alignment, Visual Passive Alignment, and Utilities for Passive Alignment. Section 1 discusses passive alignment methods that rely on the mechanical precision of materials being assembled for providing the positional alignment. Section 2 describes visual alignment, the approach generally taken by Japanese companies, in which parts are passively aligned using visible marks on the parts in a passive way similar to photolithographic mask positioning. Finally, Section 3 discusses developments in the utilities for passive alignment that would aid both the mechanical or visual methods discussed in the first two sections.

Section 1 begins with the earliest form of passive alignment, which is used in connectors, and then has two chapters covering silicon waferboard technology, the most important basic technology supporting passive alignment. Solder technology is covered next because of the application of solder bumps for mechanical positioning and because of the need for fixing parts once aligned. "Jitney" passive alignment is described because of its application to very low cost plastic optical interconnects for computers. Finally, the section shows how mechanical passive alignment can be applied out of plane into three-dimensional optical benches using free-space passive alignments and ganged alignments of arrayed devices.

Section 2 describes key achievements for visual passive alignment. The chapters show its application toward building modules including planar lightwave circuits, low-cost plastic packaging, and surface mount packaging.

Section 3 completes the book by describing the key development of large spot lasers, vertical-cavity surface-emitting lasers, and other devices that relax the alignment tolerance for the assembly. In addition, the Monte Carlo analysis is showcased in its ability to evaluate the manufacturing potential for any passive alignment method. The methods in this section thus serve as utilities that assist the passive alignment design.

About the Editors

Robert A. Boudreau is manager of advanced OE packaging at Inplane Photonics, in South Plainfield, New Jersey where he has integrated high power lasers to planar lightwave circuits. He holds a PhD in chemistry from the University of New Hampshire, and an MBA and a BS in chemistry from the University of Illinois. For 20 years he has pioneered opto packaging at Verizon Laboratories, Tyco Electronics, and Corning Inc., including fundamental work in passive-alignment and passive-active alignment. He holds 32 patents and has authored over 100 publications. Currently he is associate editor of the *IEEE Journal of Advanced Packaging*. He founded the IEEE LEOS Subcommittee on Packaging, Manufacturing, and Reliability and was past Chair of the New England Chapter of IEEE LEOS, during which it won the Best Chapter award. He is a member of the American Chemical Society, a Senior Member of the IEEE, and a member of Sigma Xi.

Sharon M. Boudreau is a secondary chemistry/biology teacher for the Corning Painted Post School District in Corning, New York. She holds a PhD in chemistry from the University of New Hampshire, and a BA degree from Wheaton College, Norton, Massachusetts. Her research interests include crystal growth, x-ray crystallography, battery chemistry, solar energy and commercial packaging. In addition to teaching at the University of New Hampshire, she was assistant professor of chemistry at Wheaton College, where she taught inorganic, physical, and analytical chemistry. She has published in numerous journals including the *Journal of Chemical Education*, and is a member of Sigma Xi.

Contributors

Yuji Akahori
NTT Electronics Corporation
Ibaraki-ken, Japan

Nagesh Basavanhally
Lucent Technologies
Murray Hill, New Jersey

Robert A. Boudreau
Inplane Photonics
South Plainfield, New Jersey

Sharon M. Boudreau
Department of Chemistry
Corning Painted Post
  High School
Corning, New York

John V. Collins
BT Laboratories
Martlesham Heath
United Kingdom

Hongtao Han
Digital Optics Corporation
Charlotte, North Carolina

Tsuyoshi Hayashi
NTT Electronics Corporation
Ibaraki-ken, Japan

Werner Hunziker
Harting Mitronics AG
Luterbach, Switzerland

Masataka Itoh
Ibiden USA
Torrance, California

Kuniharu Katoh
NTT Electronics Corporation
Ibaraki-ken, Japan

James Kevern
AMP Tyco Electronics
Harrisburg, Pennsylvania

Kazuhiko Kurata
NEC Corporation
Ibaraki, Japan

Yung-Cheng Lee
University of Colorado
Boulder, CO

Karen Matthews
Corning Incorporated
Corning, New York

Alan Plotts
AMP Tyco Electronics
Harrisburg, Pennsylvania

Qing Tan
Optical Platform Division
Intel Corporation
Newark, California

Songsheng Tan
Infotonics
Canandaigua, New York

Kimio Tatsuno
NTT Electronics Corporation
Tokyo, Japan

Hideki Tsunetsugu
NTT Electronics Corporation
Ibaraki-ken, Japan

Werner Vogt
ETH
Zurich, Switzerland

Randall B. Wilson
BioArray Solutions Ltd.
Warren, New Jersey

Yasufumi Yamada
Department Director, Design
  Department
Planar Lightwave Circuit Group
Photonics Business Group
NTT Electronics Corporation
Tokyo, Japan

Contents

SECTION 2   *Visual Passive Alignment*

SECTION 3   *Utilities for Passive Alignment*

1 Overview of Passive Optical Alignment

Robert A. Boudreau, Sharon M. Boudreau, and Karen Matthews

CONTENTS

1.1 DEFINITIONS OF TERMS

Passive optical alignment is the method for aligning optical elements, including lasers, optical semiconductor devices, or lenses, without having to power the system to find the positions. If the system is powered to find the alignment, then the alignment is called an active alignment. Active alignment is commonplace but can be expensive to use. The term passive alignment usually applies to tiny optical elements ordinarily found in fiber optic components and systems, planar lightwave circuits (PLCs), or free-space micro-optical systems. In recent years passive alignment has gained importance in reducing the manufacturing cost of fiber optic components, in particular, and in enabling the assembly of more complex optical systems that do not easily lend themselves to active optical alignment. Probably the oldest and simplest form of passive optical alignment is the alignment that takes place when snapping together an optical cable connection. Chapter 2 will discuss this type of passive alignment in detail.

One variation of passive alignment is called optical self-alignment. In this case the component geometry is designed in such a way that there is a force

1

inherent in the system that pulls the parts together into proper alignment. This is discussed in some detail in Chapter 5, where it is shown how the surface tension of solder can be used to perform this type of passive alignment.

The newest variation of passive alignment is called passive-active alignment. In this approach passive alignment is used to achieve close alignment for bonding, and then the structure is bent into precise alignment using active alignment. Bonding is performed using a bendable flexure made of a material of low yield strength such as Kovar™ or gold. More detail is covered below under the section on the Suss Diebonder, a bonder currently used in manufacturing with this approach.

There are two major approaches to achieving passive optical alignment. One is to use mechanical passive alignment, in which parts are made so precisely that they properly fit into alignment, and the other approach is to use a camera for visual passive alignment, to guide the manipulation of parts into position. These approaches are discussed in the first two major sections of this book, but for both of them it is common to use silicon as the substrate material on which to build the optical system.

Silicon has a lot of advantages as a platform for assembly because it can be processed in a semiconductor fabrication facility to produce very high precision optical targets or mechanical features for the alignment, and it enables many different options for integration of other elements. When used as an assembly platform it has many names. Chief among these are Silicon Optical Bench, SiOB, Silicon Waferboard, and Silicon Motherboard.

One of the most common and challenging alignments for passive optical alignment is the alignment of single mode (SM) optical fiber to a semiconductor laser, where the optical fiber typically has a core of about 8 microns. This is shown in Figure 3.20. If a lensed tipped fiber is used to maximize optical coupling, nearly half the light is lost if its position is misaligned only 1 micron out of position. This level of passive alignment control is hard to achieve and is the focus of most of this book. Optical coupling graphs such as in Figure 3.20 are often used to analyze the performance of an optical system and to set engineering targets for the precision of the passive alignment system. Shown in the graph, the sensitivity to misalignment in this system can be reduced by defocusing the lens at the expense of peak optical coupling. Another way to reduce the sensitivity is to put a mode transformer on the laser device, known as a Large Spot device, as discussed briefly below and in Chapter 10, but this is done at the expense of building a larger and more complicated semiconductor device.

A more relaxed tolerance passive alignment system is available when multimode optical fiber is used. For this type of system, a large-cored fiber is used, typically about 50 microns or more in size, but the data rates or distances traveled are much less than with a single-mode fiber (SMF), so this only works for certain applications. Multimode systems are common in datacom, vehicle, and audio systems. Passive alignment in these systems can often be achieved using plastic molded parts, and the emphasis is on very low cost. An example of this system is shown in Chapter 6.

Issues of the mode shape and device selection enter into the design of the passive alignment system because they affect both the ultimate coupling possible

and the sensitivity to misalignment. There are two basic emitting device geometries to couple to: the edge emitting device and the surface emitting device. An
edge emitting device such as the common Fabry Perot laser [1] has an emitting
spot that is usually elliptical in shape. The near field shape at the facet is wider
than it is tall, and the far field pattern projected is taller than it is wide. Some
optical coupling is usually lost because the mode shape inside an optical fiber is
round, and that shape has to be matched. Chisel-shaped lensed fiber tips are
sometimes used to overcome this. In contrast, the VCSEL, or vertical cavity laser,
presents a round mode to couple to and is better matched to optical fiber. VCSELs,
however, do not deliver as much optical power as edge emitters.

The remainder of this chapter will highlight an overview of places in which
passive optical alignment is used, and the rest of the book will provide detailed
examples and data for the application.

1.2 ALIGNMENT OF ARRAYS

One of the earliest implementations of passive alignment was for the alignment
of arrays, and this introduced the use of the silicon optical bench. Multichannel,
high-throughput, low-cost, compact interconnection modules are of particular
interest to the telecommunications and datacom industries. In the rapidly growing
telecommunications area, 1 Tbit/sec class-throughput switching systems and multimedia Tbit/sec servers may soon be required. Asynchronous transfer mode
(ATM) switching systems with a throughput of more than 1 Tbit/sec are expected
to provide high-speed digital communication services such as video, data, and
high-definition television (HDTV). Optical interconnection is a promising technology for high-speed switching systems because it can eliminate the bottleneck
of electronic interconnection, caused by its limited signal bandwidth and number
of input/output (I/O) connections. Adopting optical interconnects to interboard,
intraframe, and interframe connections is a key to achieving these levels of performance. Optical interconnection modules that are easy to fabricate, low cost,
and compact are needed.

In many multichip module applications, data distribution to and from the
memory chips can be one of the key bottlenecks in high-performance system
implementation. Vertical cavity surface emitting lasers (VCSELs), optoelectronic
(OE) devices, modulators, and detectors are developed to the point that they can
enable both high density and high speed. The integration of these devices thus
needs to be developed. A packaging architecture and associated technologies must
also be developed to integrate the OE devices and optical components in a way
that is compatible to conventional electronic packaging.

Packaging of a 40-channel parallel optical interconnection module with over
25 Gbit/sec throughput has been realized in ParaBIT, a parallel, interboard, optical
interconnect technology [2]. A major feature of this module is the use of VCSEL
arrays as cost-effective light sources. The ability to use the same packaging structure
for the transmitter and receiver is enabled via the VCSEL arrays. High-density
multiport bare fiber (BF) connectors are developed for the module's optical interface

to achieve supermultichannel performance. These bare fiber connectors are novel in that they do not require a ferrule or a spring, ensuring physical contact with excellent insertion loss (< 0.1 dB per channel). Here, a polymeric optical waveguide film with a 45 degree mirror for coupling to the VCSEL and photodiode (PD) arrays by passive optical alignment was also developed. The optical array chips were aligned and die bonded to a substrate using a die bonder that looks for fiducial marks. The technique referred to as transferred multichip bonding (TMB), shown in Figure 1.1, can be used to mount optical array chips on a substrate with a positioning error of less than ±6 μm. This bonding process used an alignment template that was removed from the four VCSEL/PD array chips and aluminum nitride substrate after cooling. This is a form of template passive alignment.

Another form of template passive alignment is shown in Figure 1.2 [3,4], described by the authors as the Index Method. Here it was being applied to arrays. The process went as follows: First the fiber array was held in a Si V-groove array block that was marked with fiducial marks. These marks were then aligned to marks on a template that has vacuum holding ability to pick up the block. The template, while still holding the fiber array block, was then used to pick up the laser array bar, aligning to its own separate set of fiducial marks on the template. At this point the fiber array and laser array were then aligned to each other and stuck to the template. The aligned assembly was then brought to a bed of molten solder pads on a product substrate, and the solder was cooled to harden. The template was then released to be reused again, leaving the aligned parts bonded to the solder.

In many multichip module applications, data distribution to and from the memory chips can hinder high-performance system implementation. This type of performance-limited data path can be replaced by using a free-space optics interconnect (FSOI) module with a crossbar connectivity between point-to-point interconnects [5]. Fully packaged FSOI systems for multichip interconnection have been demonstrated. A conventional printed circuit board (PCB)/ceramic board is populated with Si and OE chips mated to a FSOI layer, assembled separately. The OE module has a ringed PCB chip registration board with alignment marks used to place the VCSEL and detector arrays. Fiducial marks match pinholes in the optical modules, and the chips are transported via a lithographic chip transfer technique. A plastic FSOI module snaps to the OE module via mechanical pins. These parallel OE free-space interconnects were demonstrated with link speeds of up to 400 Mbit/sec per channel.

A surface-mountable, low-cost, parallel-fiber optical link that employs passive mechanical alignment of its optical elements was also demonstrated. A consortium named PONI, led by HP Laboratories with DARPA support, demonstrated the parallel-optical subassembly (POSA). The POSA could be used as stand-alone transmitters and receivers for compact 12-channel optical links that carry data at 1.25 to 2.5 Gbit/sec [6]. Separate receiver and transmitter modules, comprised of 12 or 8 channels operating at a maximum rate of 1.25 Gbit/sec per channel, defined the optical links. Materials and techniques originally developed for tape ball grid array IC packages were used to construct the electro-mechanical platform. Electrical IO was provided via a 6×10 solder ball array on 50 mil centers. Passive alignment

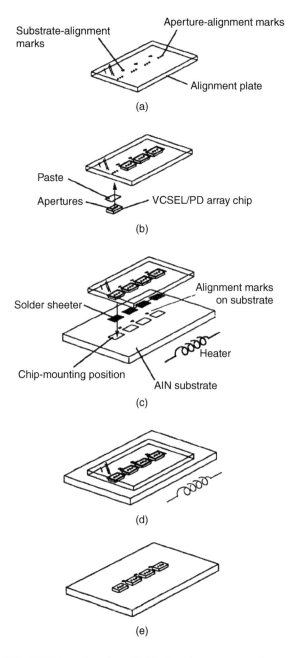

FIGURE 1.1 "ParaBIT" transferred multichip bonding process using template passive alignment. From reference [2].

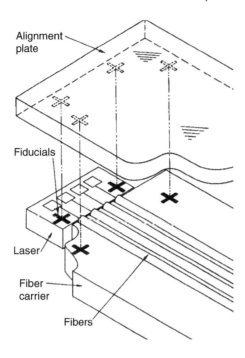

FIGURE 1.2 Template passive alignment using the index method. From reference [3].

involved accurately positioning the OE arrays with respect to alignment data in the base. A die attach fixture incorporating a MT ferrule with guide pins located the substrate. A tool maker's microscope measured the position of the diode centers in relationship to the base data and fed this information back to control the nudge micrometers. Die attach position accuracy averaged 5.2 μm.

NEC demonstrated another example of the use of arrays in optical interconnects with a 622 Mbit/sec per channel 12-channel optical transmitter and receiver designed with large-tolerance optical fiber [7]. This design incorporated a high-speed 12-channel transmitter and receiver large scale integration (LSI) with automatic decision threshold control; a 12-channel, 1.3-μm laser diode (LD) array; 12-channel In GaAs PIN PD array; low-power consumption; and passive optical alignment technology, as shown in Figure 1.3. The optical coupling scheme of the optical transmitter uses LDs coupled to graded index GI-62.5 multimode fiber (MMF) spaced at 250-μm intervals. The assembly process starts with a mechanically constructed optical coupling unit on a silicon substrate followed by photolithographically patterning alignment marks on the arrayed LD and Si substrate. The LD array is mechanically positioned and bonded with AuSn solder. Finally, the 12-channnel MMF is passively aligned on a Si-V groove with ±0.5 μm accuracy in each direction. The optical coupling scheme of the receiver module uses flip-chip bonding. The MMF array optical output signals are reflected vertically from the metallized Si-V grooved end walls and are coupled to the light-receiving ends

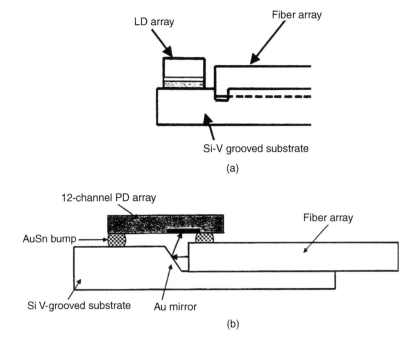

FIGURE 1.3 Passive alignment schemes for the (a) transmitter and (b) detector of a 12-channel multimode fiber transceiver array. From reference [7].

of PIN-PD array. AuSn solder was deposited with a micro-punch and die technique. The PIN-PD array chip was placed in the Si-groove substrate, and the Si submount was heated to melt the solder for bonding. Positioning accuracy was reported to be within ±0.5 μm in both directions. In this demonstration, more than an 890 psec eye opening at the receiver was obtained over the 0 to 70°C temperature range, indicating the applicability of the link for intra- and intercabinet interconnections in broadband communication systems.

Optical interconnect modules are required to be compact and low cost. The typical approach to their construction is to actively align the LD or PD array to an array of lensed fibers while individually controlling 6 degrees of freedom for alignment. This process, however, is a painstaking and time-consuming one. To simplify this process, passive alignment techniques using V grooves on Si mother-boards and flip-chip bonding using self-aligning effects of solder bumps were investigated. Accurately etched V grooves in the Si motherboard were used to guide fibers to the LD or PD arrays. However, the alignment accuracy in the vertical direction depended on the accuracy of the LD thickness in the case of conventional epi-up or junction-up LD bonding. Flip-chip bonding achieves accurate self-alignment in the horizontal direction within less than 1 μm, but accurate vertical alignment is difficult because of complicated control of solder volume to suit the weight of the LD or PD. In lieu of this information, NTT developed a passive alignment technique between a four-channel LD array and a hemispherically lensed MMF array (GI50/125 µm), and a four-channel PD array and a slant polished MMF (GI50/125 µm) [8]. A stair-shaped Si platform was used to hold the PD array, and capillaries were bonded to 250-µm pitch Si V grooves on an upper plate, which was used for alignment in the angular and x directions. The Si V grooves helped to guide the fiber array. Alignment in the x direction was achieved passively by fiber insertion into the capillaries. The y alignment was controlled by the capillaries' outer diameter. Angular alignment was controlled parallel to the LD stripes by inserting the fibers into the capillaries, and a microscope aided in spacing the fiber and LD arrays. The fiber alignment to the PD array was similar. Overall, alignment time was reduced to less than 1/20 the time of conventional active alignment by simply positioning the fiber array through capillaries in V grooves in the Si motherboard. High coupling efficiencies of 3.4 ± 0.5 dB for LD submodules and 0.4 ± 0.1 dB for PD submodules were achieved via these techniques.

Researchers at the French Institute of Microtechniques proposed a mechanical microconnector to obtain precise optical self-alignment of parallel optical integrated circuit (OEIC) to optical ribbon fibers. Groups of waveguides were made on a wafer, in addition to alignment marks on the waveguide substrate. The alignment marks enabled precise positioning of Ni pins, which were electrodeposited on the OEIC, using photolithographic processes. A silicon platform was fabricated with V grooves to position the optical fibers and to provide precise openings for insertion of the Ni alignment pins. During assembly, the waveguide was first locked to the platform, and then the optical fibers were slid to their contact with the waveguide. Preliminary experimental results showed excess optical losses on the order of 3 dB [9].

1.3 ALIGNMENT OF TRANSMITTERS AND RECEIVERS

The wide expansion and variety of communication media, such as Internet, mobile telephones, cable television, and video transmissions, has lead to a large demand for optical communication and transmission networks for multimedia services [10]. The next-generation communications networks will require low-cost, fast, compact, and low-power consumption modules. Both subscriber and trunk line networks require more economical and smaller optical transmitters, receivers, and transceivers.

A wavelength division multiplexing (WDM) transceiver module was demonstrated that used surface-mountable 1.3-µm laser diodes and 1.55-µm PDs passively aligned to an embedded fiber [11]. A single fiber was used for both transmission and reception. In this design, a Si motherboard was used as a platform to attach the fiber in a V groove and align to the transmitting LD. The receiving PD was surface mounted on an insulating substrate that had a SMF embedded in it for both input and output. Inserted in the SMF was a 20-µm-thick polyimide WDM filter, which was used to direct the incoming 1.55-µm light to the detector and to transmit the 1.3-µm light into a slit that transverses the fiber at an angle 30 degrees normal to the substrate surface. Index-matching epoxy was used to both hold the fiber in place and surface mount the PD using a microbump process. The index-matching epoxy eliminated all of the reflective surfaces in the PD portion. Solder was used to mount the LD. The output power at 1.3-μm wavelength was found to be greater than 1.5 mW, with a coupling efficiency of 12% to the fiber. The PD responsivity was 0.85 A/W at 1.55 μm wavelength.

One way to reduce module size is to use multichip module (MCM) technology in the electrical part of the transceiver. Researchers at NEC developed a packaging technology that passively aligned the LD to a fiber stub in the package, using an LSI technique [12]. By integrating the passive alignment of the LD with the IC mounting techniques, a low-cost and high-density packaging technology was demonstrated. Miniaturization of the optical transmitter (Tx) and receiver (Rx) pair was achieved via an optical coupling unit on a silicon motherboard. This structure made it possible to reflow solder in the same way as conventional surface-mounted electrical components. A receptacle structure consisting of a glass ferrule with a short length of fiber and hooked parts inserted allowed fiber pigtails to be temporarily detached from the housings during the board assembly. The Tx/Rx pair measured $10 \times 26 \times 3$ mm^3 each. These researchers also demonstrated a high thermal conductivity package that operated from $-40°$ to $+85°C$ ambient temperature.

A 2.5 Gbit/sec optical receiver module that used a one-chip receiver IC for an ultrabroadband optical access system was demonstrated [13]. The IC provided reshaping, regeneration, and retiming functions required for the receiver. The IC and a superlattice avalanche PD (SL-APD), used for detection, were packaged in a compact ceramic housing, using passive alignment technology, as shown in Figure 1.4. The AuSn electrodes, V groove, and alignment patterns were patterned on the silicon substrate. An automated machine positioned the SL-APD by overlapping the Si substrate, followed by attachment to the AuSn electrodes with flip-chip mounting. The optical signal from the V-groove mounted fiber was reflected by the Au mirror and injected into the SL-APD. The module's dimensions were

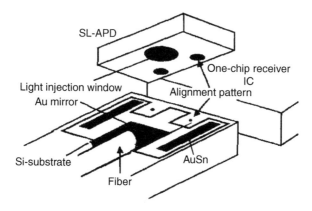

FIGURE 1.4 Passive alignment scheme between an optical fiber and a superlattice avalanche photodiode. From reference [13].

14×19 mm. The power consumption was 450 mW at a +3.3 V supply voltage and showed a -31.7 dBm receiver sensitivity.

Another attempt to demonstrate a low-cost, manufacturable module was made by researchers at Fujitsu. The module was produced using hybrid integration mounting techniques including passive alignment of the optical devices, plastic molded packaging techniques, and a receptacle interface [14]. The module demonstrated sufficient reliability for trunk line applications. A mean time to failure of 20 years under the conditions of 45°C and 50% relative humidity was achieved with this module.

The combination of light emitting diode-passive alignment carrier (LED-PAC) optical subassembly technology and conventional multilayer thick-film MCM-ceramic (C) circuit technology was used to demonstrate viability for high-performance military-aerospace digital fiber-optic transmitter applications. A hermetically sealed fiber-optic transmitter module designed for use in harsh application environments and based on a SiOB subassembly, using multilayer thick-film interconnect technology, was demonstrated in 1999 by researchers at Boeing [15]. Thick-film gold metallized precision injection-molded aluminum nitride LED submounts were bonded to an anisotropically etched Si substrate, which enabled passive alignment of the LED to a 100/140 µm MMF. The MMF was solder bonded to the SiOB and hermetically sealed inside a Kovar™ dual in-line (DIL) package. This module met Boeing 777 avionics local area network requirements over a -40 to $+10$°C temperature range. The transmitter average output power was above -14 dBm.

New passive-alignment carrier (PAC) optical subassemblies for military-aerospace fiber-optic transmitter and receiver applications were soon developed [16]. The goal was to manufacture similar rugged LED-PAC and PIN-PAC optical subassemblies followed by integration into the hermetically sealed hybrid MCM-C receiver and transmitter. The PIN-PAC design included a back-illuminated PIN PD, a reflective turning mirror, a V groove, a wire-bondable thin-film interconnect trace, and solder-bonded optical fiber bond joint. The large core/cladding 100/140 mm optical fiber pigtail configuration achievd a 0.8 A/W detector responsivity at 1.3 µm wavelength. Fabrication of the PIN-PAC and LED-PAC substrates was accomplished using multistage photolithography, silicon anisotropic etching, thin-film deposition, and patterning and dicing processes. The SiOB-based LED-PAC and PIN-PAC were fabricated together on 5-inch-diameter Si wafers.

Researchers at LG Corporate Institute of Technology developed photolithography on a three-dimensional surface, including multistep silicon anisotropic etching for SiOB manufacturing, precise control of optical active chip alignment, and fluxless, pressureless die bonding in an effort to demonstrate a fully passively aligned system of optical components [17]. Photolithographic processing on a three-dimensional surface was developed to get proper feature sizes for the electrode, solder, and mirrors. Figure 1.5 shows the self-alignment assembly for the PD. The precise width of the chip is designed to fit the etched trench on the silicon substrate. Precision cleaving of the LD allowed that chip to be defined with a dimensional accuracy of ±0.5 µm, while the tolerance of the Si grooves for the LD

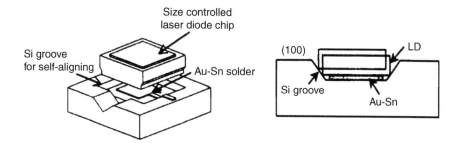

FIGURE 1.5 Self-alignment passive-alignment of a laser diode and an optical fiber's V groove. From reference [17].

and the fiber was measured to be < 0.5 μm. The fluxless, pressureless bonding technique made it possible to reduce the production cost. The LD module achieved better than 8% coupling efficiency to a cleaved end fiber and had output power of −10 dBm. The PD module responsivity was better than 0.85 A/W.

Researchers at NEC attempted to take this one step further by developing a passive alignment technique and optical coupling structure that was suitable for automated assembly [18]. Several types of optical modules, such as planar lightwave circuit (PLC) modules, LD modules, PD modules, as well as optical interfaces, were taken into account when developing this system. Packaging, optical coupling, and assembly steps used the same equipment, reducing the manufacturing costs. SiOB technology was also used to create a SMF MT-RJ Small Form Factor transceiver module [19]. Here, a shield structure was developed to reduce the electrical crosstalk in a SiOB. The shield, consisting of a metal plate, was placed in a trench groove. The groove was made by a dicing process and was located in the middle of the LD and PD position. The crosstalk in the optical subassembly (OSA) was reduced over 20 dB compared to the unshielded SiOB.

In 1999, Agilent Technologies manufactured a variety of MM and SM SFF transceivers with data rates ranging from 125 Mbit/sec to 1.25 Gbit/sec. The small form factor (SFF) transceivers offer similar functionality as SC duplex transceivers but use the mini MT-RJ optical interface to achieve twice the port density. SM SFF transceivers using a SiOB and a LD without mode expander were developed with passive alignment and attachment in mind. The transceivers achieved the required coupling power for fast ethernet [20]. The SM SFF module used silicone encapsulation for nonhermetic packaging in place of hermetic packaging. Two packaging schemes for a 2 Gbit/sec SFF transceiver using a MT-RJ receptacle were also demonstrated [21]. The package had an advantage in reliability and cost compared to standard, hermetically sealed transistor outline (TO)-canned optical devices.

A combination of Si and polymer microstructure technologies were fabricated using bichlorinated biphenol (BCB) polymers to build waveguide fiber to the home (FTTH) array components with a MT interface [22]. Passive alignment structures were used for both laser arrays and optical interfaces. The LD carrier was mounted on a polymer carrier with mounting holes, alignment trenches, and vertical alignment structures. The optical power measured from one channel was 0.135 mW at 100 mA. Total shrinkage of the structures after transfer molding was found to be only ~0.69% when measuring and comparing the structures from both mold insert and replicated carrier with a profilometer.

Passive alignment was used with hybrid integration of optical components to demonstrate a bidirectional, compact, low-cost transceiver module at NEC [23]. The LD was mounted on a PLC, resulting in a compact module of 26 × 10 × 3 mm. Alignment marks were placed on both the LD and the PLC chip such that a large-spot LD could be passively mounted on a PLC chip. The optical coupling loss was measured to be less than 4.5 dB. Both the alignment marks for the LD mounting and the V groove for fiber alignment were fabricated on the PLC chip. The output power of the Tx/Rx module was +2 dBm at a LD drive current of 45 mA. A responsivity of over 0.35 A/W, with a minimum optical receiver power of −38 dBm and dynamic range of more than 31 dB, using a 50 Mbit/sec burst signal, were attained.

A comparison of active and passive fiber alignment techniques for module pigtailing was presented [24]. The conclusion at that time was that the alignment tolerances of MM devices could be met with low-cost precision machining techniques, such as lithography-electroplating-injection-molding (LIGA), but that efficient cost utilization of precision substrates required large-volume production and the use of assembly automation. In 2001, the possibilities of using low-temperature cofired ceramics (LTCC) technology were reviewed as a means to fabricate fiber pigtailed transmitter arrays [25]. These arrays used surface-emitting sources, such as VCSELs, equipped with vertical-mounted MMF pigtails. The LTCC module was mounted vertically on a printed circuit board (PCB) to provide a small, 1-D PCB footprint. The light source was flip-chip bonded on the opposite side of the substrate, and the fiber was aligned and held in place via a hole structure through the layers, thus allowing for the use of a detachable electrical interface between the fiber optic media and applications electronics. The alignment tolerances were therefore more relaxed compared to conventional optical connections.

1.4 LARGE-SPOT LASERS

Development in telecommunications requires improvement in coupling efficiency (chip to fiber), components, and modules. Optical mode transformers (OMTs) are one of the venues used to enable increased coupling efficiencies and ease the assembly tolerances, a key requirement for all passive alignment techniques. Various taper designs have been discussed in the literature [26]. All spot size transformers are based on the volume reduction, lateral or vertical, of the active layer along the axis of propagation to decrease the confinement mode and match the geometry and width of a semiconductor mode waveguide to a SMF or another semiconductor mode waveguide [27]. Broadly speaking, there are two families of spot-size converters—lateral tapers and vertical tapers—as shown in Figure 1.6. In lateral tapers, only the width of the active area is varied. In vertical tapers, the thickness of the active area, and possibly the width, is varied. There is no universal OMT. The design of a spot-size transformer must remain compatible with the optoelectronic

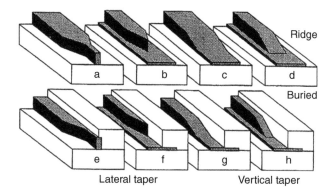

FIGURE 1.6 Various forms of waveguide tapers that produce large spot lasers that aid passive alignment by increasing the mode size of the beam exiting the laser facet. Panels (a), (c), (e), and (g) are single-core waveguides, whereas panels (b), (d), (f), (h), are double-core waveguides. From reference [26].

component manufacturing process and tends to be based on application and technological know-how in a given laboratory.

The PLC platform concept has been largely exploited to develop photonic hybrid circuits. A reduction in beam divergence from 15 to 13 degrees and a reduction of coupling loss from 11 to 4.6 dB for uncoated cleaved fiber was demonstrated [26]. The coupling loss to a cleaved SMF (core diameter = 8 µm) was reduced to 4.8 dB. The coupling loss to a lensed fiber (beam radius = 2.3 µm) was reduced to 1.6 dB. The best results, using nontapered devices, were 9.5 and 4 dB, respectively. Coupling loss figures for a lensed fiber and an antireflection (AR)-coated DH waveguide were reduced to less than 1 dB.

A large-spot laser diode, two waveguide detectors, and a 1.3/1.55-µm thin film filter were hybrid integrated on a Si motherboard, resulting in a −32-dBm sensitivity at 156 Mbit/sec [26]. The output power was 0 dBm at 60 mA. A coupling loss of only 1 dB, laser to standard SMF, was achieved with tolerances of ±2 µm in both vertical and horizontal directions and without the use of micro-optic elements [28]. In this case, both Fabry-Perot (FP) and distributed feedback (DFB) lasers were fabricated. The longitudinal SM emission of the FP laser was stabilized with an external fiber Bragg grating (FBG), which was directly butt-coupled to the laser. High-efficiency coupling (~80%) and high optical power in the fiber (5 mW) were demonstrated. Single longitudinal mode emission with narrow line width with high side-mode suppression was also demonstrated.

Other sources report a high coupling efficiency of 47%, with a −1 dB alignment tolerance of ±2.1 µm for both vertical and horizontal alignments being demonstrated between LS lasers and SMF [29]. To achieve this, both active and tapered passive waveguide regions were simultaneously grown by selective area metal organic chemical vapor deposition (MOCVD) growth on a patterned InP substrate. A high coupling efficiency of 85% was demonstrated when a cruciform cylindrical lens was used to replace the corrective lens in a confocal optical system. This is a 3 dB improvement over the conventional confocal system of SMF coupled to a 980-nm pump laser [30]. More examples of the use of large-spot devices as an aid for passive alignment can be seen in Chapter 10.

1.5 MICRO-OPTICS

Micro-optics silicon V-groove structures have been used for passive alignment of optical fibers with microrod optics and lenses [31, 32]. V-groove structures are easily fabricated in Si using anisotropic etchants, as described in Chapter 3. Diffractive optics can also be fabricated in silicon via photolithography and etching techniques. Diffractive optical elements (DOEs) can provide many optical functions that refractive optics cannot, such as beam reshaping, beam splitting, and beam homogenizing (diffusing). Digital Optics Corporation has developed a manufacturing technique for fabricating diffractive optics on the end of microrods [33]. The combination of microrods with optical fibers has been used to either modify the output optical beam from the optical fibers or reshape the optical beam launching into the optical fiber. Refractive lenses, desirable for high numerical aperture applications, can be fabricated on silicon using photolithography, photoresist reflow, and various etching techniques. The integration of N fiber arrays bonded into silicon V grooves with a $1 \times N$ micro lens array composed of 2^{n}-phase-level diffractive optics ($n > 1$) has been demonstrated [33], as shown in Figure 1.7. Here, a 1×6 fiber array was used with a lens array having 16 phase level diffractive optics, resulting in a total insertion loss for a pair of lens/fiber arrays of only 1.5 to 2.0 dB/channel.

Wafer-scale manufacturing techniques for making both micro refractive optics and multiphase level diffractive optics offer compact size, light weight, and low manufacturing cost. Integrated micro-optical systems (IMOS) containing a VCSEL array, edge-emitting lasers, detector arrays, and fiber arrays for opto-electronic and fiber-optic applications have been demonstrated [34]. Several examples of IMOS integration were shown. One example was a compact integrated optical sensor containing an edge emitting laser, diffractive optical elements, refractive elements,

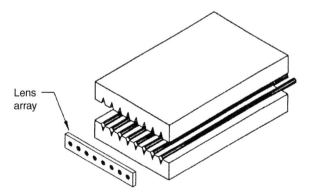

FIGURE 1.7 Passive alignment scheme for mating a lens array with an array of optical fiber. From reference [33].

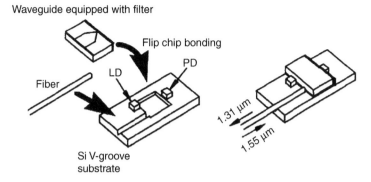

FIGURE 1.8 Simple passive alignment scheme for wavelength division multiplexing used in a bidirectional link. From reference [38].

a three-element signal detector, a turning mirror, and a monitor detector. Another example was an integrated optical sensor, which was composed of diffractive optical elements, a VCSEL array, and a detector array. The third example was a fiber-optic collimated array assembly, which contained a fiber array and a micro lens array. Collimated lens arrays were fabricated with experimental total round-trip insertion loss of 1.5 to 2.0 dB/channel [35]. Integration was also used to produce a low-cost light source [36]. Here the combination of an Axicon lens with a microcavity organic light emitting diode (OLED) was used.

WDMs are key components for optical access networks. Hybrid integrated WDM optical modules usually consist of a dielectric multilayer filter and an optical waveguide, but the assembly of a film filter and an optical waveguide can be difficult. For ease of assembly, and thus reduction in manufacturing cost, a fluorinated polyimide [37, 38] polymer optical waveguide integrated with a filter was fabricated. Here, the Ta_2O_5/SiO_2 dielectric multilayer filters were directly deposited onto the end face of the polymer waveguides by an ion beam sputtering process. This waveguide-filter combination acts as a wavelength separation device. This technology, combined with silicon optical bench, was used for a new type of WDM module [38]. All optical devices were mounted on the silicon V-groove substrate by flip-chip bonding. The waveguide-filter combination was then placed on the silicon bench, thus achieving a passive alignment process, as shown in Figure 1.8. This approach decreased the assembly cost and size of the WDM module.

1.6 LIGA AND HIGH-ASPECT STRUCTURES FOR PASSIVE ALIGNMENT

Lithography-Electroplating-Injection-Molding (LIGA) has been applied to the passive alignment of micro-optics [39]. Micro-optics is a very fast growing field driven by requests from the optical data communication and sensor fields. Years ago, devices made by integrated optics based on silicon and other semiconductor material and polymers were the dominating technology. Today, optical systems and devices based on free-space optical set-ups fabricated by micromachining technologies are beginning to enter the field.

High-aspect ratio structures (HARS) have requirements that include anisotropy, high etch rate, good selectivity to masking material, and compatibility with other processes [31]. The etch rates of conventional dry-etching processes such as reactive ion etching (RIE) and electron cyclotron resonance (ECR) are usually too low for deep and through-hole etching. In addition, the selectivity is also too low, requiring additional time to deposit thicker masking layers and creating difficulty in maintaining the masks during the long etch.

The LIGA process is characterized by fabrication of microstructures with high aspect ratios (> 50), vertical and smooth sidewalls (roughness < 30 nm), precision in the lateral position (< 1 μm), structural dimensions in the micrometer range, heights up to millimeters, and tolerances in the submicrometer range. Thus, the LIGA process is viable for use when building micro-optical and micro-electro-optical-mechanical (MEMS) devices. Fabrication of micro-optical benches with alignment structures for hybrid and passive integration of optical elements has been demonstrated [40].

Many MEMS applications require multilevel microstructures having two or more layers that need to be aligned to one another during processing. A passive alignment system based on mechanical registration via reference posts has been demonstrated for MEMS [41,42], as shown in Figure 1.9. For this, an alignment accuracy of ±5 μm between two layers was achieved. This alignment method was applied to almost all kinds of mask membrane materials, with no extra processing steps to the substrate.

Passive alignment of optical fibers, detectors, and light sources in micro-optical benches was achieved with submicron precision using transparent polymer elements containing three-dimensional-positioning structures and planar optical elements made by surface structuring [42]. Here, a fabrication process that combined the

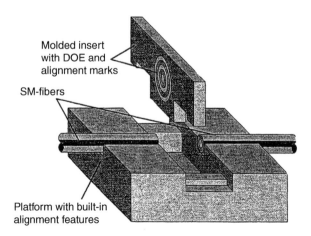

FIGURE 1.9 Lithography-electroplating-injection-molding (LIGA) structures used to passively align diffractive optical elements with single-mode fiber. From reference [42].

mechanical precision of the LIGA process with the optical functionality offered by DOEs was used to make polymer inserts in micro-optical benches. Low-cost replication technology was used to fabricate polymer bench inserts with imaging properties. LIGA-fabricated metal frames with smooth vertical and side walls of 500 microns thickness were used to fabricate injection mold cavities with integrated three-dimensional positioning features and surface relief structures for optical purposes. Injection molding was also used to make lens components that are arranged in an array with a 250-micron pitch to be used with fiber ribbons. A measured fiber-to-fiber coupling efficiency of 27% for SMF fiber was obtained.

1.7 THE SUSS DIEBONDER

The first commercially available diebonder that approaches passive-alignment performance in its die placement accuracy was introduced by Suss Microtec, formerly known as Karl Suss in 1998 [43]. This bonder, known as the FC 150, relied on precision optics and staging to achieve a 1 to 2 micron placement accuracy. It is shown in Figure 1.10, on the left.

The FC-150 uses an optic to simultaneously image the bottom of a die to be placed as it is held by a vacuum pick-up tool, as well as the surface of the substrate to be placed upon. The operator aligns these images, and then the optic is removed and the parts come together to be bonded. This approach to alignment had been used by other bonders before, but has been implemented with higher precision in the FC-150. This diebonder relies on the precision of the mechanics and optics because the actual bond takes place at a different place and time than what is observed in the viewer. In other words, a translation is necessary.

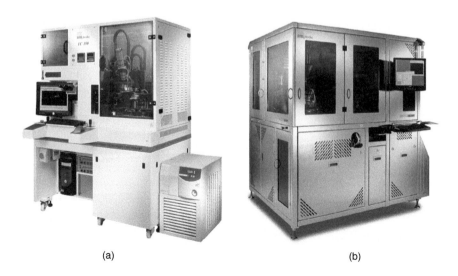

(a)   (b)

FIGURE 1.10 Commercially available diebonders that have pioneered optical passive alignment, including the (a) FC-150 and (b) Triad 0.5 AP. Courtesy of Suss Microtec.

To make the translation work, the bonder relies on the liberal use of Invar™, a near-zero thermal coefficient of expansion iron alloy, 36% Ni 64% Fe, that won the 1920 Nobel Prize in physics for Charles-Edouard Guillame, its discoverer. Other materials used include Zerodur™, a synthetic transparent ceramic with very low expansion, similar to Corning ultra-low-expansion (ULE™) glass. The viewer on the bonder can zoom to 400 × magnification and can resolve down to 0.6 microns in a 300 × 400 micron field of view. One of the great advantages of this bonder is that it enables automation, and a model FC 250 diebonder was produced that is fully robotic for production. Aligned diebonds could be made at a rate of one every 30 seconds or so.

Development of the bonder has taught a number of things. First, the bonder can place to a precision approaching 0.5 microns, but the bonding process itself causes movement, resulting in the final placement accuracy of 1 to 2 microns. Both thermocompression bonding and solder reflow bonding are available, but the solder reflow bonding has the highest placement accuracy because it requires the least force. Downward force can cause deflections that can misalign die. Placement accuracy also improves as the quantity of bonded die increases because it may take some anticipated "scrap" production before everything stabilizes. A final issue to be resolved was the placement height of the die. The viewer only observes the die in two dimensions, but for laser die, in particular, it is critical to set the height above the substrate. For this, both focal point and bump feeling techniques were explored, but bump feeling was the only approach not affected by chip thickness. These techniques rely on using a force gauge to sense contact with the surface to be bonded to, and then the machine backs off a set distance while the solder is melted.

This type of visual passive alignment works best for alignment of large spot lasers, or die that feature mode expanders to bring the alignment tolerance into the 1 to 2-micron range for single-mode alignments. More details concerning these mode expanded devices will be described in Chapter 10.

Another place where these bonders excel is in their application toward an assembly technique known as passive-active alignment. With this approach, the optical system of the product being assembled could be designed such that all but one of its elements needs to be aligned to a precision of 1 to 2 microns, leaving one element to be actively aligned. The active alignment of this one element then results in the alignment of the entire system in the product. This passive-active alignment approach is illustrated in Figure 1.11, as performed by Axsun Corporation [44]. In this case, the optical system consists of an assembly of free-space imaging optics, separately placed and diebonded, where the final alignment is performed on a previously diebonded structure that is purposely designed to be able to flex. This flexing is done after bonding, and after power is turned on. The final alignment is thus an active alignment, which steers the beams of the rest of the optics into alignment. This passive-active alignment requires that the initial alignment is nearly aligned, and that the flexure used to bring the optics into alignment has the right flexing characteristics for it to be bent and stay put. The Axsun flexure is produced using the LIGA MEMS process and provides the right

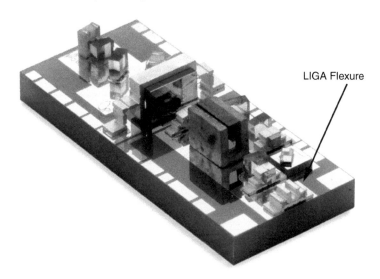

LIGA Flexure

FIGURE 1.11 Passive-active alignment using a lithography-electroplating-injection-molding micro-electro-mechanical systems-flexure (LIGA-MEMS) to peak the alignment of an otherwise all passively aligned assembly. Courtesy of Axsun Incorporated.

mechanical characteristics to provide desired low-yield strength flexing properties. This means it can be flexed to a position with a minimal amount of spring back.

The Suss bonder continues to be advanced, however, for visual passive alignment. At the time of this writing, a second-generation bonder called the Triad 05 AP, shown in Figure 1.10, on the right, was introduced. With this bonder, an improved air-bearing stage is provided, along with better optics, a better bonding arm, and new pulse heating to minimize heating of bonder mechanics and speed bond times. Most notable, however, is that for the first time there will be an automatic feedback loop. With this feature, the bonder automatically does a visual inspection of the bonded die right after it is bonded, to measure the offset from the correct position, and then inserts a correction offset into the bonding routine to compensate. The bonder therefore automatically compensates for various complex drifts and process aberrations no matter where they come from and hones in on the target alignment position. This produces a very tight alignment possibility for high-volume manufacturing, where many die are to be bonded in a repeatable way. Early results for this bonder indicate a bonding placement accuracy approaching 0.2 microns [45], which may eliminate the need of a mode expander on the laser device to be bonded.

1.8 FINETECH'S FINEPLACER LAMBDA BONDER

This very impressive bonder, shown in Figure 1.12, provides 1 micron placement precision for visual passive alignment *and* provides the ability to do mechanical passive alignment in the same versatile machine. A low cost, rugged table top unit, about the size of a wirebonder, it is very simple to use and maintain, owing to its

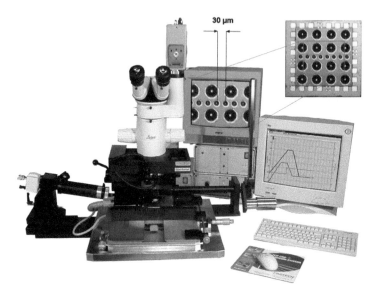

FIGURE 1.12 The Fineplacer Lambda bonder is capable of 1 micron visual passive alignment and also can be used for mechanical passive alignment. Courtesy of Finetech Incorporated.

intuitive manual mechanics that eliminate expensive electronics and actuators. The bonder achieves its impressive 1 micron optical placement precision by keeping its alignment beam splitter fixed, thus eliminating the beamsplitter translation error found in other bonders. It also places components using a pivot arm, thus eliminating the translation error of a bond head slider. Other unusual features include a dead weight lever arm for setting the bond force rather than a fragile strain gauge, the use of a frictionless air bearing table to quickly preposition parts manually, and a second high powered side viewing camera that can be used to hover components just above contact with a substrate. This last feature is unusual and especially valuable, because in addition to observing the bond event, it also enables the use of this bonder for mechanical passive alignment. For example, when a component is in hover position, the substrate can be translated, thereby bumping the held component's edge against a mechanical feature on the substrate, thus registering the component position to that feature prior to applying the bond force. Bonding can be by thermocompression, solder reflow, or adhesive. The stage has low thermal mass and is radiatively heated below its surface, enabling rapid temperature ramps.

1.9   SELF-ALIGNMENT

This type of alignment is a variation on passive alignment, where, in addition to having a means for finding the location of the position, there is a force present that automatically pushes the optics into alignment. The most common version of this is to use the surface tension of molten solder to provide the force. The solder can be very precisely defined in terms of location and shape, using wafer deposition techniques. There is a geometry before melting and a predictable geometry after melting. An example of self-assembly by this approach was shown for the alignment of MEMS optics [46]. In this system the solder was used to raise hinged plates up to a vertical position from their initial position flat on the substrate where they were formed. This enabled the passive alignment self-assembly of free-space optics such as lenses. In all cases, using solder this way, it is extremely important to make sure it is totally free flowing and not interfered with by oxidation or contamination.

The most common self-alignment approach is to have an array of solder bumps on one plate and a corresponding array of wettable pads on another plate [47]. When the two plates are placed together and the solder is reflowed, there is a force trying to minimize the surface area of the solder that pulls the arrays into alignment. The design of this system can be optimized for distance between the plates and alignment forces, depending on the number and sizes of the solder bumps. In addition, there is the option of using the solder bumps to push or pull the plate to one side into a mechanical stop, blocking the full motion, so that the plate is located against a known stop position. Both the hinged plate and solder bump alignment approach are shown in Figure 1.13. A variation of this is using solder bumps to

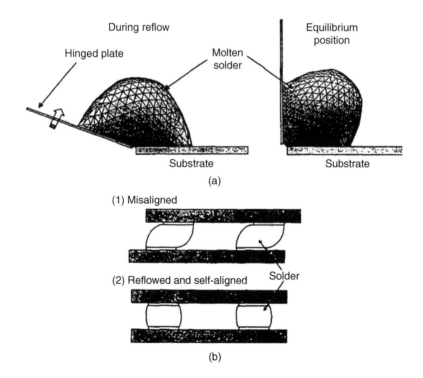

FIGURE 1.13 Self-alignment passive-alignment of (a) vertical plates and (b) horizontal plates using solder surface tension. From reference [47].

pull chips down into pockets [48] and registering edges with the pocket walls to find a position, as shown earlier in Figure 1.5. This approach has an alignment precision of ±3 microns, making it useful for detector and multimode fiber alignments, but marginal for standard semiconductor lasers.

Another form of self-alignment is to provide a trench structure with built-in features surrounding the trench that provides a force on the object that is to be aligned in that trench. This has been demonstrated for both the self-alignment of optical fiber and chips [49]. In the case of optical fiber, the silicon V grooves used to hold the fiber have an overlaying layer of "fingers" that cover the V groove. When a fiber is pushed into the V groove, these fingers provide a spring force that holds the fiber in the V groove and against the precision passive alignment side walls of the groove. The fiber is automatically centered in the groove. In the case of self-aligning a chip, a socket is etched into the substrate that has fingers surrounding the socket walls that are crushed when pushing the chip into the socket. The opposing forces to this crushing action provide the self-alignment and centering of the chip.

1.10 PASSIVE ALIGNMENT IN PLCS

One important arena for passive alignment is its application toward planar lightwave circuits (PLCs). PLCs route signals around on a plate using optical waveguides that are patterned into the plate. They can be constructed of silicon, glass, or plastic materials, and it is often easy to include the passive alignment structures because they can be formed by the same processes that form the optical waveguides. Methods to passively align chips, filters, and micro-optics, as well as optical fiber, have been developed.

One way to align chips to the PLC is to rely on the reflections of markers as seen in the facets of the chips [50], as shown in Figure 1.14. With this approach, a camera is used to inspect a laser diode chip as it is manipulated in front of waveguide facet on a PLC, and the reflections of the marker spots are observed in the facet of the laser chip. This approach applied to a glass on silicon PLC eliminated errors associated with imaging a chipped and damaged edge chip. The approach also eliminated the effect of variation in thickness of the laser die on its epi side and was a sensitive indicator to position because when the laser diode is rotated a given amount, the reflected markers rotate by twice that amount. Taken together, the researchers were able to obtain a three-sigma alignment tolerance of 0.9 microns and 0.21 degrees.

Direct writing of multimode PLC waveguides that includes passive fiber alignment grooves was performed in polymethyl methacrylate acrylic (PMMA) plastic [51]. Low-loss, smooth sidewall guides were obtained by slow cutting or multiple-pass cutting. The basic approach involved the cutting of grooves for the guides and backfilling with liquid core material followed by an ultraviolet cure. Fibers were placed in wider grooves set in front of the guide grooves, and the voids were back filled with liquid core. Placing fiber in grooves like this tends to have a self-centering

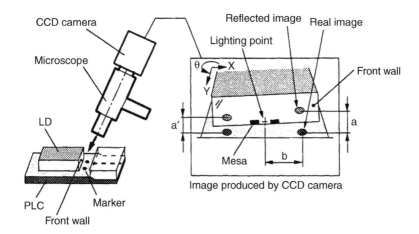

FIGURE 1.14 Use of optical reflections reflected from device facets to achieve passive alignment between a laser diode and a planar lightwave circuit. From reference [50].

effect that aids in the passive alignment because the grooves can be made slightly narrower than the fiber diameter to provide some compression and self-alignment.

Fiber-to-waveguide passive alignment was performed with plastic polymer PLC circuits that have been spin-cast onto silicon wafers [52]. Alignment precision for this SMF system has been estimated to be better than a micron, with a coupling loss of about 0.7 dB. To achieve this high-precision alignment, the V-groove structure and the waveguide structure were fabricated with the same mask set. In this system, the optical fiber was held in a silicon V groove, and the end face of the polymer waveguide was defined by a saw cut. In general, it is much easier to obtain a high-optical quality, low-loss facet on the end of a polymer waveguide than it is on a glass waveguide because the cutting of polymer material and the feature of back-filling with index-matching fills make for a good optical interface. In comparison, although glass waveguides are lower loss than plastic waveguides, it is harder to achieve an all-glass end facet of high optical quality by saw cutting or etching because polishing is normally required. More examples of passive alignment in PLC circuits can be found in Chapter 8.

1.11 MISCELLANEOUS APPLICATIONS

Most of this book discusses the use of passive alignment for use with transmitters and receivers in various forms, but the technology has shown broader application as more forms of micro-optical devices emerge. One such application is in optical switching, and specifically an example that required a fan-out geometry for Si V grooves [53]. A fan-out V groove of very high precision was required, rather than the usual parallel V groove. Very high precision fan-out V-groove arrays can be produced if appropriate mask compensations are used because the etch rate rises as the angle of etch goes farther off the crystal axis. This type of Si crystal etching is discussed in detail in Chapter 3. Using these V grooves, it was possible to achieve a single-mode optical mechanical switch with an insertion loss of only 0.64 dB. In this case, the passive alignment was actively aligned to a moving mirror actuator, but the array geometry passively positions the alignment of all the fiber ends with a single alignment. Another example of a passively aligned switch is one that makes use of MEMS. This switch [54] uses a tall micromirror that is formed with thick photoresist technology and is moved by electrostatic forces. The insertion loss for this switch was 4 dB.

Another interesting application for passive alignment is optical accelerometers [55]. In this system, a MEMS structure was made that could wobble a beam holding a mirror. The mirror movement was then sensed in two dimensions by two Fabry-Perot cavities, as shown in Figure 1.15. The resolution of such an accelerometer was shown to be submicron, just as with the capacitance type, except that the optical type has the additional advantages of immunity to electromagnetic interference and possible operation in hazardous environments.

One other area that has seen some rise in the use of passive alignment is in the area of scanners and tunable filters [56]. The scanner was achieved by encapsulating a MEMS beam structure inside a vacuum capsule and driving it with two Helmholtz coils. A translucent glass package allowed for the deflection of a laser beam. In contrast, the tunable filter was achieved by building a set of V grooves and then

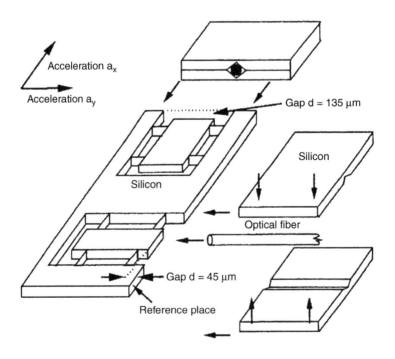

FIGURE 1.15 Passive alignment used in the assembly of a two-dimensional optical accelerometer. From reference [55].

etching a pocket that held a filter array vertical with respect to the beam path. The micromachining enables the assembly from fabricated chips.

REFERENCES

1. J. LaCourse, "Laser primer for fiber optics users," *Circuits and Devices,* 8, pp. 27–32 (1992).
2. K. Katsura, M. Usui, N. Sato, A. Ohki, N. Tanaka, N. Matsuura, T. Kagawa, K. Tateno, M. Hikita, R. Yoshimura, and Y. Ando, "Packaging for a 40-channel parallel optical interconnection module with an over-25-Gbit/sec throughput," *IEEE Transactions on Advanced Packaging,* 22(4), pp. 551–560 (1999).
3. M. S. Cohen, M. F. Cina, E. Bassous, M. M. Oprysko, and J. L. Speidell, "Passive laser-fiber alignment by index method," *Photonic Technology Letters,* 3, pp. 985–987 (1991).
4. M. S. Cohen, M. J. Defranza, F. J. Canora, M. F. Cina, R. A. Rand, and P. D. Hoh, "Improvements in index alignment method for laser-fiber array packaging," *IEEE Transactions of Components, Packaging, and Manufacturing Part B,* 17, pp. 402–411 (1994).
5. X. Zheng, P. J. Marchand, D. Huang, and S. C. Esener, "High speed parallel multi-chip interconnection with free space optics," *Applied Optics,* 39(20), pp. 3516–3524 (2000).
6. P. Rosenburg, K. Giboney, A. Yuen, J. Straznicky, D. Haritos, L. Buckman, R. Schneider, S. Corzine, F. Kiamilev, and D. Dolfi, "The PONI-1 parallel-optical link," *1999 Electronics Components and Technology Conference,* pp. 763–769 (1999).
7. N. Watanabe, T. Shine, and T. Nagahori, "Data-format-free 622-Mbit/s/ch 12 channel parallel optical transmitter and receiver," *Proceedings of SPIE: Optoelectronic Interconnects VII: Photonics Packaging and Integration II,* 3952, pp. 84–97 (2000).
8. H. Takahara, N. Tanaka, and Y. Arai, "Passively aligned LD/PD array submodules by using micro-capillaries," *IEEE Transactions on Advanced Packaging,* 23(2), pp. 323–327 (2000).
9. N. Kaou, V. Armbruster, J. C. Jeannot, P. Mollier, H. Porte, N. Devoldere, and M. de Labachelerie, "Microconnectors for the passive alignment of optical waveguides and ribbon optical fibers," *IEEE-MEMS Conference,* pp. 692–697 (2000).
10. A. P. McDonna and B. MacDonald, "Optical component technologies for FTTH applications," *Proceedings of the 45th Electronics Components and Technology Conference,* pp. 1087–1091 (1995).
11. G. Tohmon, T. Uno, T. Nishikawa, M. Kito, T. Yoshida, and Y. Matsui, "Passively aligned WDM transmitter/receiver module using fiber embedded circuit," EP-Vol. 19-1, *Advances in Electronic Packaging,* 1, pp. 779–783 (1997).
12. A. Kawatani, H. Fujimi, K. Shuke, K. Kurata, and R. Nagaoka, "Packaging technology for surface mountable, low profile fiber optic transmitter and receiver," *1998 International Symposium on Microelectronics, SPIE,* 3582, pp. 353–358 (1998).
13. M. Soda, S. Shioiri, T. Morikawa, M. Tachigori, I. Watanabe, and M. Shibutani, "A 2.5-Gbit/sec one chip receiver module for gigabit-to-the-home (GTTH) system," *IEEE 1998 Custom Integrated Circuits Conference,* pp. 273–276 (1998).

14. N. Yamamoto, T. Kojima, T. Watanabe, S. Sasaki, A. Mesaki, F. Suzuki, H. Hakogi, and K. Miura, "Surface mount type LD module with receptacle structure," *1999 Proceedings of 49th IEEE Electronics Components and Technology Conference*, San Diego, pp. 1129– 1134 (1999).

15. M. W. Beranek, E. Y. Chan, K. W. Davido, H. E. Hager, D. G. Koshinz, C. L. Larson, C. J. Moore, H. P. Soares Jr., and R. L. St. Pierre, "Hermetically sealed fiber-optic transmitter based on silicon micro-optical bench optical subassembly and multi-player thick film interconnect," *1999 International Symposium on Microelectronics*, pp. 90–91 (1999).

16. M. W. Beranek, E. Y. Chan, C. C. Chen, K. W. Davido, H. E. Hager, C. S. Hong, D. G. Koshinz, M. Rassaian, H. P. Soares Jr., R. L. St. Pierre, P. J. Anthony, M. A. Cappuzzo, J. V. Gates, L. T. Gomez, G. E. Henein, J. Shmulovich, M. A. Occhionero, and K. P. Fennessy, "Passive alignment optical subassemblies for military/aerospace fiber optic transmitter/receiver modules," *IEEE Transactions on Advanced Packaging*, 23(3), pp. 461–469 (2000).

17. K. Song, J. Bu, Y. Jeon, C. Park, J. Jeong, H. Koh, and M. Choi, "Micromachined silicon optical bench for the low cost optical module," *SPIE Conference on Miniaturized Systems with Micro-optics and MEMS, SPIE*, 3878, pp. 375–383 (1999).

18. K. Yamauchi, K. Kurata, M. Kurihara, Y. Sano, and Y. Sato, "Automated mass production line for optical module using passive alignment technique," *IEEE 2000 Electronics Components and Technology Conference*, Las Vegas, may 2000, pp. 15–20.

19. M. Iwase, T. Nomura, A. Izawa, H. Mori, S. Tamura, T. Shirai, and T. Kamiya, "Single mode fiber MTRJ SFF transceiver module using optical sub assembly with a new shielded silicon optical bench," *IEEE 2000 Electronics Components and Technology Conference*, Las Vegas, 24(4), pp. 419–428 (2000).

20. M. Owen, "Agilent Tech. Single mode small form factor (SFF) module incorporated micromachined silicon, automated passive alignment, and nonhermetic packaging to enable the next generation of low-cost fiber optic transceivers," *IEEE Transactions on Advanced Packaging*, 23(2), pp. 182–187 (2000).

21. Y. Sunaga, R. Takahashi, T. Tokoro, and M. Kobayashi, "2 Gbit/sec small form factor fiber optic transceiver for single mode optical fiber," *IEEE Transactions on Advanced Packaging*, 23(2), pp. 176–181 (2000).

22. T. Ericson, G. Palmskog, P. Eriken, P. Lundstrom, M. Granberg, L. Backlin, K. Frojd, and C. Vieider, "Precision passive alignment technologies for low cost array FTTH component," *IEEE Lasers and Electro-optics Society Conference*, Kauai, Hawaii, August 2000. pp. 115–116.

23. N. Kimura, K. Kurata, N. Kitamura, M. Funabashi, A. Goto, T. Kanai, H. Ando, and T. Tamura, "Low cost bi-directional optical transmitter/receiver module for subscriber system," *Proceedings of SPIE*, 3952, April 2000, pp. 354–361.

24. P. Karioja, J. Ollila, V. P. Putila, K. Keranen, J. Hakkila, and H. Kopola, "Comparison of active and passive fiber alignment techniques for multimode laser pigtailing," *IEEE 2000 Electronic Components and Technology Conference*, Las Vegas, May 2000, pp. 244–249.

25. M. Karppinen, K. Kautio, M. Heikkinen, J. Hakkila, and P. Karioja, "Passively aligned fiber-optic transmitter integrated into LTCC module," *IEEE 2001 Electronic Components and Technology Conference*, Orlando, FL, May/June 2001, pp. 20–25.

26. B. Mersali, A. Ramdane, and A. Carenco, "Optical-mode transformer: a III-V circuit integration enabler," *IEEE Journal of Selected Topics in Quantum Electronics*, 3(6), pp. 1321–1330 (1997).

27. I. Moerman, G. Vermeire, M. D'Hondt, W. Vanderbawhede, J. Blondelle, G. Coudenys, P. Van Daele, and P. Demeester, "III-V semiconductor waveguiding devices using adiabatic tapers," *Journal of Microelectronics*, 25, pp. 675–690 (1994).

28. B. Hubner, G. Vollrath, R. Ries, C. Greus, H. Janning, E. Ronneberg, E. Kuphal, B. Kempf, R. Gobel, F. Fiedler, R. Zengerle, and H. Burkhard, "Laser diodes with integrated spot-size transformer as low-cost optical transmitter elements for telecommunications," *IEEE Journal of Selected Topics in Quantum Electronics*, 3(6), pp. 1372–1383 (1997).

29. T. J. Kim, J. K. Ji, Y. C. Keh, H. S. Kim, S. D. Lee, A. G. Choo, and T. I. Kim, "Monolithic Integration of laser and spot size converter using selective area MOCVD growth," *Pacific Rim 1999 CLEO Conference*, Seoul, Aug./Sept. 1999, pp. 1219–1220.

30. S. Y. Huang, C. E. Gaebe, K. A. Miller, G. T. Wiand, and T. S. Stakelon, "High coupling optical design for laser diodes with large aspect ratio," *IEEE Transactions on Advanced Packaging*, 3(2), pp. 165–169 (2000).

31. P. J. Hesketh and J. D. Harrison, "Micromachining, the fabrication of microstructures and microsensors," *The Electrochemical Society, Interface*, 3(4), pp. 21–26 (1994).

32. H. Han, J. D. Stack, J. Mathews, C. S. Koehler, E. Johnson, and A. D. Kathman, "Integration of silicon bench with micro optics," *SPIE Conference on Photonic Packaging and Integration, SPIE*, 3631 pp. 234–243 (1999).

33. H. Han, J. Mathews, J. Stack, and B. Hammond, "Packaging of integrated micro optical systems," Lasers and Electro-Optics Society 1999 12th Annual meeting LEOS '99. *IEEE*, San Francisco, Nov. 1999, pp. 90–91.

34. M. Feldman, H. Han, J. Stack, and J. Mathews, "Integrated micro-optical assemblies for optical interconnects," Parallel Interconnects, Proceedings, 6th International Conference, Anchorage, AK, October, 1999, pp. 141–144.

35. H. Kopola, J. Hiltunen, J. Hakkila, P. Karioja, T. Kololuoma, A. Karkkainen, J. Passo, and J. Rantala, "Design, fabrication and packaging of micro-optical components and systems," Optical MEMS, IEEE/LEOS International Conference, Kavai, HI, August, 2000, pp. 105–106.

36. T. Matsuura, S. Ando, S. Sasaki, and F. Yamamoto, "Low-loss heat resistant optical waveguides using new fluorinated polyimides," *Electronic Letters*, 29(24), pp. 269–271 (1993).

37. M. Ukechi, et al., *Technical Report of IEICE, EMD 98-9*, p. 19 (1998).

38. M. Ukechi, T. Miyashita, Y. Komine, T. Mase, A. Takahashi, T. Nishimura, R. Kaku, S. Hirayama, N. Uehara, and K. Ito, "A new concept for the WDM module using a waveguide equipped with filter," Optical Fiber Communication Conference, Baltimore, MD, March, 2000, 2, pp. 97–99 (2000).

39. W. Ehrfeld, P. Bley, F. Gotz, P. Hagmann, A. Maner, J. Mohr, H. O. Moser, D. Munchmeyer, D. Schmidt, and E. W. Becker, "Fabrication of microstructures using the LIGA process," *Proceedings of the IEEE Micro Robots and Teleoperators Workshop*, Hyannis, MA, 9–11 November, 1987.

40. J. Mohr, "Free space optical components and system based on LIGA technology," Optical MEMS, IEEE/LEOS International Conference, Kauai, HI, August, 2000, pp. 147–148.

41. Z. Ling, K. Lian, and J. Goettert, "Passive alignment and its application in multi level x-ray lithography," *Materials and Device Characterization in Micromachining III Proceedings of SPIE*, 4175, pp. 43–49 (2000).

42. H. Schift, J. Sochtig, F. Glaus, A. Vonlanthen, and S. Westenhofer, "Fabrication of replicated high precision insert elements for micro-optical bench arrangements," *SPIE Conference on Microelectronic Structures and MEMS for Optical Processing IV*, 3513, pp. 122–134 (1998).

43. K. A. Cooper, R. Yang, J. S. Mottet, and G. Lecarpentier, "Flip chip equipment for high end electro-optical modules," Proceedings of the 48th ECTC Electronic Components and Technology Conference, Seattle, WA, 27 May, 1998. pp. 176–180.

44. P. Whitney, "Hybrid integration of photonic subsystems," Proceedings of the 52nd ECTC Electronic Components and Technology Conference, May 2002. pp. 578–582.

45. G. Lecarpentier, J. S. Mottet, O. Pizzirusso, and K. Cooper, "Optoelectronic packaging: the need for sub micron post-bond accuracy device bonding," Proceedings of the Opto 2003 Conference, Bethlehem, PA, 7–10 October 2003.

46. K. F. Harsh, V. M. Bright, and Y. C. Lee, "Study of micro-scale limits of solder self-assembly for MEMS," IEEE Electronic Components and Technology Conference 50th Proceedings, Las Vegas, May 2000, pp. 1690–1695.

47. K. F. Harsh, P. Kladitis, M. Adrian Michalicek, J. Zhang, W. Zhang, A. Tuantranont, V. Bright, and Y. C. Lee, "Solder self-alignment for optical MEMS," Lasers and Electro-Optics Society 12th Annual Meeting LEOS'99, San Francisco, Nov 1999, 2, pp. 860–861.

48. M. H. Choi, H. J. Koh, E. S. Yoon, K. C. Shin, and K. C. Song, "Self-aligning silicon groove technology platform for the low cost optical module," IEEE Electronic Components and Technology Conference, San Diego, June 1999, pp. 1140–1144.

49. Y. Backlund, "Micromachining in silicon for passive alignment of optical fibres," *SPIE Micro-Opto-Electro-Mechanical Systems*, 4075, pp. 118–123 (2000).

50. Y. Nakamura, H. Komoriya, T. Hirahara, and T. Koezuka, "Automated optical passive alignment technique for PLC modules," *SPIE Conference on Machine Vision Applications for Industrial Inspection,* 3652, pp. 78–85 (1999).

51. T. Klotzbucher, M. Popp, T. Braune, J. Haase, A. Gaudron, I. Smaglinski, T. Paatzsch, H. Bauer, and W. Ehrfeld, "Custom specific fabrication of integrated optical devices by excimer laser ablation of polymers," *Proceedings of SPIE, Laser Applications in Microelectronic and Optoelectronic Manufacturing*, 3933, pp. 290–298 (2000).

52. R. Moosburger, R. Hauffe, U. Siebel, D. Arndt, J. Kropp, and K. Petermann, "Passive alignment of single-mode fiber integrated polymer waveguide structures utilizing a single-mask process," *IEEE Photonics Technical Letters*, 11(7), pp. 848–850 (1999).

53. H. Han, B. Caron, W. Lewis, S. Tan, J. Mathews, C. Drabenstadt, R. Boudreau, T. Bowen, and D. Murray, "Refractive plate optical switches using off-axis V-groove array for fiber positioning," *Advances in Electronic Packaging,* 19–1, pp. 765–771 (1997).

54. Y. Kato, K. Mori, T. Mase, A. Takahashi, O. Imaki, and R. Kaku, "Development of 4 × 4 MEMS optical switch," Optical MEMS 2000 IEEE/LEOS International Conference, Kauai, HI, August, 2000, pp. 95–96.

55. G. Schropfer and M. Lebachelerie, "Comparison between an optical and a capacitive transducer for a novel multi-axial bulk-micromachined accelerometer," SPIE Conference on Micromachined Devices and Components IV, Santa Clara, CA, September 1998. 3514, pp. 199–209.

56. H. Fujita, "Magnetostrictive 2-D scanners and pig-tailed tunable MEMS filters studies on vacuum packaging and interconnection of optical MEMS," Optical MEMS, 2000 IEEE/LEOS International Conference, Kauai, HI, August, 2000. pp. 5–6.

Section 1

Mechanical Passive Alignment

2 Passive Alignment in Connectors and Splices

James Kevern and Alan Plotts

CONTENTS

2.1 INTRODUCTION

Optical fiber connectors represent one of the earliest and most widely installed examples of passive alignment. Interconnection of the various components of a fiber-optic system is a vital part of system performance. Connection by splices and connectors couples light from one component to another with as little loss of optical power as possible. Throughout a link, a fiber must be connected to sources, detectors, and other fibers. This chapter describes the basic considerations involved in fiber-optic interconnections and provides several examples of actual connectors and splices. To simplify the discussion, we emphasize connecting one fiber to another.

By popular usage, a connector is a separable device that connects a fiber to a source, detector, or another fiber. It is designed to be easily connected and disconnected many times. A splice is a device used to connect one fiber to another permanently. Even so, some vendors offer separable splices that are not permanent and can be disconnected for repairs or rearrangement of circuits.

The requirements for a fiber-optic connection and a wire connection are very different. Two copper conductors can be joined directly by solder or by connectors that have been crimped or soldered to the wires. The purpose is to create intimate contact between the mated halves to maintain a low-resistance path across the junction. Connectors are simple, easy to attach, reliable, and essentially free of loss.

The key to a fiber-optic interconnection is precise alignment of the mated fiber cores (or mode fields in single-mode fibers) so that nearly all the light is coupled from one fiber across the junction into the other fiber. Contact between the fibers is not even mandatory. The demands of precise alignment on small fibers create a challenge to the designer of the connector or splice.

The following is a list of desirable features for a fiber-optic connector or splice [1]:

- **Low loss:** The connector or splice should align the fibers such that loss of optical power is minimized.
- **Easy installation:** The connector or splice should be easily and rapidly installed without need for extensive special tools or training.
- **Easy operation:** The connector or splice should be easily actuated without excessive insertion/withdrawal forces.
- **Repeatability:** A connector should be able to be connected and disconnected many times without changes in loss.
- **Economical:** The connector or splice should be inexpensive, both in itself and in special application tooling.

It can be very difficult to design and manufacture a connector that meets all the requirements. A low-loss connector may be more expensive than a high-loss connector, or it may require relatively expensive application tooling. Although low losses are always desirable, other factors clearly influence the design choices.

2.2 HISTORICAL PERSPECTIVE

The origin of fiber optic connectors can be traced back to the early 1980s. These early connectors had large tolerance ranges associated with the critical components necessary for good fiber–fiber alignment. In addition, compared to current state of the art, early fibers were not as well controlled dimensionally, particularly in outside diameter and core-to-clad eccentricity. Typical tolerances associated with these early fibers were ±3 μm (.003 mm) on the fiber diameter (typically 125 μm) and ±3 μm on the fiber core-to-cladding concentricity.

TABLE 2.1
Typical Manufacturing Process Capability—Early to Mid-1980s

Material	Manufacturing Method	Typical Tolerance (μm)		
		Ferrule Outside Diameter	Hole Size	Concentricity
Polymer (thermoplastic)	Molded	7	3	5
Metal (stainless steel)	Machined	4	2	3
Ceramic (alumina)	Extruded/machined	3	3	< 2

The precision components used in the manufacture of early optical connectors consisted primarily of three materials: plastic (molded), metal (machined), and ceramic (formed, fired, and ground). The values in Table 2.1 illustrate the precision of the "state-of-the-art" manufacturing processes used to produce early fiber-optic connector components. Typical accepted values of insertion loss for these early connectors were in the range of 1.5 to 2.0 dB for multimode and 1.0 to 1.5 dB for single mode.

The manufacturing imprecision was further compounded by the measurement uncertainty inherent in the equipment of that day. Recent work done by the Telecommunications Industry Association, F06.3 Committee [2,3], and work done by the National Institute for Standards Technology to develop standard reference materials have gone a long way to eliminate sources of measurement error, both random and systematic.

The inability of these manufacturing processes to produce ferrules to the desired tolerance range on a consistent basis led manufacturers to rely on sorting to segregate parts by tolerance grade. The more tightly toleranced parts were then sold as single-mode components. This was particularly true with ceramic ferrules, which were usually reserved for single mode because of their higher cost.

Over time, vast improvements have taken place in the manufacturing methods and materials used to produce these components. Measurement precision and accuracy have also improved dramatically [4]. Process capability and material improvements now allow connector manufacturers to achieve single-mode-toleranced

TABLE 2.2
Typical Tolerances Achievable in Today's Connectors Are Significantly Better Than Those of Early Connectors

Material	Manufacturing Method	Typical Tolerance (μm)		
		Ferrule O.D.	Hole Size	Concentricity
Polymer (thermoplastic)	Molded	2	2	1.4
Metal (stainless steel)	Machined	1.5	1.5	2
Ceramic (zirconia)	Extruded/machined	1	1	< 1

components in virtually all of the standard materials. Now the connector material is chosen more by its suitability for a particular application than by the tolerance that can be achieved. Applications that require a low insertion loss but not a high degree of ruggedness or exposure to temperature extremes may now use an all-plastic connector, whereas in years past the choice would have been limited to a connector that used ceramic for the precision components. For today's connectors, typical values of insertion loss of < 0.5 dB for multimode and < 0.2 dB for single mode are obtainable.

2.3 FERRULE-BASED CONNECTORS

As in any optical fiber coupling, the primary objective of a connector is to control all six degrees of freedom; namely, transverse, angular, and axial offsets. The manner in which this objective is achieved falls into two main categories: those that directly align the fibers with each other, and those that mount the fiber permanently to another member (ferrule), which is then aligned. Connectors are generally intended for repeated mating and unmating, whereas splices join two fibers more or less permanently. Some splices are differentiated from connectors by only the design intent.

Strictly speaking, a ferrule is "a ring or cap usually of metal put around a slender shaft to strengthen it or prevent splitting" (*Merriam Webster Online Dictionary*) In common practice, it has come to mean any part in which the fibers are mounted. Ferrules have traditionally been cylindrical, but other shapes have been used as well (e.g., spherical and rectangular). Use of a ferrule creates additional features, whose tolerances must be managed to achieve good alignment, but provides protection for the fiber, enhances durability, and generally makes the fiber easier to handle.

2.3.1 SINGLE-FIBER CONNECTORS

The capability of fabrication processes and the relative sensitivity to offset combine to make transverse offset the most difficult to control. Two major issues contribute to effecting good alignment. First, the core of the optical fiber must be located precisely within the ferrule. The true position of the fiber hole is specified with respect to datum features. For cylindrical ferrule connectors, the datum feature is the outside ferrule diameter. Second, the ferrules are aligned to each other, usually with the same datum features from which the fiber position is determined. Therefore, alignment loss has two main contributors: first, deviation of fiber location from its true (ideal) position with respect to the ferrule datum, and second, misalignment between the datum features of the two ferrules (or between the ferrule and device optics for active component interfaces). It is primarily in this second aspect that connector designs differ. Regardless of the method used to align the ferrules with each other, all must first locate the fiber core to a precision adequate for the intended application.

Most often, the fiber is positioned within the ferrule by means of a precision diameter hole, which itself is precisely located with respect to features that align the ferrules with respect to each other. The core is centrally located within the fiber to tolerances specified by the fiber manufacturer. The diameter of both the fiber and the hole determine the tightness of fit. Unless the fit is extremely close (< 1 micron or so), fibers usually tend to be biased toward one side, even though theoretically they can be located anywhere within the hole. For very tight fits, capillary action creates a self-centering effect. Certain epoxies are also formulated in such a way as to promote this self centering.

The following dimensions (and their tolerances) are the primary contributors that control the precision with which the core is located. Note that both the fibers and the preponderance of connectors to date are based on cylindrical features; therefore, the mathematical treatment here is based on cylindrical coordinates. (The z-axis is aligned with the axis of the fibers and is treated separately in Section 2.3.1.3.) The choice of coordinate system for offset analysis will depend on the character of the alignment geometry. For example, the guide pin approach of the MT ferrule described in Section 2.3.2 is more easily described in Cartesian coordinates.

Because the individual offsets occur randomly, the net offset vector's azimuth angle will be uniformly distributed between 0 and 360 degrees. Therefore, the total transverse offset is defined probabilistically.

Referring to Figure 2.1 and denoting vectors in **bold**, first define $\boldsymbol{R}$ as a randomly oriented unit vector

$$\boldsymbol{hc} = (r_h - r_c)\boldsymbol{R} \tag{2.1}$$

$$\boldsymbol{bf} = (r_b - r_f)\boldsymbol{R} \tag{2.2}$$

The offset of the fiber core from the alignment bore is then simply:

$$\boldsymbol{O} = \boldsymbol{bf} + \boldsymbol{h} + \boldsymbol{hc} + \boldsymbol{cc} \tag{2.3}$$

We can then write the general equation for fiber to fiber alignment as:

$$A = |\boldsymbol{O1} - \boldsymbol{O2}| \tag{2.4}$$

where $\boldsymbol{O1}$ and $\boldsymbol{O2}$ represent the offsets of the launch and receive fibers, respectively.

Most single-fiber ferrule connectors rely on cylindrical ferrules aligned by a bore that is either solid or resilient. Resilience is used in both the radial and axial directions to minimize the effects of tolerances. The drawback to any approach that employs resilience is that stability against outside mechanical disturbance is diminished. For this reason, most connector designs use a "floating" optical subassembly composed of two ferrules and an alignment sleeve. This subassembly is contained in a cavity and supported by springs (Figure 2.2).

2.3.1.1 Cylinder in Bore Style Connectors

2.3.1.1.1 Solid Cylinder–Solid Bore

One of the earliest commercially successful incarnations of this type is the FSMA (see Figure 2.3), which is derived from connectors designed to terminate coaxial copper cable with a hole drilled in the center to accommodate a fiber. With one

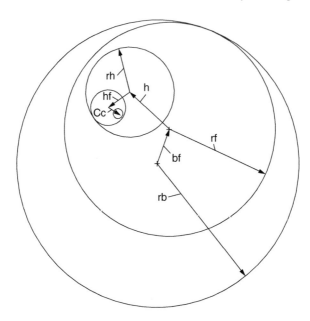

FIGURE 2.1 Offset in a cylindrical ferrule described as a vector combination of individual offset parameters. r_c = fiber cladding radius; cc = offset vector resulting from the fiber core to clad offset tolerance; r_h = ferrule hole radius; h = offset vector resulting from the ferrule hole location tolerance; r_f = ferrule radius; hf = offset vector resulting from the clearance between fiber cladding and ferrule hole; r_b = adapter or coupling bore radius; bf = offset vector resulting from the clearance between bore and ferrule.

exception (see Figure 2.4), alignment in FSMA connectors is controlled simply by the tolerances to which the parts could be fabricated, and Equation 2.3 applies.

Because both the bore and the ferrule are essentially rigid bodies, there must never be an interference fit. Therefore:

$$r_b \geq r_f \tag{2.5}$$

and the maximum transverse and z-axis offset is controlled simply by the precision with which the parts can be fabricated. Because of the rigid nature of the design, if the ferrules touch, there is no mechanism by which the force at the fiber to fiber interface will be limited. Therefore, there must always be some z-axis gap between the fibers to avoid damage to the glass surface. This results in these types of connector designs always having a "loss floor" because fresnel reflection will always be present. It also makes them unsuitable for laser-based systems in which back reflections must be minimized. (Therefore, the design in Figure 2.5 is generally preferred.)

Angular offset is somewhat of a concern primarily because of the fabrication methods used. If the ferrules are machined, the fiber hole very often has a short aspect ratio, increasing the angular uncertainty (see Figure 2.6).

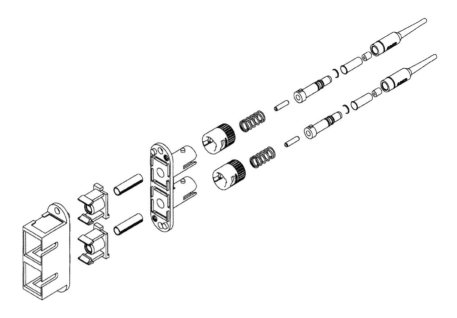

FIGURE 2.2 Illustration of common connector construction. Resilience (springs) absorb *z*-axis tolerance, ensuring positive contact. Resilience in the alignment sleeve absorbs ferrule radius tolerance. (Courtesy Tyco Electronics.)

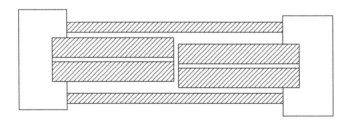

FIGURE 2.3 FSMA connectors are made in two configurations. This one has no resilience anywhere. In this version, members are rigid, and tolerances control both transverse and *z*-axis alignment.

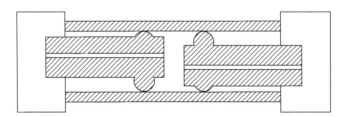

FIGURE 2.4 Another FSMA variant from Tyco Electronics incorporates resilience by way of deformable protrusions on a polymer tip.

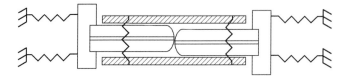

FIGURE 2.5 Schematic representation of a solid cylinder-resilient bore-style connector with spring loading to exert normal force. Most common single-fiber connectors use this alignment concept.

2.3.1.1.2 Solid Cylinder–Resilient Bore

One way to minimize the effects of some tolerances is to make one of the alignment features resilient. Typically this involves making the alignment bore from a resilient material or, more commonly, by using a resilient structure. Split alignment sleeves made from ceramic, metal, or plastic are the predominant form of alignment in optical connectors today. The connector shown in Figure 2.2 is a commercial implementation of this concept, shown schematically in Figure 2.5. The resilient bore is the most common type of single-fiber connector.

2.3.1.1.3 Resilient Cylinder–Solid Bore

Because motion is relative, from a physical standpoint, tolerance effects are also minimized by making the cylinder (ferrule) resilient. In either case, the fit between the cylinder and bore is a slight interference. The stiffness of the resilience is a trade-off between low insertion forces (low stiffness), tolerances of bore and cylinder (the greater the tolerance, the lower the stiffness must be), and stability. To prevent misalignment caused by external loads, a high stiffness is desired. For example, projections on the ferrule can deform when inserted into a rigid bore (see Figure 2.4).

2.3.1.1.4 Tapered Cylinder/Bore (Biconic)

When both alignment surfaces are tapered, the aligning surface is attempting to control all 3 degrees of freedom (transverse, angular, and axial) or all 5 degrees

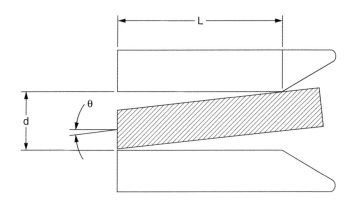

FIGURE 2.6 The aspect ratio (L/d) of the fiber hole influences angular tilt.

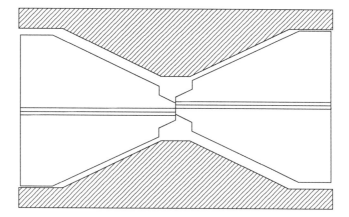

FIGURE 2.7 Biconic alignment requires precise matching of polish length to the cone spacing of the adapter.

(2 transverse, 2 angular, and axial; rotational is not controlled). If parts are not fitted precisely, control in one or more directions is lost. For instance, if the parts bottom out in the cone before contact is achieved, an end gap results. However, if parts bottom out on the ends first, both transverse and angular alignment are compromised (Figure 2.7). Slight resilience in the form of local deformations at the point of contact between the conical parts helps to expand the operational tolerance range.

2.3.1.2 Tuning

Tuning is employed to improve the performance of passive optical connectors. Technically speaking, it is not passive alignment but, rather, an active "prealignment." Parts are mated and optical performance is measured. Then slight adjustments in the fiber position are achieved through various means to minimize the optical loss.

Although early cylindrical ferrule connectors allowed random angular orientation of the ferrules about the z-axis, connectors are commonly keyed to improve repeatability. Often the keying feature can be assembled to the ferrule in various orientations, usually every 60 or 90 degrees. This allows the rotation to be "clocked" to achieve minimum loss. The relative offset between the fibers as a function of clocking angle is governed by Equation 2.6 and shown in Figure 2.8. Observe that variation resulting from clocking is controlled by the best of the two connectors. A connector pair in which the core is perfectly centered in one ferrule (i.e., $r_i = 0$) will have zero clocking variation independent of the radial offset of the other ferrule. For cylindrical connectors, in general terms, the offset of a given fiber core is described by Equation 2.3.

Describing the transverse misalignment of each fiber core as simply r_1 and r_2 for fiber 1 and fiber 2, respectively, the resulting clocking offset from (2.4) becomes:

$$A(\theta,\phi) = \sqrt{(r_1\cos\theta - r_2\cos\phi)^2 + (r_1\sin\theta - r_2\sin\phi)^2} \qquad (2.6)$$

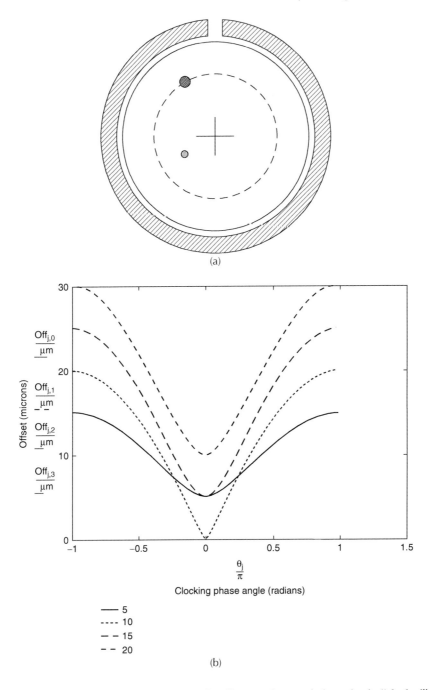

FIGURE 2.8 (a) Paths of launch and receive fibers as they are independently "clocked" around the alignment axis. (b) Fiber offset as a function of phase angle between the launch and receive fibers ($\phi = 0$).

where θ and ϕ represent the clocking angle of fibers 1 and 2 respectively. Observe from Figure 2.8b the change in shape as the offset radii become equal.

2.3.1.3 Z-Axis Alignment

Early connectors, such as FSMA, shown in Figure 2.3 and Figure 2.4, clamped flanges on the ferrule body to rigidly fix the relative location in the axial direction. This results in a gap between the fiber ends, the size of which is limited by tolerances on the ferrule length and the adapter. To minimize loss caused by beam spreading, the gap should be as small as possible. In addition to beam spreading, loss occurs because of reflection from each of the two air-to-glass interfaces. For large gaps (> 20 wavelengths), the loss caused by this Fresnel reflection is approximately 0.3 dB, and the resulting back reflection is −11 dB. For very small gaps, however, constructive and destructive interference make the loss and reflection a strong function of both the gap and the wavelength. The span of this regime in which coherency is a factor depends on the coherence of the light-emitting source. The relationship between transmission efficiency and gap for a Gaussian source is given in terms of the time delay for propagation across the gap [5] by:

$$T = 1 - 2R + \frac{2R\sin(\pi v_0 \Delta\tau)}{v_0 \Delta\tau} \text{gauss} \ (\Delta v \bar{\tau})\cos(2\pi v_0 \bar{\tau}) \qquad (2.7)$$

where R = the single surface reflection coefficient, τ_0 = the round-trip time delay across the gap for normal incidence, τ_1 = the round-trip time delay across the gap for maximum propagation angle, v_0 = the optical frequency of the source, Δv = the source spectral width, and $\Delta\tau = \tau_0 - \tau_1$.

Even in systems using LEDs, which are normally considered to be incoherent, gaps small enough to be within the coherence length can occur (Figure 2.9). Note that when very small gaps are present, strong temperature effects can be seen (Figure 2.10). This phenomenon has also been used to explore the thickness of what is known as the damage layer — a layer of elevated refractive index produced by polishing operations [6].

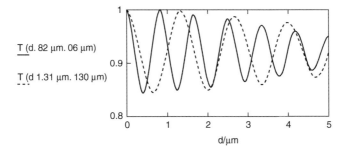

FIGURE 2.9 Transmission efficiency across a small gap illustrating the dependence on wavelength.

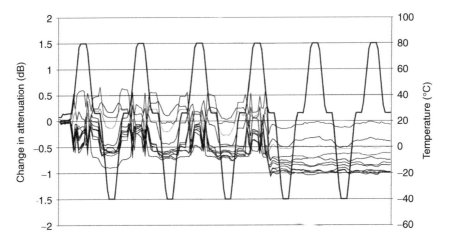

FIGURE 2.10 Thermal cycling performance of an MT style connector having small air gaps. Note how attenuation stabilizes after index matching gel is applied two-thirds of the way through exposure.

At one time it was thought that contact between fibers should be avoided (Figure 2.11), however, more recent connectors are generally designed to achieve positive contact between the fibers, thereby assuring that there will be no z-axis misalignment (Figure 2.12) [7]. Positive contact designs must be controlled to

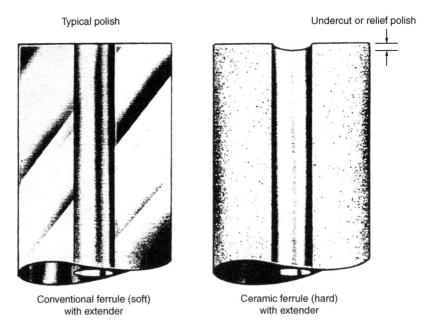

FIGURE 2.11 Early connector designs purposely avoided contact between the fibers themselves. (Courtesy Buehler.)

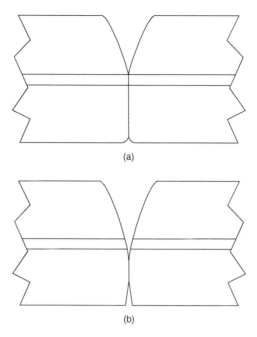

(a)

(b)

FIGURE 2.12 (a) Positive contact eliminates Fresnel reflection and loss resulting from *z*-axis misalignment (gap). (b) Offset of the point of contact can create small air gaps that degrade performance.

ensure that variations resulting from normal manufacturing tolerances do not allow deviations from perfect geometry such that a gap occurs between the fiber ends. The most common deviations are nonperpendicularity, also known as vertex or dome offset [8,9], and fiber recess caused by overpolishing ferrules that have a hardness rating higher than that of the glass fiber (e.g., ceramic). Fiber recess can be accomodated by deformation of the ferrule material under spring pressure [10]. Even though the ferrule material is typically quite hard, the fiber recess is often limited to tens of nanometers, such that deformation caused by hertzian stresses allows the gaps to close [11].

The tolerances are interrelated, such that ferrules with smaller radius tips can tolerate higher nonperpendicularity and so forth [12].

2.3.1.4 Splices

Although some splices are essentially separable connectors using ferrules and alignment sleeves, a number of splice products align the fibers directly, thereby eliminating the tolerances associated with positioning the fiber within a ferrule. This capitalizes on the essential nature of splices in that they are permanent connections having no requirement for durability when repeatedly mated.

One of the simplest and most widely used forms of splicing is to align fibers in a V-groove. Fibers are biased into the V-groove by pressure, usually generated

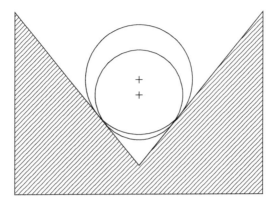

FIGURE 2.13 V-groove misalignment is primarily caused by fiber diameter tolerance.

by slight bends in the fiber. X–Y alignment tolerance is governed simply by the properties of the fibers themselves, provided the fibers and V-grooves can be kept free of contamination (Figure 2.13). Z alignment is achieved by butting the fibers with light force, which again is usually achieved by bending the fiber. Once fibers are in position, they are locked in place either by mechanically clamping or by potting with an index-matching adhesive such as an ultraviolet-cured adhesive.

2.3.2 ARRAY CONNECTORS

Array connectors precisely position the fibers within a block of material (also called a ferrule, even though it is not cylindrical) and then align the blocks with respect to each other. Array ferrule types differ in both how their transverse alignment is achieved (i.e., selection of datum features) and in fabrication techniques. Fibers are normally placed in the block in one of two ways: the blocks are fabricated with holes in which the fibers are fastened or bonded, or the blocks are fabricated in two halves, one or both of which has V-grooves to align the fibers.

The ferrules are aligned with each other by datum features, which are in turn aligned by some intermediary surface. As with cylindrical connectors, z-axis alignment is achieved through positive contact [13,14].

2.3.2.1 Alignment by Planar Features

The first obvious approach is to select two orthogonal outside surfaces as datum features (Figure 2.14a) and to precisely control fiber position with respect to them. Because the fiber locations are different distances from the datum features, this approach is limited to manufacturing processes with excellent dimensional control such as etched silicon V-grooves. It does not lend itself to processes such as molding, where variation in material properties such as postmold shrinkage contribute systematic deviations in position.

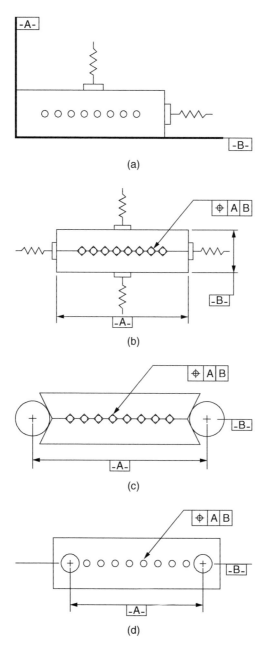

FIGURE 2.14 Array connector alignment is accomplished by selection of specific features to align ferrules to each other. (a) External orthogonal surfaces; (b) centered on external features; (c) external V-grooves; (d) guide pins and holes.

2.3.2.2 Alignment by Center of Features

One way to minimize misalignment tolerances is to align the ferrules by features rather than planes. The general principle here is to place the center of the alignment features in such a way as to minimize the distance between the datum and the furthest fiber location (see Figure 2.14b, c, d). Thus, the distance from the alignment datum to the fiber is controlled such that manufacturing variations and differential thermal expansion effects are minimized. However, the size of the datum features must also be tightly controlled to avoid misalignment between the ferrules themselves, in the same way that ferrule diameter must be controlled in cylindrical connectors.

Some array connector systems align the ferrules using V-grooves along the outsides of the ferrules coincident with the axis of the fibers (Figure 2.14c). This locates the datum at the center of the fiber array pattern. The ferrule alignment is achieved through guide pins that are pressed into the V-grooves. Tolerance variations in the size of the ferrules (i.e., distance between V-groove vertices) and included angle provide an opportunity for transverse misalignment analogous to that induced by variations in ferrule diameter in cylindrical ferrule alignment.

2.3.2.3 Guide Pin/Hole Ferrules

Probably the most widely deployed array technology to date is the MT [15,16,20] shown in Figure 2.15. The MT ferrule alignment is achieved by mounting precision-diameter guide pins in precision holes in the ferrule (Figure 2.14d). Guide pins can be fabricated much like needle bearings and can therefore be mass produced

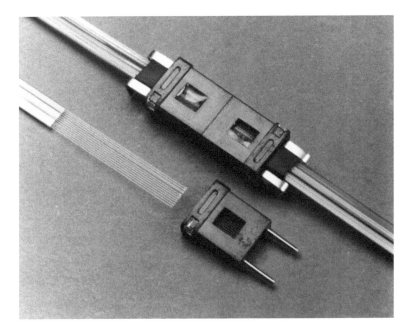

FIGURE 2.15 MT array splice.

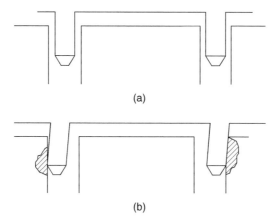

(a)

(b)

FIGURE 2.16 Deformation of the guide pins and the ferrule material surrounding the holes absorbs slight variation in pin/hole spacing. (a) Perfect matching of pin/hole spacing rarely occurs; (b) typical situation—slight interference causes local deformation, which provides a "self-centering" effect.

with tight tolerances. Similar to the V-groove array ferrules, the guide pin holes are positioned such that the point halfway between the guide pin holes is the desired center of the fiber hole pattern. In theory, slight clearance between the guide pin and the hole, caused by diameter tolerances, results in misalignment analogous to a rigid cylindrical ferrule in a rigid bore (Section 2.3.1.1), This situation can only occur, however, when the guide pin holes in both ferrules are positioned exactly the same distance apart (Figure 2.16a). Because this rarely occurs in practice, slight material deformation occurs, which serves to absorb the pin/hole diameter tolerance and provides self-centering behavior (Figure 16b).

MT ferrules have been designed into a number of connector systems. In these designs, the ferrules are spring-loaded to attain positive contact between fibers [19]. Guide pins are retained, usually in one of the connector plugs, although "hermaphroditic" configurations have also been postulated in which each plug contains one pin. Guide pin location depends on connector polarization and the type of fiber interconnection required (e.g., some applications require fibers to "cross over" in each cable section, wheras others require "straight through" connectivity). It is especially useful in applications where high fiber density is required. MT ferrules are available in fiber counts up to 72 (6 rows × 12 columns).

2.4 FERRULE-LESS CONNECTORS

One way to reduce both cost and tolerance effects is to align the fibers directly. This might be thought of as using splice technology for separable connectors. To accomplish this, the fibers are aligned using a single bore or V-groove. Fresnel reflections are minimized through the use of index gel, positive contact, or angled end faces. These connectors usually feature a moveable shutter to protect the exposed fibers when unmated. Often the shutter also serves as an optical barrier

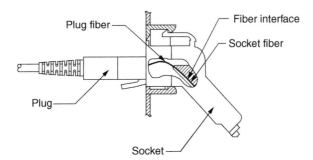

FIGURE 2.17 Volition™ VF-45 plug inserted into the socket. The fibers are engaged and guided by the V-grooves to the fibers in the socket. (Courtesy 3M Corporation.)

for eye safety; however, whether this function actually meets legal requirements depends on how easy it is to actuate (in most jurisdictions it must require a tool), and whether the material is opaque at the wavelength in question. Many materials that are opaque in the visible spectrum provide much less attenuation at the longer wavelengths typical in optical communication systems.

An example of a ferrule-less connector is shown in Figure 2.17. The Volition™ VF-45 connector uses V-groove fiber alignment technology and is used for premise applications. In the installed position, the plug fiber is bent in a bowed position to provide stored energy, which both seats the fiber in the V-groove and provides a downward compressive force on the resident cable plant fiber to establish and maintain optical continuity. The V-groove provides a simple convenient structure to align two fibers by using the precision geometry of the fiber to effect alignment of the fiber cores (see Figure 2.13).

The cable plant fibers are polished and clamped in one end of the V-groove. The plug fibers are typically factory terminated and are suspended in free space within a protective shroud, with the plug fiber ends beveled and polished optically. As the plug is inserted into the socket, the beveled plug fiber end slides along the V-groove to optically mate with the cable plant fiber clamped to the V-groove. The beveled fiber end face removes sharp edges of the glass that can shatter, reduces skiving of material along the V-groove, and provides a place for debris to accumulate (Figure 2.18).

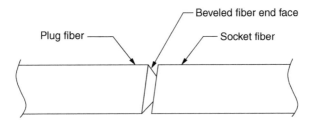

FIGURE 2.18 Cleaved and beveled fiber end face reduces shatter of fiber end face, reduces skiving of the V-groove, and provides a place for debris to accumulate. (Courtesy 3M Corporation.)

In the installed position, the coated fiber is bent to provide stored energy, which both seats the fiber in the V-groove and provides light normal force (approximately 0.1 N) between the mated fibers.

2.5 BACK PLANE CONNECTORS

Fiber connectors continue to play an increasing role in connecting both telecommunications and computer networks. Once the fiber reaches an electronic box, however, it enters into the realm of electrical package interconnects. To keep step with increasing signal and data rates, manufacturers are placing active optical devices on their printed circuit boards right next to their electrical counterparts. Copper back plane connectors, made by many connector manufacturers, have existed for years and address the need to quickly and reliably mate line cards with larger, stationary back planes. Although the art of developing this type of electrical interconnection is quite mature, the optical counterparts of these connectors are embryonic by comparison.

Optical connectors have also been designed to operate on printed circuit boards next to larger, less precise electrical connectors. Their existence on the printed circuit board must be totally transparent to the user; that is, no special procedures or care must be added by their use. To accomplish this, the optical connector must operate within the parameters of the electrical connector. Typically, the tolerances associated with mating two electrical connectors together are orders of magnitude greater than those required to mate two optical connectors together. One way around this is to make the optical connector an integral part of the electrical connector. By doing this, all of the components can be referenced to each other through a common datum point. Then when the connector halves come together, the tolerances between the mating components (line card and backplane) are reduced. An example of a connector using this principle is the (Tyco Electronics) 2-mm HM Z-PACK™ electrical back plane system. Here a module containing the optical section of the connector is tied directly to the electrical section. When the mating of the two connectors occurs, a single alignment feature guides the electrical and optical pins together simultaneously (see Figure 2.19 and Figure 2.20).

This is the ideal solution to designing a back plane system from scratch. In many cases, the use of fiberoptic connectors is an afterthought and is added to an already existing design to extend the capability of an installed system. This presents an altogether different set of circumstances that requires a unique solution set. When electrical and optical connectors are rigidly mounted on the same daughter card, they must be located precisely in relation to one another. Typically, a true position of 0.1 mm (.004 in.) is a rule of thumb for locating several components on the same board. If manufacturing tolerances or other factors prevent this location tolerance from being obtained, then other design options must be considered. One of these options is to "float" the optical connector with respect to the board that it occupies.

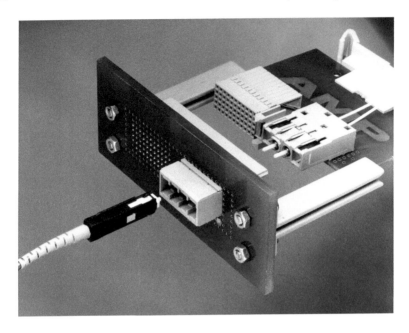

FIGURE 2.19 A feature is provided to provide mechanical "prealignment" between the line card and back plane before engagement of the electrical and Tyco Electronics Z-PACK MSC™ fiber optic connector. (Courtesy Tyco Electronics)

One example of this connector is the Tyco Electronics LIGHTPLANE™ back plane connector (Figure 2.21). The line card connector is attached to the PC card using shoulder screws. This allows the connector to float on all three axes, which helps to ease the tolerance restrictions. The down side to using a connector in this

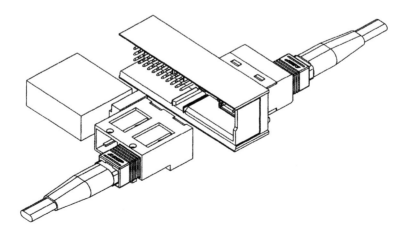

FIGURE 2.20 LIGHTRAY MPX™ Fiber array connector integrated with Tyco Electronics 2-mm HM Z-PACK™ electrical contacts. (Courtesy Tyco Electronics)

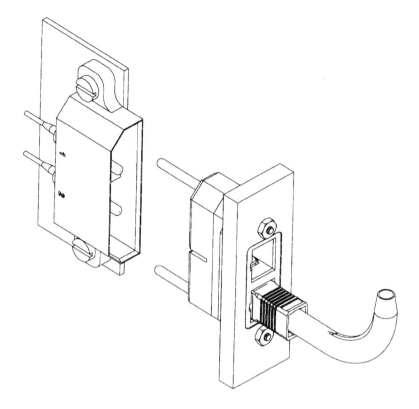

FIGURE 2.21 Back plane connector housing alignment pins and industry-standard SC connectors. Note use of shoulder screws to enable line card connector to float, which enables alignment pins to guide the optical connector into engagement. (Courtesy Tyco Electronics)

manner is that it takes up additional board real estate and requires the use of mounting hardware.

2.5 CONCLUSION

Although optical connectors take on a wide variety of appearances, in general the alignment methodology employed involves one of a handful of basic schemes. By understanding the alignment mechanism underlying each design, it is possible to recognize similarities and differences in designs that influence the ability to achieve the precise alignments required for communication grade optical fibers. Obviously larger fiber technology, such as might be used for sensors, medical applications, "light pipes," displays and so forth provides more design latitude with regard to how alignment is achieved [17,18].

REFERENCES

1. Sterling, *Technician's Guide to FIBER OPTICS,* Second Edition, Delmar Publishers, Albany, NY, 1993.

2. Telecommunications Industry Association, EIA-455-134, FOTP-134, "Measurement of Connector Ferrule Hole Inside Diameter" ANSI/Telecommunications Industry Association/EIA-455-134-1996, Arlington, VA, January 1997. More about Telecommunications Industry Association Fiber Optic Standards can be found online at http://www.tiaonline.org/standards/.

3. Telecommunications Industry Association, EIA-455-135, FOTP-135, "Measurement of Connector Ferrule Inside and Outside Diameter Circular Runout" ANSI/Telecommunications Industry Association/EIA-455-135-1996, Arlington, VA, January 1997.

4. Young, Hale, Mechels, "Optical Fiber Geometry: Accurate Measurement of Cladding Diameter," *Journal of Research of the National Institute of Standards and Technology,* Vol 98, No 2, March–April 1993, pp. 203.

5. Wagner and Sandahl, *Applied Optics,* Vol 21, No 8, 15 April 1982, pp. 1383

6. Shah, Young, Curtis, Optical Fiber Communication Conference, 19–22 January 1987 Session TUF4, Optical Society of America.

7. Shintaku, Nagase, Sugita, "Connection Mechanism of Physical-Contact Optical Fiber Connectors with Spherical Convex Polished Ends," *Applied Optics,* Vol 30, No 36, December 1991, pp. 52–60.

8. IEC 61300-3-15 (1995-02) Fibre optic interconnecting devices and passive components—Basic test and measurement procedures—Part 3–15: Measurements—Eccentricity of a convex polished ferrule endface. Listings of IEC documents can be found online at http://www.iec.ch/.

9. IEC 61300-3-16 (1995-05) Fibre optic interconnecting devices and passive components—Basic test and measurement procedures—Part 3–16: Examinations and measurements—Endface radius of spherically polished ferrules.

10. Deeg, "Effect of Elastic Properties of Ferrule Materials on Fiber-Optic Physical-Contact (PC) Connections," *Amp Journal of Technology,* Vol 1, November 1991, p. 29.

11. Bolhar and Deeg, "Contact Zone and Hertzian Stress in Fiber-Optic Connections with Spherical or Ellipsoidal Fiber Endfaces," *AMP Journal of Technology,* Vol 2, November 1992, p. 29.

12. Reith, Grimado, Brickel, "Effect of Ferrule-Endface Geometry on Connector Intermateability," *National Fiber Optic Engineers Conference Proceedings,* June 1995, p. 635.

13. Kevern, Harper, Knight, Knasel, Satake, "Multifiber Connector Endface Attributes for Optimal Connector Performance," *Electronic Components and Technology Conference Proceedings,* Orlando, FL, 1996, pp. 936.

14. Matsuura, Ueda, Honjo, Yamanishi, "Development of 16-Fiber Push-On Type Optical Connector," *International Wire & Cable Symposium Proceedings*, 1994, pp. 768.

15. IEC 60874-16 (1994-09) Connectors for optical fibres and cables—Part 16: Sectional specification for fibre optic connector—Type MT.

16. IEC 61754-5 (1996-12) Fibre optic connector interfaces—Part 5: Type MT connector family.

17. Miller, *Optical Fiber Splices and Connectors,* Marcel Dekker, 1986.

18. Senior, *Optical Fiber Communications,* Second Edition, Prentice Hall, 1992.
19. IEC 61300-3-30 (2004-01-23) Polish angle and fibre position on single ferrule multifiber connectors.
20. Satake, Tatsuno, Ouchi, Knasel, Knight, Lundberg, Keller, "Single-Mode Multi-fiber Connector Design and Performance," Electronic Components and Technology Conference Proceedings, 1996 p. 494.

3 Mechanical Passive Alignment I

Songsheng Tan, Hongtao Han,
and Robert A. Boudreau

CONTENTS

3.1 INTRODUCTION

Silicon waferboard refers to the technology of a silicon platform that enables passive alignment and other features to reduce cost in manufacturing optoelectronics. In particular, the platform is based on a mechanical passive alignment technique, as originally developed at GTE Laboratories (now Verizon Laboratories) and further developed at AMP–Tyco Electronics. It is suitable for component construction based on hybrid or monolithic integration.

The mechanical passive alignment technique for silicon waferboard creates reactive ion-etched features in the form of blocks that have been micromachined into the silicon surface, providing stops to position the semiconductor die on the surface. This technique relies totally on the precision of the parts to be aligned and not on an optical vision system, so the assembly equipment can operate quickly and be inexpensive.

3.2 ADVANTAGES OF SILICON WAFERBOARD

The silicon waferboard technology for packaging has a number of advantages, as shown in Table 3.1. The most important advantage is the passive alignment, which will be discussed in the next section, but there are a number of other important features that are advantages. These are discussed next, starting with flexible manufacturing.

Flexible manufacturing is a term applied to a system of manufacturing that is able to rapidly and inexpensively change from one product to another without a lot of retooling and design engineering. This is possible with silicon waferboard

---

TABLE 3.1
**Strategic Advantages of Silicon Optical Bench
or Silicon Waferboard**

Assembly and Test Automation	Physical
On-wafer die-bonding	Environmental stability
On-wafer burn in	Miniature size
On-wafer testing	Heat sinking
Step and repeat handling	High-speed electrical transmission
Design for Manufacturing	Integration
True passive alignment	Hybrid
Flexible manufacture	Monolithic
Rapid prototyping	Overmolding compatible
Applies electronics manufacturing to optics	Microelectromechanical systems
	Planar lightwave circuit
	Micro-optics
	Electronics
Simplicity	
Readily available materials	
Low cost materials	
Reliability (few parts)	

because the micromachined structures are fabricated with soft tooling, otherwise known as photomasks. Photomasks can also be built up from a library of previously debugged standard cells of structures in a manner similar to electronic chip fabrication. Artwork can be quickly generated for a new product, and then the soft tooling can be fabricated by a photomask shop before being used to fabricate the new product, using standard processing steps.

Hybrid integration is extended to optics using silicon waferboard. This occurs in a variety of ways. Photonic devices that efficiently emit light in the infrared region used by fiber-optics cannot be made of silicon at the present time. For this reason, InP- and AlGaAs-based materials are bonded to the silicon surface to provide the required optical device function. Beyond this, hybrid structures of passive optical devices, such as lenses and optical fiber, can be mounted on the surface.

On-wafer assembly, die-bonding, burn in, and testing refer to wafer scale processes made possible because of the known positions of repeated product cells on a silicon waferboard. The repeated pattern enables repetitive and automated processes, not possible with conventional fiber-optics manufacture. Testing of optics can be done using equipment developed for the electronics industry, such as automatic probers, by using the monitor detectors behind the laser chips as a means to provide the electrical signal to track the laser output for a given applied bias.

Monolithic integration refers to the growth or etching of silicon waferboard such that all the functional features are formed into the wafer itself rather than bonded to it. The alignment features and metallizations are made into the substrate, but it is also possible to include circuit elements such as bi-pass capacitors, as well as resistors. Furthermore, growth may include thick glass layers that can be fabricated into dielectric insulators or optical waveguides.

Silicon waferboard is considered more environmentally stable than conventional packaging techniques because there are fewer parts and fewer interfaces. The silicon is also more closely expansion matched for the optics materials than if the mounts were made of metal. Conventional fiber-optic packaging uses bulk mechanical sub assemblies of metal, lenses, and fiber blocks and contact blocks. These small parts are largely eliminated with silicon waferboard, reducing the handling as well as reducing the number of interfaces, which aids reliability.

The materials used to fabricate silicon waferboard are inexpensive. The amount of materials used to make these parts is small, and the silicon wafer cost divided by the number of product sites per wafer is also small. A typical 100-mm wafer might yield from 600 to 1000 product die to use for assembling 600 to 1000 products.

Silicon waferboard manufacturing leverages off of conventional electronics manufacturing equipment for wafer handling, processing, cleaning, and testing. The only special equipment needed would be for measuring critical mechanical dimensions. Dimensions need to be measured to better than 0.1 μm, but measured over distances of several hundred microns. Throughout this chapter it will become self evident that another key advantage of silicon waferboard is that it provides for miniaturization. All the functions of a transceiver can be provided by bare die bonded to the silicon surface and interconnected with wirebonds. There are no large parts to the assembly unless they are involved with interfacing with the outside world. Both electrical and optical interconnect interfaces could exceed the size of everything else on the silicon waferboard.

Heat sinking is an advantage for silicon waferboard because the thermal conductivity of silicon is fairly high—about half that of copper. Silicon waferboard can also serve as a platform for microelectromechanical systems (MEMS). These MEMS structures can have many functions. They can serve as mechanical members that redirect light in switches, and they can encode signals onto a light path.

3.3 MECHANICAL PASSIVE ALIGNMENT

The mechanical passive alignment discussed below is shown in Figure 3.1. It consists of a very small structural assembly because the chips involved are also very small: on the order of 300 to 500 μm. Figure 3.1 shows the basic silicon waferboard substrate with its V-groove and alignment pedestals and the basic laser chip with its notch designed to engage with a side pedestal. The front-to-back positioning of the laser chip is controlled by the position of the forward pedestal. The side-to-side positioning is controlled by the laser notch and the position of the side pedestal. The vertical height of the chip above the V-groove is controlled by the stand-offs. Beneath the laser chip is the solder layer used to bond the laser chip.

The figure shows a mechanical passive alignment because the positional registration of the chip is done entirely by the mechanical precision of the etched surfaces that are to be in contact with it. As can be seen elsewhere in this chapter, the etching of silicon can be done controllably and to high precision. The laser notching can also be done to high precision, and the location of the active light emitting spot on the laser is known to high precision because the thickness of the layers grown on the laser are known. The mechanical passive alignment scheme is thus fairly simple and straightforward because all the qualities needed to align the parts are built into the parts themselves.

This provides some key advantages. First, the die-bonder used to place the die can be inexpensive and relatively simple because it does not need a vision system or piezo electric stage for fine positional adjustment. The die-bonding can

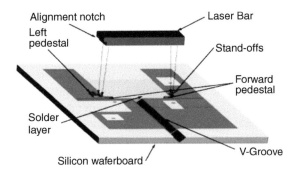

FIGURE 3.1 Mechanical passive alignment structure.

be done quickly because there is no hunting for the optimum optical alignment of a vision target on the die.

In contrast, vision-system passive alignment requires patterned feducial targets on the substrate and on the chips as location references. The alignment requires a special imaging algorithm and fine stage motions to center the targets for positioning the die. This is in part because the alignment tolerance requirement is smaller than the wavelength of light, so improved alignment performance requires a statistical model for studying the fuzzy images to that level of precision. Mechanical passive alignment also provides for a means of setting the vertical placement of the die, using a hard stop, whereas the vision system methods do not. There are no vertical placement positioning feducials to control the height of the chip with a vision system.

3.4 WAFER FABRICATION

To provide for a mechanical passive alignment assembly, it is critical that the parts be made to high precision. Figure 3.2 shows a schematic of the basic fabrication sequence, which uses etching to create mechanical alignment features. Details of the etching are found later in this chapter.

The basic wafer fabrication scheme is to use photolithography to provide the high degree of dimensional control needed. The photolithography is used to pattern

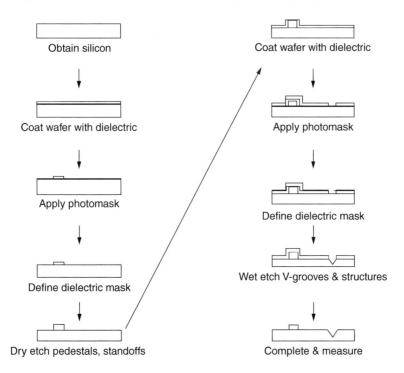

FIGURE 3.2 Silicon waferboard micromachining process.

photoresist, which is then used to pattern dielectric masks that are subsequently used to mask the etching processes. The resist masks by themselves are not resistant enough to the etching processes to survive, so their image is transferred to the dielectric layer. Much of the development work done in creating this process was in investigating how processing parameters affected the replication of the masked image to the actual part. Processes were optimized to minimize undercut or overcut, which would either shrink or enlarge the dimension of the part being etched. Also, parameters were optimized to provide a good etching rate so more parts could be made per unit time.

Most of the micromachining process time consists of etching. Dry etching is used for the standoffs and side pedestals, and wet etching is used for the V-grooves. The wet etch is used because it makes a deep V-groove structure in the silicon wafer that is capable of holding an optical fiber. The regular dry etching process, in contrast, is only capable of etching to a depth of about 12 μm. A special dry etching method known as Bosch etching, developed in recent years, can go much deeper but requires a special reaction that alternates between etching and a mask protection every few seconds. The Bosch process should be considered on a case-by-case basis depending on mechanical requirements. The advantage of the dry-etch process is that it offers vertical side walls to the etched structures, but the wet-etching process is generally less expensive to do, so wet etching should be used wherever possible.

3.4.1 SILICON ISOTROPIC DRY ETCHING

The advancement of microelectronics requires faithful pattern transfer, as microelectronic feature sizes become smaller and smaller. Micromechanics and micromachining are technologies used for forming high-aspect ratio microstructures. Having a high aspect ratio means that the depth or height of the feature is much greater than the size and precision of the width of the feature. These needs cannot be met by conventional wet chemical etching, which typically shows significant undercut and isotropy. Dimensional control, fine pattern transfer, and high-aspect ratio structures have made plasma etching technology more and more attractive. These requirements are also the major driving forces for the development of plasma-based anisotropic etching techniques [2, 3]. Plasma etching technology, also commonly called dry-etching technology, has attracted lots of attention in the field of microelectronics, micromechanics, MEMS, and micro-opto-electro mechanical system [4, 5]. Another attractive feature of plasma-based dry etching is its compatibility with other vacuum-based processes such as chemical vapor deposition (CVD) and molecular beam epitaxy (MBE) [2]. Plasma-etching technology has many years of history. There are numerous articles, review papers, and books talking about the fundamentals of plasma-etching technology [2–17]. We do not intend to emphasize these fundamentals here but encourage the readers to read these references for better understanding.

3.4.1.1 Plasma Etching

Plasma etching technology uses a gas glow discharge to dissociate and ionize gas species (relatively stable molecules), forming chemically reactive and ionic species, which react with the materials to be etched to form volatile products,

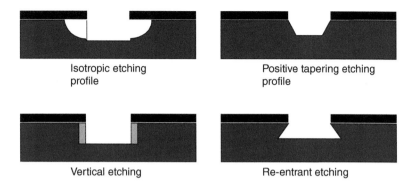

FIGURE 3.3 Typical plasma etching profiles.

such volatile products are pumped away by a vacuum system. Plasma etching may also form energetic ion beams to physically sputter off and remove the material from the substrate.

One of the major advantages of plasma-based etching is that it can provide a considerable degree of profile control. This includes negative tapering profile (also called reentrant profile), vertical profile, positive tapering profile, and isotropic profile (see Figure 3.3), depending on the etching system configuration, etching chemistries, and etch parameter settings. Generally speaking, there are three types of plasma-etching configurations: plasma etching (PE), ion beam etching (IBE), and reactive ion etching (RIE).

PE, or radical etching [3, 18], is a purely chemical etching. It features isotropic etching with minimal ion bombardment. Such systems are usually used for photoresist stripping and other applications in which high selectivity and low radiation damage are key requirements, and the isotropic nature of the etch is not a problem — or may even be an advantage. Normally, the plasma-etching mode is operated at relatively high pressure (say 1 torr), where plasma potential and resulting ion bombardment will be low.

Ion beam etching [19–21], or ion milling, is primarily physical etching resulting from a directed energetic ion flux. The selectivity is low, almost 1:1. In some applications (e.g., fabrication of a refractive lens), the 1:1 selectivity is desired. To prevent redeposition of etched products, the etching is usually operated at very low pressure (< 1 mtorr). Because of this low pressure, a plasma is normally created in a remote, small, high-pressure chamber, called a Kaufman source, and the ions are extracted from this plasma with electrostatically controlled grids and directed to the substrate to be etched [3]. Because of the nature of physical etching, the etch rate of ion beam etching is usually low, typically 1 to 30 nm/min. To increase the etch rate, an extra reactive feed gas is introduced into the reactor. Such a process is often called chemical-assisted ion beam etching or chemical assisted reactive ion beam etching. In terms of the etching profile, the IBE features a positive tapering profile [22].

Reactive ion etching, often called RIE, is one of the most popular plasma etching configurations [2, 23]. The wafer to be etched is placed on the RF-driven

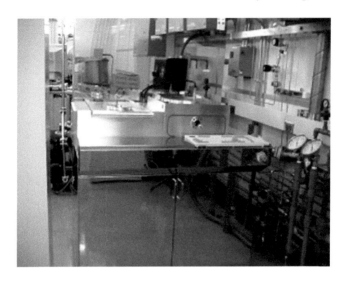

FIGURE 3.4 A typical RIE etching system.

electrode. Figure 3.4 illustrates the schematic of typical RIE system. Because the mobility of electrons in the discharge is much greater than that of the ions, more electrons initially reach the wafer electrode. In the meantime, the coupling capacitor prevents any DC current flow, and a negative charge is accumulated on the wafer electrode [2]. The RIE configuration uses a negative self-bias DC voltage developed between the plasma and the wafer electrode to accelerate ions from the plasma to the wafers. Etching is attributed to both the chemical reaction at the wafer surface and the physical removal of the material by ion bombardment (sputtering) [11]. The great feature of RIE etching is the anisotropy. Although the selectivity may not be as high as plasma etching, and the surface radiation damage may be a concern because of the high self-bias voltage (typically 300 to 500 V), it has been widely used because of its etching anisotropy and its profile control. The typical operation pressure in RIE mode is typically in the range of 10 to 200 mtorr [2].

In recent years, because of the high etching throughput and low radiation damage requirements for microelectronics, and the high aspect ratio and deep vertical microstructures for MEMS applications, high-plasma density etching techniques have advanced dramatically [24]. Among these techniques, electron cyclotron resonance (ECR) and inductively coupled plasma (ICP) are widely used [25, 26]. In the ECR system, a plasma discharge is generated by microwave excitation (commonly 2.45 GHz). ECR is operated in a low-pressure regime (< 10 mtorr). Compared to conventional RIE and an ICP system, the ECR system is very complicated. It consists of microwave generator, waveguide, dummy load, recirculator, and so on. To tune the microwave, a three-stub tuner is usually used. Because of the complexity of an ECR system, it costs more to build it. In addition, because the chamber walls for the ECR system are made of metal material (typically stainless steel), it leads to a diverging field issue (i.e., the electrons could be grounded to the chamber walls, which results in an etching nonuniformity). Finally, the ECR system is difficult to scale with the newer larger wafer sizes. In comparison, an ICP system has its inductive source mounted on a standard RIE system platform and consists of a modified chamber top, which contains the inductively coupled plasma source. In addition to the source, there is a RF power supply and an impedance-matching network. The RF frequency powering the coil is adjustable from 1.7 to 2.1 MHz. This frequency is kept low to reduce the RF impedance of the coil, allowing for a larger current to flow, but one above the average ion transit frequencies to avoid direct acceleration of ions by the RF energy. The lower electrode is powered as in a conventional RIE system at 13.56 MHz to allow for independent control of the substrate direct current bias voltage. Compared to an ECR system, the ICP system has a simpler source design, lower-cost components, fully automatic impedance matching, higher effective pressure range of operation, and more flexible source designs.

3.4.1.2 RIE Etching Chemistries

For RIE etching of silicon, the etchant gas has to meet two requirements. First, it has to chemically react with the material being etched, and second, its etching product has to be volatile and can be pumped away. Hydrogen-based and halogen-based (i.e., F, Cl, Br) plasma can be used for the RIE of silicon, and the etch products SiH_4, SiF_4, $SiCl_4$, and $SiBr_4$ are volatile. Halogen-based gases such as CHF_3, CF_4, SF_6, Cl_2, HBr, and mixtures of these gases with O_2, H_2, Ar, and He have been used for the etching of Si and SiO_2 [1]. Generally speaking, the fluorine-based plasmas are used for isotropic etching. To achieve the anisotropic profile, an etching inhibitor such as oxygen is often introduced ($SF_6/O_2/CHF_3$) [3, 27]. In fluorine chemistry, the chemical reaction of F atoms with Si surface atoms is spontaneous. In contrast, Si and SiO_2 are not etched spontaneously by chlorine atoms [14, 28, 29] or bromine atoms [30]. The chlorine- and bromine-based RIE is an ion-induced etching, where accelerated ions modify the surface reactions in one way or another (e.g., chemical sputtering, chemically enhanced physical sputtering, or lattice damage) [2, 3, 18] to make the radicals react with Si and form volatile compounds $SiCl_4$ or $SiBr_4$. The masking materials are very important for the dry-etching process. SiO_2 and metals such as Cr, Au, Ag, and Al can be used as etching masks depending on the etching chemistries used. In general, an etching mask will affect the silicon trench etching profile because the mask is retarding when its profile is not fully vertical (see following section).

The etching of silicon is a fluorine-based chemistry and normally results in large undercut of the mask. With Br-based RIE [31] the selectivity to masking material SiO_2 is higher [32–34], but it is a toxic and corrosive process and is difficult to handle. Chlorine-based RIE [35–37] still has some safety and handling concerns. In the following section, we will report about our silicon RIE etching using chlorine-based chemistry.

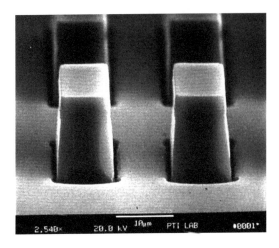

FIGURE 3.5 SEM micrograph of RIE etched silicon posts.

3.4.1.3 RIE Etched Silicon Structures for Passive Alignment

The advancement of silicon micromachining and MEMS technology contributed to the study and development of silicon dry-etching techniques. Unlike planar processes in conventional VLSI or ULSI technology, bulk micromachining processes deal with real three-dimensional microstructures. Generally speaking, the vertical etch depth of bulk micromachining is in the range of from tens of micrometers to submillimeter size, depending on the application. In addition, the etch profile is a major concern, where anisotropic etching (i.e., vertical structures) is desirable for many applications including silicon waferboard technology. Researchers have been studying a variety of plasma system configurations [27, 38].

In this chapter, we report our results based on the conventional RIE etching technique. As Figure 3.1 suggests, etch anisotropy is extremely important in achieving submicron accuracy for single-mode passive alignment. Figure 3.5 is a scanning electron micrograph (SEM) of typical RIE-etched silicon pedestals showing a slightly tapered profile using chlorine-based chemistry. High-precision single-mode passive alignment of optical fibers to devices such as semiconductor lasers requires a more vertical profile. The taper was not caused by any under-cutting, as the etching is an ion-induced process. In fact, we found that the profile of the masking material produced the taper because of its limited etch selectivity. Figure 3.6 is an SEM of the SiO_2 mask, which is used for silicon etching. The mask profile was effectively transferred to the silicon, producing the profile shown in Figure 3.5. Once the profile of the mask was improved, silicon posts with excellent etch profiles were readily obtainable and are demonstrated in Figure 3.7.

Because the chlorine chemistry is very sensitive to moisture, the RIE system we developed for the etching processes has a load lock (Plasmatherm SLR 720 shuttle lock RIE etching system). The etching parameters are Cl_2 flow rate, 30 sccm; pressure,

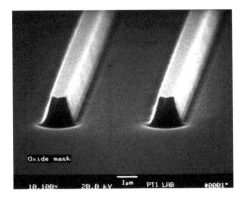

FIGURE 3.6 SEM micrograph of SiO_2 mask used for silicon etching having poor vertical side walls.

15 mtorr; RF power, 175 W. At this etching condition, the Si etch rate is about 0.1 μm/min, and the selectivity to silicon dioxide is about 10:1. The etching mask we used is thermally grown SiO_2 with thickness of 1.2 μm. Because the etching selectivity of silicon to silicon dioxide is not very high (10:1), the etching profile of the masking

FIGURE 3.7 SEM micrograph of RIE etched silicon microstructures.

material (SiO_2) is extremely important. To ensure the anisotropy of SiO_2, we used chrome as an etching mask instead of a typical photoresist. Figure 3.6 shows the bad result of using photoresist as an etching mask for SiO_2 etching. Note that the side walls are not vertical in the formed SiO_2 mask. Chrome is a better mask for this than photoresist. To pattern chrome, Cl_2/O_2 chemistry-based RIE etching technique was implemented (Plasma Therm SLR 720 shuttle lock RIE etching system). After the patterning of chrome, we used fluorine-based chemistry (CHF_3/O_2) to etch the thermal oxide using chrome as the etching mask. The system we used for the oxide etch is a Plasma Therm 700 series PE/RIE system. The parameter settings are CHF_3/O_2 flow rate, 47.5/2.5 sccm; etching pressure, 20 to 25 mtorr; and RF power, 180 watts.

In addition to the improvements we made in the silicon etch profile, we eliminated the use of polyimide standoffs because of reliability concerns, process yield, and height control. Polyimide was originally used to set the height of the chip above the silicon waferboard. Through additional masking steps, both the alignment pedestals and the standoffs could be formed into the silicon substrate. Figure 3.8 shows bilevel silicon posts, where the 5-μm-high post is the chip height standoff, and the other, 10-μm, posts are for lateral alignment of the chip. The formation of multilevel silicon structures using RIE may have many other applications in the micromechanical and micro-optic fields.

3.4.1.4 Passive Alignment Structures and Accuracy

To develop a process for product manufacturing, the etching repeatability and uniformity are the most important issues. Wafers were patterned and separately etched to determine etch rate repeatability and etch uniformity of the RIE etch

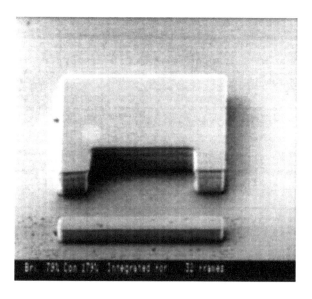

FIGURE 3.8 SEM micrograph of RIE etched bi-level silicon microstructures.

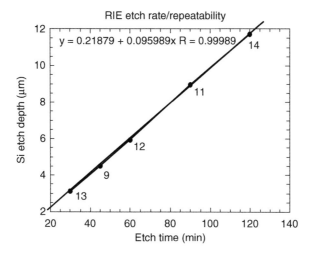

FIGURE 3.9 Plot of etch depth vs. etch time.

process. Figure 3.9 shows the RIE etch depth vs. etch time for five wafers. Note the linear relationship indicating a constant etch rate of 0.1 μm/min over the timescale shown.

Figure 3.10a to Figure 3.10c show the uniformity of depth control across 3-in. wafers for etch times of 30, 45, and 60 min, respectively. The corresponding

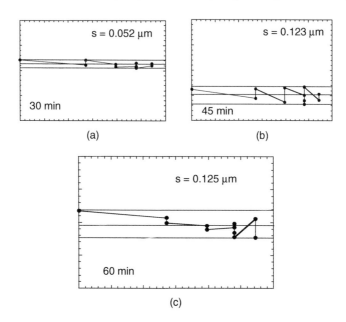

FIGURE 3.10 Plots of silicon post height vs. distance across wafer for etch times of (a) 30 minutes, (b) 45 minutes, and (c) 60 minutes.

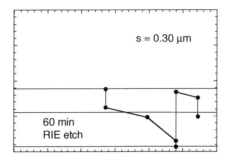

FIGURE 3.11 Plot of undercut from nominal feature width vs. distance across a 3-in. wafer.

standard deviations for these times in the etch depth uniformity across each wafer are 0.05, 0.12, and 0.13 μm, respectively. The nominal standoff height used in our silicon waferboard process is 5 μm; thus, the RIE-etched silicon standoff uniformity exceeds the requirement for single-mode passive alignment.

RIE-etched silicon forward-posts and left-posts, as shown in Figure 3.1, are the planar locators for the edge of the laser bar in the lateral plane. Single-mode passive alignment requires that these structures be vertical to match the notch on the laser. Figure 3.11 shows the variation in width of silicon alignment posts resulting from the RIE etch process across the wafer.

The average width change across the wafer was about 0.5 μm, with a standard deviation of about 0.3 μm. This result indicates that the alignment control of 0.25 μm was achieved because only one edge of the alignment pedestals are used for mechanical registration.

3.4.2 SILICON ANISOTROPIC WET ETCHING

Anisotropic etching of silicon is a ubiquitous process in micromachining. Perhaps the most prolific application of silicon anisotropic etching is the formation of silicon V-grooves for optical fiber devices, as described by Kurt E. Petersen [39]. Silicon V-grooves fabricated by anisotropic etching are especially ideal for precise placement of delicate, small-diameter optical fibers, and their etching is one of the most critical processes for passively aligning optical fibers with edge emitting lasers in silicon waferboard technology (SWT) or silicon optical bench.

Although the exploration of V-groove etching technique was started in 1969 [40] and the utilization of V-grooves for precise alignment of optical fiber was started in 1970 [41–44], the study of the V-groove etching technique is still ongoing in attempts to improve reproducibility. Various projects related to the silicon anisotropic etching have been studied during the last 30 years, including the development of anisotropic etching solutions [41, 45, 46–81], the effect of the crystal orientation of silicon [82–94], the etching inhibition on boron heavily doped silicon [95–100], the corner undercut and compensation [40, 52, 101–105], etching defects [52, 68, 69, 106–112], and so on. The following sections will deliberate the topics, which are closely related to the V-groove etching used in SWT.

3.4.2.1 Etchants for Silicon Anisotropic Etching

Three categories of etchant systems are available for silicon anisotropic etching: ethylenediamine, pyrocatechol, and water (EPW) solutions; inorganic aqueous alkaline solutions including KOH, NaOH, NH_4OH, and so on; and tetramethyl ammoniumhydroxide (TMAH) solutions. The proposed mechanism of silicon etching with all the above etchants assumes that OH ions are the prime reactants at the silicon surface. The fundamental reaction process is the oxidation of silicon to form hydrated silica, followed by the dissolution of the oxidation products.

3.4.2.2 EPW Solutions

The EPW etchant system was first disclosed in 1962 [46], using hydrazine, catechol, and water solutions. Later on, the etchant system was modified to use ethylenediamine instead of hydrazine because of the higher stability and lower toxicity of the chemical. Because hydrazine NH_2NH_2 and ethylenediamine $NH_2(CH_2)_2NH_2$ belong to the same chemical group of amine, their behaviors toward silicon etching are similar. The major advantages of this etchant system are the very slow etching rate for the SiO_2 mask and that the etchants are compatible with an integrated circuit process. The ratio of etching rates of (100) oriented Si wafers with SiO_2 masks in an EPW solution at 90 °C can be as large as 9000:1, according to the data published by Seidel et al. [66].

Finne and Klein [47] were the first authors to study the chemical reaction of silicon etching in EPW solutions. Their experimental results indicate that water and amine are the necessary components for the etching process. Both components react with each other to produce OH ions. The hydroxyl ions, together with water, then oxidize the silicon to form hydrous silica $Si(OH)_6^=$ with the evolution of hydrogen. There is a peak observed in the etch rate curve as the function of the etchant concentration. The maximum was observed for the etch composition corresponding to a mole ratio of water to ethylenediamine of approximately two. The pyrocatechol acts as a complexing agent to create chelation reaction in the process. The hydrous silica is converted to the complex pyrocatecholate, which is soluble in the amine solution, causing an increase in the etch rate by a factor of three. The etching behavior of EPW solutions as a function of trace concentrations has also been examined by Reisman et al. [58]. It was found that the trace quantities of pyrazine $C_4H_4N_2$ inside ethylenediamine led to an increase of the etch rate. The enhancement effect of the etch rate at a trace concentration range for pyrazine up to a few thousand parts per million is very powerful, whereas above this range it becomes more gentle. The etch rates as a function of pyrazine fit well to a linear relationship in a log–log plot. It was suggested that the pyrazine might function as a catalyst in the reaction. The addition of pyrazine into EPW solution up to 6 g/l of ethylenediamine was suggested to make the solution less susceptible to small variation in pyrazine content. It was also observed that the addition of pyrazine into EPW would improve the surface smoothness and the reproducibility of the silicon etch rate and would provide an excellent resistance to oxygen exposure. The effects of additives were also reported by others [72, 76]. Several recipes were proposed for the EPW etchant system, as shown in Table 3.2.

TABLE 3.2
Composition of Different Solutions Published by Finne and Klein [47], Reisman et al. [58], and Bassous [55]

Type,	S [58]	F [58]	B [55]	T [47]
Water (ml)	133	320	320	470
ED, l	1.0	1.0	1.0	1.0
Pyrocatechol, g	160	320	160	176
Pyrazine, g	6	6		

Adapted from H. Seidel et al. [66].

The S solution was developed for silicon etching at slower etch rates or at lower temperatures. The F solution was developed for use where high etch rates were required. The B solution was presented by Bassous, and the T solution was presented by Finne and Klein. In general, the etching temperatures of single-crystal silicon in EPW solution range from 100 to 115°C. A glass or quartz reflux bath is necessary, with a boiler at the bottom and a condenser in the top. A small amount of Ar or N_2 flow through the solution during the etching also protects the solution from an "aging" effect.

During deep etching of silicon in EPW solution, there is a tendency for undesired residue formation on the silicon surface, which inhibits achieving smooth etching surfaces, as reported by several authors [40, 47, 58, 64]. Two kinds of precipitations appear during silicon etching in EPW solutions. One is a reaction product that is soluble in water, with a crystalline appearance. The other is a residue that is insoluble in water and that has the appearance of a white spongy substance. The occurrence of residues depends on the composition of the solution, on the saturation level of the solution with silicon, and on the temperature of the silicon etching.

Reisman et al. [58] suggested that a higher oxidation rate compared to the complex-forming rate might cause the residue formation. The authors observed a tendency for residue formation with increasing pyrazine content, and a reverse tendency, eliminating residue formation, with the addition of more pyrocatechole in the solution. Wu et al. [64] did measure the solubility of silicon in EPW solution as a function of temperature and found that a maximum amount of 8 g silicon could be etched without producing residue in the F solution containing 1 l ethylenediamine at the temperature of 95°C. It was also found that the oversaturated EPW solution could still etch silicon, but the white residue would appear in addition to the reaction product. A chemical analysis of the residues indicates that they consist mainly of SiO_2 mixed with a trace amount of reaction products. The tendency of residue formation increases with the reduction of the reaction temperature, as the solubility of silicon in EPW solution is lower with a reduced temperature.

One of the benefits of using EPW solution is that it achieves a very flat etched surface [40, 52, 67, 64], which is necessary for processing a high-quality silicon diaphragm in the microsensor industry. It has been observed that the quality of the etched surface might be affected in two ways: first, by the formation of pyramidal hillocks, and second, by the formation of a wavy-textured structure. The quality of etch surfaces was dependent on the temperature and the solution concentration. Declercq et al. [54] observed both hillocks and wavy surfaces in a half-half volume etching solution of water and hydrazine at any temperature. Less water in the etching solution at 100°C improves the flatness of the etched surfaces. Their study indicates that no hillocks and no wavy surface were achieved when etching in a solution with a water concentration of 30% or less at a temperature of 100°C. EPW is a good etchant for producing silicon V-grooves with smooth and mirror-like sidewalls, except for the hazardous nature of the chemicals.

3.4.2.3 Inorganic Aqueous Alkaline Solutions

Inorganic aqueous alkaline solution is also an anisotropic etchant for crystal silicon. In general, the etching temperatures of single-crystal silicon in aqueous alkaline solution range from 65 to 85°C, which are lower than the temperatures used for silicon etching in EPW solutions. This etchant system was presented early in 1959 [45]. Among the aqueous alkaline solutions, the one most frequently used is based on KOH solutions. Price [50] did a systematic study in ternary mixtures of KOH, H_2O, and isopropyl alcohol (IPA) systems. The KOH and water are the necessary components of the etching process in this etchant system. In general, the amount of IPA is oversaturated in the KOH etchant for process simplicity and forms a layer on top of the solutions. Because the addition of IPA causes a decrease in the etch rate, IPA can not act as a complexing agent but, rather, as a moderator [40] or as a softener [103] of the reaction. In diluted KOH solution the etch rate of silicon increases with the raising of KOH concentration. For a high-concentration solution, the etch rate of silicon decreases with the raising of KOH concentration. There is a broad peak in the etch rate curve as a function of KOH concentration. On the basis of the discussion of the mechanism of silicon etching process in KOH solution, Seidel et al. [66] presented an assumption that four water molecules were necessary for the dissolution of one silicon atom. Using this ratio, a curve fit for the silicon etch rate over the full range of KOH concentration was obtained by taking $R \sim [H_2O]^4[KOH]^{1/4}$. The peak of the etch rate curve with KOH concentration was found to be about 20% wt for (100) and (110) silicon wafers, which was consistent with Kendall's results [88]. With the addition of IPA, the etch rate peak shifts to a concentration of around 30% wt for (100) silicon etching and to a concentration of over 45% wt for (110) silicon etching [50]. The major purpose of the addition of IPA in aqueous alkaline solution is to smooth the etched surface and to reduce the undercut of the etched pattern, such as the sidewalls of V-grooves.

Strandman et al. [75] did compare the roughness of the etched groove wall by measuring the top-to-top value in KOH/IPA and KOH solutions. The roughness of the groove wall was about 10 µm when etching, with approximately 51% KOH solutions without IPA. The roughness reduced to about 0.3 µm when etching in the same concentration of KOH solution with IPA. Furthermore, in aqueous KOH/IPA etchant solution, the surface roughness of the sidewalls was reduced with decreasing KOH concentration, but the risk for hillock formation increased. Findler et al. [71] investigated the surface roughness Ra of (100) silicon etching in KOH without IPA solutions. It was found that the best results of surface roughness were obtained for 30% KOH solution. As an anisotropic etchant, the smallest etch rate is always in the (111) orientation, whether it is with or without IPA. In aqueous KOH solution with IPA, the (100) etch rate is faster than the (110) etch rate. In the case of KOH solution without IPA, the etch rate of (110) becomes larger than that of (100). The aqueous KOH solution without IPA is prefered during deep groove etching on (110) silicon wafers [48, 83, 85, 94], as the large differential etch ratio between (110) and (111) permits deep, high-aspect ratio grooves with minimal undercutting of the masks.

Although the aqueous alkaline solution was introduced earlier as an anisotropic etchant of silicon and the chemicals are not hazardous, EPW solution was the first solution widely adopted in the industry to produce micro-silicon sensors [59–62], micro-ink jet [56], and other micromechanical parts [39, 58, 63]. Several concerns for the inorganic alkaline solutions have been discussed in the literature. One is related to the etch rate of SiO_2 in alkaline solutions. This is important because SiO_2 is often used as an etching mask.

Seidel et al. [66] compared the SiO_2 etch rate in both EPW and KOH solutions. It was found that the SiO_2 etch rate in KOH solutions was about two orders of magnitude larger than that in EPW solutions. The etch rate ratio between (100) silicon wafers and SiO_2 becomes about 420 in 20% KOH solution at 80°C. For highly concentrated KOH solutions at temperatures typically used for etching, this ratio would reduce to about 100 to 200, which is usually not enough for the deep etching [39, 56]. In the case of prolonged etching, the LPCVD Si_3N_4 film is a preferred masking material for KOH. Another concern is the flatness of the etching surface in KOH [71, 80]. Wavy-textured structure is frequently seen on the (100) etched surface, even in a KOH solution with IPA, when the etch hillocks are disappeared. Another concern might be the compatibility of KOH etching with the IC process [64, 69]. With the improvement of etching techniques, more people are using aqueous KOH solution to perform silicon anisotropic etching, especially in V-groove etching.

Aqueous ammonium hydroxide NH_4OH solution is a unique etchant in inorganic alkaline solutions. The same type of solution with the addition of H_2O_2 is broadly adopted as a cleaning solution of silicon wafers in IC process. Schnakenberg et al. [67, 68] investigated the compatibility of NH_4OH silicon etching with the IC process. The etch rate of thermally grown SiO_2 in NH_4OH was found to be one magnitude less compared with the values published for KOH solutions. Aluminium metallization would not be attacked in NH_4OH solution containing dissolved silicon with concentrations larger than 0.1 g/l. The addition of H_2O_2 in a concentration range between 0.65×10^{-2} and $1.84 \times 10^{-2} M$ would increase the etch rate of (100) silicon to 75 μm/h, which was about a factor of 2.5 higher compared with the NH_4OH solution without H_2O_2 addition.

3.4.2.4 TMAH Solutions

TMAH as a silicon anisotropic etchant was developed during the 1990s [69–71, 78, 81]. The motivation for this came from the integration of sensors and actuators with IC devices. TMAH is of great interest because of the absence of alkaline metal ions, the very low etch rate of SiO_2, its nontoxic character, and the possibility of passivating aluminum metallization. This allows anisotropic silicon etching with TMAH to be an IC-compatible process that can be employed in IC fabrication lines. In fact, TMAH was already used as a developing solution for positive photoresist in the IC industry. Tabata et al. [69] investigated the anisotropic etchant of TMAH within the concentration range of 5 to 40 wt% over a temperature range of 70 to 90°C. It was found that the etch rates of (100) and (111) decreased with increasing concentration linearly in the above concentration range, and that the etch rates increased with the raising of the temperature. At the same time, Schnakenberg et al. [70] reported that a maximum (100) silicon etch rate of 39 μm/h was observed when the TMAH concentration ranged up to 2 wt%. The authors also measured the dependence of SiO_2 etch rates on concentration and temperature and found that the ratio of etch rate of (100) Si to SiO_2 was about 5000, calculated from the measurement results. The compatibility of TMAH etchant with conventional IC processes was also investigated [70, 78].

The etch rates of various dielectric layers including LPCVD silicon nitride, PECVD silicon oxide, PECVD silicon nitride, and thermally grown SiO_2 in TMAH etchant were measured. The results showed that the etch rates of dielectric films increased with increasing temperature and with decreasing TMAH concentration. The LPCVD Si_3N_4 had a negligible etch rate, whereas the PECVD oxide had the highest etch rate within the investigated masking materials. The etch rate ratio of (100) silicon to PECVD oxide was about 1000. That meant that the PECVD oxide was still a good masking material in TMAH solutions, even compared to the thermally grown SiO_2 used as a masking material in KOH solutions. It was also found that the TMAH solution successfully passivates the aluminum metallization when the solution is doped with silicon.

Flatness of etched surfaces is one criterion of good etchants for silicon anisotropic etching. Findler et al. [71] presented measurement results for the surface roughness, Ra, of (100) silicon planes during anisotropic etching in both KOH solutions and TMAH solutions. The researchers found that the preclean process for silicon wafers had a large effect on the roughness of etched surfaces. Dipping silicon wafers into a buffered HF solution would make etched surfaces smoother. Their results also indicated that the TMAH-based etchant would provide a superior-quality silicon surface as compared to KOH. As discussed above, the flatness of etched surfaces is affected by two factors: formation of hillocks and surface roughness. It is a unique characteristic of TMAH etchant that the hillock density and surface roughness decrease with the raising of the TMAH concentration [69, 71, 78, 110]. TMAH is therefore supposed to be an ideal etchant for silicon anisotropic etching, including V-groove etching, for producing mirror-like etched surfaces.

3.4.2.5 Crystallographic Orientation Effect

Anisotropic etchants, by definition, exhibit a strong dependence of their etch rates on the crystallographic orientation of the single crystal silicon. Numerous works have engaged in the study of this dependence [40, 63, 66, 77, 82, 84, 90–92]. All measurement results indicated there was a sharp minimum at silicon (111) planes in addition to two relative minima around silicon (100) and (110) planes. However, the fabrication of three-dimensional microstructures relies on the knowledge of etch rates of the low-index planes as well as of those of the higher-index planes. The etch-rate ratio for the three main silicon planes, (100), (110), and (111), becomes a feature of general interest for many practical applications of silicon anisotropic etching. This value varies from lecture to lecture, depending on the etchant solutions, the etching temperatures, the etching conditions, and the tolerance of the measurements.

Seidel et al. [66] investigated the anisotropic etching behavior of silicon, in EPW solutions as well as in aqueous KOH solutions, extensively, based on the measurement of wagon-wheel patterns. In EPW solution type S, the etch-rate ratio of silicon (100), (110), and (111) was found to increase from 30:30:1 near the boiling point of the solution to about 100:150:1 at 50 °C. A similar behavior was observed for KOH solutions. The value for this ratio was found to vary from 30:50:1 at 100 °C to about 100:160:1 at room temperature. Further, when isopropyl alcohol was added to a KOH solution, a decrease in the etch rates of about 20% for (100) and 90% for (110) was observed. As a result of the stronger decrease of the etch rate on (110) planes, the etch-rate ratio of (100):(110) was reversed. Earlier, Kendall [83, 85, 88] did a series of studies on silicon anisotropic etching with (110) wafers in KOH solutions while producing narrow grooves with vertical walls in silicon. On the basis of the measurement of the mask undercut to groove depth, it was found that the etch rate ratio of (110) to (111), as high as 400:1, could be obtained in 44 wt% KOH solution at 85 °C, which was much larger than reported above. An anisotropic etch ratio between (100) and (111) of 50:1 was reported in a TMAH solution with a concentration of 2 wt% and a temperature of 80 °C [70]. A reverse temperature effect on the anisotropic ratio in KOH and TMAH solution was suggested by Clark et al. [65] and Tabata et al. [69], based on their activation energy measurements. Their results will be discussed in the following section, which discusses the temperature dependence of etch ratio. It is obvious that there is a large discrepancy in the discussion of anisotropic ratio.

Because there is a sharp minimum of silicon etch rate at the {111} planes, the etch rate in the vicinity of the {111} planes is extremely sensitive to small angular misalignments. The misalignment of mask patterns with crystallographic orientation has a serious effect on deep-groove etching in (110) silicon wafers and on V-groove etching in (100) wafers [74, 83, 84, 88, 89, 93, 94]. Some studies of the misalignment effect on V-groove etching have been conducted in our lab during the course of SWT development.

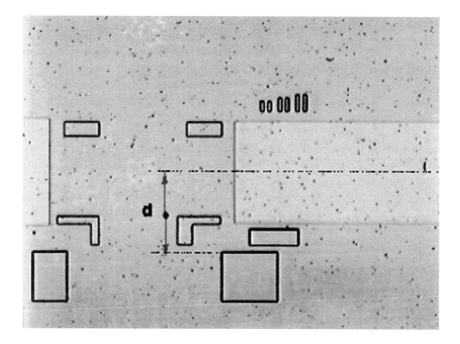

FIGURE 3.12 The layout near the front of the V-groove The distance d represents the critical distance from the emission point on the laser to the front side pedestal.

The silicon substrates are often adopted as carriers to package and passively align laser diodes (LD) and fibers with silicon waferboard technology (SWT). There are alignment side pedestals for laser diode positioning, and V-grooves for fiber placement. The distance (d) between the V-groove centerline and the side pedestal (as shown in Figure 3.12) is one of the critical parameters for the passive alignment of SWT, which should be controlled with alignment tolerances of ±1 μm [113]—the same accuracy as the fiber height control.

Various factors affect the control accuracy of this parameter, such as the size reduction of side pedestals during photolithography, the amount of undercut during dry etching of side pedestals, the alignment accuracy of V-groove patterns, and the V-groove wet etching. Here the discussion will be limited in the topic of the V-groove etching. Obtaining uniform undercuts during V-groove etching is one of the critical processes for SWT. Various nonuniform undercuts have been observed during V-groove etching that are related to the crystal orientation misalignments. Two sources could cause the crystal orientation misalignments: the V-groove being off of the [110] crystal direction, and the wafer orientation being off the {100} crystal plane.

The principle of V-groove etching in silicon is that {111} surfaces are attacked at a much slower rate than all other crystal planes in the anisotropic etchants. When rectangular strip openings are aligned along the [110] direction of the (001) wafer coated with silicon nitride, V-shaped grooves will be created by anisotropic etching. The lateral sides of the V-grooves are enclosed by {111} crystal planes, which have the lowest etch rate. The width of the V-grooves can be well controlled, as can the fiber height placement. In normal processing, misalignment of the V-groove mask compared to the crystal orientation is not perfectly avoidable. When this happens, a nonuniform undercut is observed along V-grooves. This phenomenon is caused by the fact that the sidewalls of the V-grooves will conform to the {111} crystal planes during etching. This makes the centerline of the V-grooves twist on an angle compared to the silicon nitride mask and changes the controlled distance between the V-groove centerline and the side pedestal. This change reduces the coupling efficiency between the lasers and the optical fibers when used for passive optical alignment. In the case of V-grooves with both ends closed, the twist is around the center point of the V-groove. In this case, the equivalent variation of the distance (Δd) between the V-groove centerline and the side pedestal depends on the length of the V-grooves (l) and the amount of misalignment (θ).

$$\Delta d = 1/2 \times \theta \times l \tag{3.1}$$

Figure 3.13 shows that the equivalent deviation (Δd) in this distance is directly proportional to the misalignment angles of V-grooves with the V-groove length of 2 mm. As shown in Figure 3.13, only a 0.06 degree misalignment causes a 1-µm shift in the V-groove centerline on the front side. Because a prealignment method can accurately align the mask to the crystallographic orientation within a precision better than ±0.05 degrees [89, 93], the misalignment caused by the V-groove being off the [110] crystal direction can be reduced.

The situations become more complicated if the wafer surfaces are off the (001) crystal planes. Figure 3.14 schematically shows this effect. AB and CD are the edges of V-groove sidewalls on the wafer surface of the (001) plane. These two lines are parallel with each other and along the [110] orientation. If the wafer surface tilts a small angle of α along the V-groove direction, the intersections of

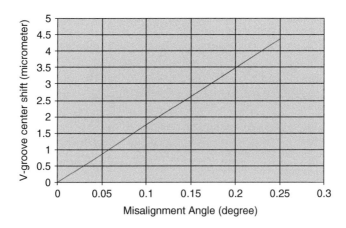

FIGURE 3.13 The deviation of the V-groove center position with the misalignment.

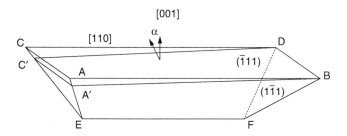

FIGURE 3.14 Schematic of V-groove with wafer tilted off the (001) direction.

the V-groove sidewalls with the wafer surface become A′B and C′D. Then these two intersections will not be parallel anymore. Simple calculation indicates that the angle between these two lines will approximate 1.4 α.

Another kind of nonuniformity of undercuts will be created depending on the alignment. Let us discuss a special case in which the V-groove window is aligned with the C′D direction. After etching, the undercuts along the back side of the V-groove will be small and uniform, and the undercuts along the front side will become larger and nonuniform along the V-groove. In that case, the fiber will not only twist, as discussed above, but will also tilt from the silicon wafer surface.

3.4.2.6 Etching Hillocks

The pyramidal defects of etching hillocks appear when (100) silicon is etched in a variety of anisotropic etchants, including EPW, inorganic alkaline, and TMAH solutions [52–56, 67, 68–69, 78, 80, 106–112]. Hillocks are usually reported as being pyramidal protrusions from {100} surfaces, bounded by {111} planes [107] or near {111} planes [108–110]. A general rule was presented for anisotropic etching that a concave surface was bounded by slow-etch surfaces, whereas a convex surface was bounded by fast-etch surfaces [114]. It was found that the corner of a mesa on silicon (100), which was originally bounded by {111} planes, would be beveled by relatively fast-etch surfaces including {211}, {212}, {331}, and {411} [40, 54, 101–105]. The fact that hillock formation can be stable during etching would appear to challenge conventional understanding of anisotropic etching. Besides the theoretical interest, the hillock formation is of critical importance in practice. Hillocks would cause a partial or total blockage of micronozzles fabricated by silicon anisotropic etching [56]. The etching hillocks created within V-grooves not only deteriorate the positional alignment accuracy but also may cause mechanical damage to the fiber [109].

Etching hillocks are pyramidal in shape, with the sides crystallographically related to the surface orientation. At low magnification, the hillocks appear as regular pentahedrons composed of four lateral {111} crystallographic planes resting on the {100} base plane. The square bottom edges of hillocks seem to be parallel to the <110> cleavage edges, which are formed by the intersection of {111} planes and the base plane. However, close examination under high magnification, using SEM, reveals a more involved defect structure. Each side is composed of two planes that are at a slight angle off the {111} family. After careful examination, Tan et al. [108] reported that these eight lateral sides were {567} planes for hillocks formed during etching in KOH/2PA solutions. Landsberger et al. [110] reported that these lateral sides were near {578} planes for hillocks formed during etching in TMAH solutions. Furthermore, they depicted the lateral sides of the hillocks as being formed by symmetrically overlapped serials of {111} planes of smaller size with intact <101> ledges as boundaries. In this scenario, the lateral sides of hillocks were composed of these symmetrically arranged <101> ledges, parallel to each other. Based on the discussion of the energy required to remove an atom from a kink and a ledge, Landsberger et al. [110] presumed that the etch-rate ratio of {101}/{100} < 1 appeared to be an indicator of hillock-producing conditions. One hypothesis of this model was that the apex of the hillock was protected by something such as impurities or defects. On the basis of the measurement of the activation energy (~1.2 eV) from an Arrhenius plot of hillock density vs. reciprocal temperature and other experimental results, Tan et al. [109] suggested that the something on the apex of the hillock might involve a regrowth mechanism. Further study is necessary for the full understanding of the hillock formation mechanism.

The prevention of etch hillocks is a primary objective of anisotropic etching processes. Various factors affect the formation of hillocks. These include etchant composition, etchant concentration, etch temperature, etch agitation, and etchant additives. In KOH/IPA solutions [109], a hillock density is strongly reduced with higher KOH concentration and lower etch temperature. The agitation of the solution also reduces total hillock density by an order of magnitude. When the KOH concentration is less than 10%, the density of etch hillock is extremely large, resulting in a very rough surface. Increasing the KOH concentration sharply decreases the hillock density, which reached zero for concentrations exceeding 45%. It was reported that the addition of fluorocarbon surfactant (FC129) into KOH/IPA solution could reduce the tendency of the hillock formation [106].

In KOH solutions, the roughness of surfaces etched in diluted solution is caused by the formation of pyramidal hillocks. The addition of potassium ferricyanide $K_3Fe(CN)_6$ at the concentration of about 18 mM into 4 M KOH solution can suppress the formation of hillocks. In this situation, the etch rate was found to increase slightly (by < 10%). The density of hillock pyramids decreases with increaseing ferricyanide concentration. However, the addition of ferricyanide over a critical level of 35 mM causes the termination of silicon etching [111]. In NH_4OH-based etchants, the addition of H_2O_2 in a concentration between 0.65×10^{-2} M and 1.84×10^{-2} M helps to suppress the formation of etch hillocks. Too much H_2O_2 causes the termination of silicon etching [68]. In TMAH etchants, the high density of hillocks is observed in solutions with low concentrations, such as 5%. The hillock density and size decrease with increasing concentration of TMAH. Hillocks will disappear when wafers are etched in the 25 wt% TMAH solution at 80°C [110]. No detailed studies on the prevention of etch hillocks in EPW solutions were published except the paper by Declercq et al. [54], which was mentioned in the EPW section.

3.4.2.7 Temperature Dependence of Etch Rates

A large effort has been made to determine the temperature dependence of etch rates on the (100), (110), and (111) crystal planes by Seidel et al. [66]. The activation energies E_a and frequency factor R_o were determined according to the Arrhenius law, where k is the Boltzmann Constant, E_a the apparent activation energy, and R_o the frequency factor:

$$R = R_o \exp(-E_a/kT)$$

Wafers with (100) and (110) orientation were used. To obtain detailed data on the crystal orientation's dependence on the etch rate, a wagon wheel-shaped masking pattern was employed, consisting of radial divergent segments with an angular separation of 1°. Wagon wheel patterns are commonly used as an efficient tool to study the etch rate dependence on the crystallographic orientation [63]. After etching, a "cloverleaf" figure appears whose shape depends on the etching anisotropy. By comparing the radial positions of minima, the visual observation could give a qualitative interpretation of anisotropy. To achieve a quantitative measurement in the neighborhood of {111} planes, where the etch rate was a very sensitive function of the angular orientation, a second, fan-shaped pattern with an angular separation of 0.1° and with a total angular spread of 4° was used. The (100) and (110) etch rates were determined by vertical etching on (100) and (110) wafers. The (111) etch rate was determined by laterally undercutting segments on (100) or (110) wafers.

In KOH solution, the nearly identical energies of 0.59 and 0.61 eV were obtained for (100) and (110), respectively, and an activation energy of 0.7 eV was obtained for (111). Seidel's result, that the activation energy of the Si (111) etch is larger than that of (100) and (110), is widely cited, especially for the discussion of the etch rate dependence on crystallographic orientation. Some of the data for the activation energies of silicon anisotropic etching are available in the lectures, especially for the (111) planes [45, 50, 65, 69]. Two results from the others challenge Seidel's results on the activation energy of (111). Clark et al. [65] investigated the (110) Si wafer etching in aqueous KOH solution, with concentrations from 9 to 54% and at a temperature range of from 23 to 80°C. From the dependence of the (110)/(111) etch rate ratio on temperature, the comparison of activation energies between (110) and (111) planes could be obtained. At low KOH concentrations, the activation energy of (110) Si was greater than that of (111) Si. With the KOH concentration increasing, the two activation energies approached each other. Finally, the (111) activation energy became greater at the highest concentration of 54%. Tabata et al. [69] investigated the (100) Si wafer etching in TMAH solutions, as mentioned earlier in the TMAH solutions section. It was found that the activation energies for (100) planes were larger than those of (111) planes within the whole concentration range investigated. Therefore, the authors suggested that the etch rate ratio of (100)/(111) would increase with rising temperature — the opposite of Seidel's results.

There was a difference in the etch mask patterns within these three works. Seidel et al. used wagon wheel patterns, but the other two did not. The etching results from wagon wheel patterns and separated V-grooves were investigated in our lab [115]. Three major sets of patterns were designed in the same mask. The first set of V-grooves was designed with angular increment of 0.05° ranging from −0.65 to +0.65°. The V-groove was designed with an open end that was 70 μm wide and 3 mm long. Each V-groove was apart from its neighbor's V-grooves, with a separation of about 1.5 mm — far larger than the V-groove width. The second set of V-grooves with open ends was designed with an angular increment of 0.5° ranging from −17 to +17°. The other parameters are the same as that of the first set. The third set of wagon wheel patterns was designed with a diameter of 25 mm and a V-groove width of 40 μm, with an angular increment of 1°.

TOPSIL, N-type FZ (float zone), 4", (100) silicon wafers with patterned Si_3N_4 as mask material were etched in 35% KOH/2PA solution at temperatures ranging from 40 to 85°C. Undercuts were measured from the second set of V-grooves and wagon wheel patterns in the same wafer. The results were plotted in the same chart in order to be compared, as shown in Figure 3.15. It was found that the undercut of the separated V-grooves was always larger than that of the wagon patterns at the same angle.

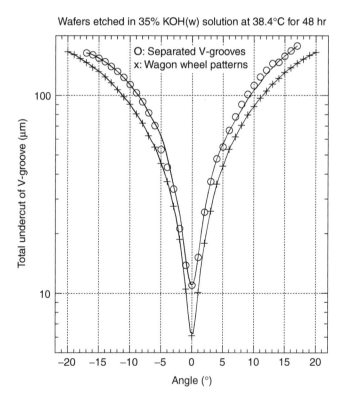

FIGURE 3.15 Comparison between separated V-grooves and wagon wheel patterns.

During chemical etching, the reacting species are continually consumed at the surface, and the concentration gradient near the reaction area will drive more reacting species toward the surface. Two processes, reaction and diffusion, are involved in the etching, and the rate controlling process is the slowest of these two. There is a general criterion [116] to distinguish the kinetics of the process. The apparent activation energy would be about 3 to 6 Kcal mole^{-1} (0.13 to 0.26 eV) for a diffusion-limited process, and about 10 to 20 Kcal mole^{-1} (0.43 to 0.86 eV) for a reaction-limited process. Because the apparent activation energy of {111} etching is larger than 0.5 eV, it appears to be a reaction-limited process, but the difference in etch rates between separated V-grooves and wagon wheel patterns indicates that the diffusion effect might still be involved, hindering the etch rate. Our early study [117] on the diffusion-limited chemical etching indicates that the etch depth in the center of the etched patterns is shallower than that at the edge, because the diffusion flux is stronger at the edge than at the center. The study also described Shaw's experimental result [118]. The larger the mask gap between the etch grooves, the higher the edge etch rate. It was very similar to the comparison of etch rates between separated V-grooves and wagon wheel patterns, as shown in Figure 3.15. Following the above discussion, it can be concluded that the undercut rates near {111} measured from the separated V-groove patterns were more accurate because of the reduced diffusion effect. The larger the distance between the separated V-grooves, the more accurate the measured {111} etch rate.

A series of temperature experiments was carried out, using the first set of patterns. The temperature dependence for the etch rates of {111} and {100} was plotted in Figure 3.16. The calculated activation energy for the {100} etch rate is 0.62 eV, which is comparable with Seidel's result of 0.61 eV in 34% KOH

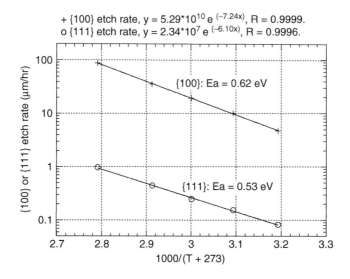

FIGURE 3.16 {100} and {111} etching rates as a function of temperature.

solution. In the same experiment, the activation energy of {111} was found to be 0.53 eV, a value less than that of {100}. The above result is similar to the results found by Clark [65] and Tabata [69].

The Arrhenius law is one of the most important equations in process kinetics, and connects the rate of a process R with the temperature T. The rate of the process R is related to both factors of the activation energy E_a and the frequency factor R_0. It is possible that the slower rate found was caused by the higher activation energy E_a. It is also possible that the slower rate is caused by the lower R_0, although E_a is lower, too. The R_0 of {100} etching was measured to be 5.29×10^{10}, which is comparable with Seidel's result of 3.10×10^{10}, but the R_0 of {111} etching was measured to be only 2.34×10^7 in our experiment.

On the basis of the above discussion, the temperature effect on the anisotropic factor could vary, depending on the pattern design. The anisotropic factor would increase with a drop in temperature where the wagon wheel pattern was used [66]. With the separation used for each pattern, the temperature effect on the anisotropic factor would tend toward zero. It has been mentioned that lowering the temperature to 70°C does not change the ratio R_{110}/R_{111} significantly [79]. Finally, if each pattern were separated far enough, the anisotropic factor would increase with a higher temperature.

3.5 CHIP NOTCHING FOR SILICON WAFERBOARD MOUNTING

A vertical mechanical surface, known as the notch, is etched into the bottom surface of the semiconductor laser or detector chip to locate the active region on the chip mechanically. Figure 3.17 shows a laser notch. The notch is especially

Epi side of laser

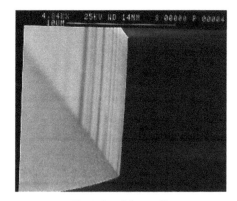

Notch sidewall

FIGURE 3.17 SEM photograph of a notched semiconductor laser. The view to the left shows the notch trough, which is along the right edge too since the die is diced in the trough. The view to the right shows a SEM of the notch edge quality. Despite expectations to the contrary, the notching process was found easy to implement and of very high yield.

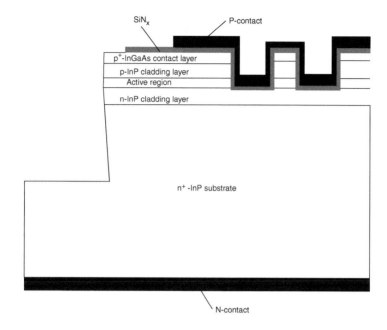

FIGURE 3.18 Drawing showing notch location as a large cut-out at the top left of the figure.

important for providing a reference location for the placement of semiconductor lasers. Lasers have to be placed and bonded to the silicon waferboard surface to within 0.2 to 0.3 μm, with respect to position.

Figure 3.18 shows the location of the notch relative to the active region. To preserve the high precision of the notch on the chip, the notch structure is formed by a self-alignment process during the chip fabrication. For a laser chip, this usually means that the laser notch structure is patterned with the same mask that patterns the active region on the chip. By doing this, there is no overlay error, as would be found trying to align different masks to each other at different parts of the process. The portion of the mask used for the notching is covered and protected during the laser device fabrication, but it is then uncovered to define it with a late-etch step. The notch is best formed by reactive ion etching followed by a quick wet etch to clean up the surfaces.

3.6   PROCESS CONTROLS

The manufacturing of micromachined silicon waferboard parts requires careful process monitoring to ensure precision control. There are different points in the process where this monitoring is done, and it is of two general types. The first type of monitoring is where the thickness of layers are checked while parts are in process, to control the process before they are completed. The second type is more like an inspection of the finished parts. In this case, if the parts fail, then the wafer is usually scrapped.

In-process checks consist of measuring the thickness uniformity of photoresist and dielectric masks used to create the structures. In addition, the etching depth of the silicon waferboard V-grooves and other structures are monitored as the etching proceeds. The in-process measurements are done with two different tools. The thickness of the dielectric layers is obtained by optical means. This is usually done by an ellipsometer, which also provides the index of refraction that helps to track the chemical composition of the Si_3N_4 dielectric. The thickness of the photoresist and the depth of the V-groove are monitored by a stylus probe, which mechanically feels the thickness by monitoring the edge of a coated layer and looking for the step.

The completed micromachined units are inspected dimensionally in two different ways, depending on whether the dimension is vertical or horizontal. The horizontal critical distances that span between registration edges are characterized as hundreds of microns in length but need to be measured to 0.1 μm precision. A custom system that uses a microscope as a pointer and a calibrated mechanical translation stage that moves the part horizontally measures these dimensions. The precision of the measurement is dependent on the operator's ability to identify the edge of an object under the microscope and a laser interferometer, which measures the translation of the stage to measure the displacement. This measurement is precise to 0.1 μm, as limited by table vibrations and judgement with respect to where the edge is as seen through the microscope. The results appear to be quite reproducible.

A stylus probe measures vertical critical distances, such as standoff height, pedestal height, or solder thickness. Just as with the horizontal measurement, it is possible to easily measure the distances to better than 0.1 μm.

3.7 DIE-BONDING ASSEMBLY

One of the greatest advantages of mechanical passive alignment is the ease of die-bonding assembly. The die-bonding process takes place quickly and requires a relatively simple die-bonder. Furthermore, it is becoming increasingly apparent that the precision of die-bonding this way may be higher than with other methods because there is more direct control over the critical nature of the parts before die-bonding, as well as more control during die-bonding.

Figure 3.19 shows a schematic of the die-bonding geometry and bonder tip. This view shows how simple the bonding can be. One stage axis is used to raise and lower a chip to be die-bonded, while another pair of stages provides translation of the Si substrate to be bonded as well as the die trays used to provide the die to the bonder. The heat source for reflowing solder can remain fixed, as well as the cameras positions used to observe the process. The location in space where the die-bonding action takes place remains fixed. With this form of die-bonding, the chip is constantly held until the bond is made, providing excellent control over the die position.

Referring to Figure 3.1, note that the positions of the substrate's side pedestals that locate the die are important. There is also a dependency on the shape of the die. A wide chip that is not very deep, as shown in the figure, can be positioned

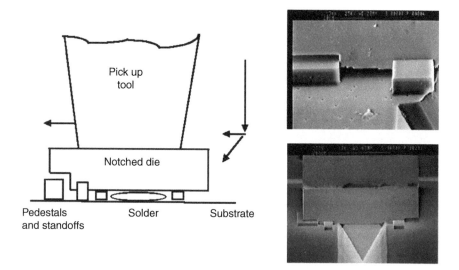

FIGURE 3.19 Schematic of diebonder tip and alignment of die.

with one side pedestal and two front pedestals. A narrow chip that is deep is best positioned with two side pedestals and one front pedestal. It is important that the "center of gravity" of the chip lie between the side pedestals or front pedestals to prevent rotation of the chip when pushing the chip up against the pedestals.

The precision of the alignment is critical but within the capabilities of the technology. A detailed discussion of this analysis is provided in Chapter 11. Figure 3.20 shows the type of alignment tolerance that is allowed with single-mode fiber coupled to a typical 1300-nm transmission laser. As can be seen, it is

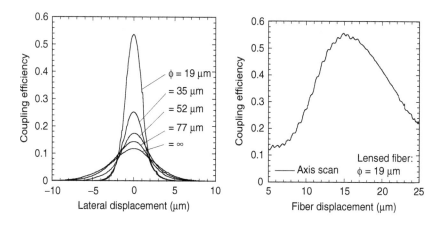

FIGURE 3.20 Single-mode optical fiber alignment sensitivities as a function of fiber tip diameter when coupled to a laser.

a trade-off between coupling efficiency and sensitivity to misalignment as provided by the diameter of the lens on the fiber. If the fiber has no lens, that is, it is flat cleaved or a diameter of infinity, then the maximum coupling efficiency possible is only about 12%; however, if a lens is present, then the coupling efficiency can be in the 70% range. For our work using passive alignment, it is necessary to have the lens on the fiber be concentric with the outside cylinder of the fiber. To date, the only lens that achieves this tolerance is one that is created by selectively etching the end of a flat-cleaved fiber in HF. In doing so, the core of the fiber tends to etch slower than the cladding resulting in a small tip, defined by the core, sitting on the fiber end. Some control over the effective radius of the lens can be attained by the degree of etching. In practice, the lens looks more like a cone in shape, and the maximum coupling efficiency with such a lens is about 50%. A typical target coupling efficiency for a passively aligned system might thus be about 30%, which is substantially better than flat-cleaved butt coupling, but less than the expensive active alignment. Figure 3.20 also shows the axial sensitivity to coupling efficiency. As shown, the peak coupling efficiency is back a distance from the laser facet. If a flat-cleaved fiber were plotted on this graph there, would be a Fabry-Perot optical interference ripple, but the cone shape of the lens suppresses that effect.

A view of a die-bonder is shown in Figure 3.21. Video cameras show the die handling and positioning on the monitors. The operation is a semiautomatic operation. The die-bonder can be programmed to bring the die close to the desired locations on the silicon wafer to be bonded. By the same process, it can also go very close to the desired location for picking up a die. The operators need only concern themselves with the fine manipulation for pick up and place, and with some practice, this can be done very quickly.

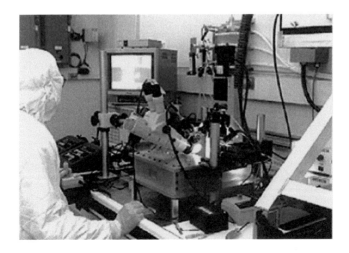

FIGURE 3.21 View of mechanical passive alignment diebonder.

There are visual cues that help with the placement of the die. If there is a gap between the side pedestals and the die, it is usually obvious, appearing as a space that looks dark. The vacuum placement tool also slides on the die surface once side pedestals are engaged. Finally, the surface of the silicon wafer is generally highly polished and acts as a mirror, providing a double image of the structure. The operator sees the die and its reflection in the wafer surface. The gap between the images provides a sense of the height of the die above the silicon surface. Because the die will rest on standoffs that set the vertical height, an expected gap between the die and its reflection will show itself.

3.8 WAFER SCALE TESTING

One of the advantages of silicon optical bench assembly practices is that they provide for the possibility of wafer scale assembly and testing if the wafer is not diced for that assembly. This is because the product die sites are in a step-and-repeat pattern across the wafer. A view of a portion of a wafer with many sites is shown in Figure 3.22. Two laser die have been mounted in this view, and other sites are vacent. Some experimental resistor traces are also shown. The step-and-repeat pattern allows the equipment designer to know the location of not only every test pad in a product site but also every test pad for all the product sites across the wafer. Furthermore, the test pattern is arrayed, so just as with electronics die wafer-scale testing, it would be possible to build a probe card for a product site and have it used to probe sequentially from die site to die site. The probe testing would be done local to the die to be tested; reducing or eliminating

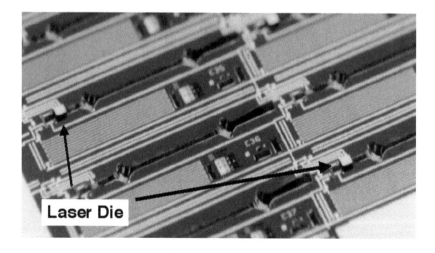

FIGURE 3.22 Wafer scale assembly and testing is made possible by known step and repeat distances. In this figure two laser die have been passively aligned while other sites are still vacant.

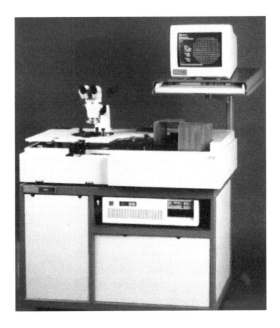

FIGURE 3.23 Pacific Western autotester.

the need for expensive and complicated testing at the wafer edge. The wafer would be left floating and the probing would be direct to the measured site.

An example of a piece of equipment that can do this testing is shown in Figure 3.23. This piece of equipment, known as an autotester, is designed for testing electronics die across a wafer. Wafers are taken in from a cassette and orientated using the wafer flat. A single probe card measures each die site for characteristics by making a connection to that individual product site's pads.

The autotester can be made to test optoelectronics by using it to test either the electrical characteristics or using it to indirectly detect the optical characteristics by measuring detectors in the optoelectronic assembly. Most semiconductor lasers are packaged in products having a monitor detector present. Monitor detectors often measure the light output at the rear facet of a semiconductor laser. By measuring the electrical response on the detector, the laser can be evaluated.

3.9 WAFER SCALE BURN IN

A side benefit of using silicon wafers for passive alignment platforms is that it enables burning semiconductor devices on a wafer scale if the assembly is done on wafer scale. This is important because the manufacturing yield of semiconductor lasers, in particular, is often less than 30% and can be much less for state-of-the-art laser devices, where specifications are pushed to the limits. If the devices can be burned in while they are very early in the manufacturing process, the bad devices can be screened out before much additional packaging and cost is invested in them. Devices frequently change their characteristics following a die-bond and burn in, so it is possible to eliminate the bad parts early by testing wafer scale following a wafer-scale burn in.

There are two types of active devices that need burning in: current-based devices and voltage-based devices. The electrical current-based devices, such as semiconductor lasers, should be burned in using a series circuit. The voltage-based devices, such as photodetectors, should be burned in using a series parallel circuit having burn-in resistors in series with the detectors.

The voltage-based burn in is easiest to implement. If the device burns out by shorting out, the series resistor will maintain the burn-in circuit for other detectors undergoing burn in. If the device burns out by open circuit, the burn in circuit is still maintained.

Current-based devices are burned in in series. For this, a row across the silicon wafer may be used. If the device burns out to a short, then the burn in continues for the rest of the row. If it burns out to open, then the burn in of other devices on the row stops until the bad device can be jumped. This is usually done by wirebonding.

3.10 FIBER ATTACH

Most of this chapter discusses passive optical alignment in terms of the optical alignment of chips to a silicon wafer, but if a fiber is used, it also must be aligned and bonded. The optical fiber is usually aligned using a precision wet-etched Si V-groove of a known depth, and the axial position of the fiber is determined using an optical microscope.

Either epoxy or solder may be used to bond the fiber. If epoxy is used, there is a drawback in terms of the stability of the adhesive, as well as the time it takes to cure. Ultraviolet-curable, low-shrinkage adhesives may be used to decrease the cure time, but an oven bake may still be required. Filling the adhesive with a powder of low-expansion materials reduces the shrinkage and movement. If solder is used, there is the drawback of creating stress in the glass of the fiber, which can lead to failures if the bonding design is inadequate. Solder also requires that the fiber be metallized to create a wettable bonding surface. The big advantage of solder is its quick bond and inherent stability over adhesives.

Figure 3.24 shows a fiber bonded with solder. The fiber is metallized to create a wettable surface for the solder. A large Si block is placed over the fiber in the V-groove, with a solder preform in between, and the solder is then reflowed to make the bond. In this case, there is metallization on both sides of the V-groove and on the bottom of the Si-block holding down the fiber, but there is no metal-lization in the V-groove. Gold-tin eutectic solder was used to bond these fibers, and it was found to be reliable when mounted with this geometry.

FIGURE 3.24 Fiber attach.

3.11 TECHNOLOGY AND PRODUCT
APPLICATIONS

The silicon wafer platform for packaging enables a lot of manufacturing and integration options other than passive alignment. Table 3.1 at the beginning of the chapter shows a listing of what may be possible. Because the parts are differentiated from one product to the next by soft-tooling changes, it is possible to produce many different products without having to change processes. The wafer-scale fabrication methods are also enabled, allowing for automation.

The most important initial product applications for this technology are small products requiring a competitive cost reduction. In this role, the features of the silicon wafer platform are leveraged to make new products that displace existing products made at higher cost. Later on, however, it is expected that the ease of hybrid integration on silicon will result in the production of integrated products. Finally, integrated products will be made of increasing complexity, such that eventually products will be made with numbers of features that can only be made by wafer fabrication methods. This is the path that electronics took and that is now being applied to optics.

3.11.1 BIDIRECTIONAL LINKS

A bidirectional link is a single optical fiber that carries information in both directions. It can be made bidirectional without interference between the downlink and uplink information by having the information transmitted in each direction with different wavelengths (wavelength division multiplexing) or the information can be transmitted each way, sharing different time slots for the transmission (time division multiplexing).

FIGURE 3.25 Bidirectional link module. Top is cut open to view contents.

The application of silicon waferboard in packaging is nicely illustrated in a project to build a bidirectional link. This project was run by a team lead by AMP Incorporated (now Tyco Electronics), with GTE Laboratories (now Verizon Laboratories), Lasertron Inc. (most recently Corning Lasertron), Digital Optics, the University of Colorado, and Broadband Technologies, and was partially funded by DARPA. The completed packaged link module made by this program is shown in Figure 3.25. The assembly parts are shown in Figure 3.26. As the figure shows, the hybrid integration packaging has reduced a very complex module down to a collection of just a few parts.

A number of fresh new approaches to packaging were used in this module. The silicon optical bench enabled the assembly of a miniature beam splitter to provide for the bidirectional function. This beam splitter is shown in Figure 3.27. This splitter consists of a slab of glass that has, half-buried in it along the surface,

FIGURE 3.26 Bidirectional link assembly parts.

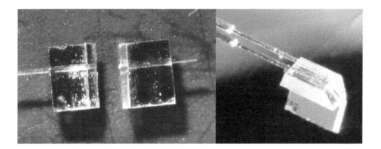

FIGURE 3.27 Micro beam splitter.

an optical fiber. The slab is cut simultaneously with its embedded fiber, thus exposing end faces, and is then polished and optically coated on the end faces before being reassembled along a V-groove. The keying to the V-groove is possible because half the circumference of the cylinder of the fiber is exposed and mates with the 111 faces of the V-groove. From a passive-alignment point of view, the assembly operates like a conventional fiber passive alignment, yet the extra feature of the beam splitter is present. Optical coupling efficiency is quite high — near 80% — because the mated pair of embedded fiber in the slab is perfectly matched, having previously been contiguous. In use, the function is also defined by the nature of the optical filters on the fiber ends. The filters are engineered to pass one band of wavelengths and to reflect the other. Outgoing light from the laser goes straight through the filter and out the fiber, while incoming light is reflected at the filter interface and passes to the receiver detector. The detector is looking for very weak signals from far away yet is very close to the intense outgoing laser light. The beam splitter blocks much of the outgoing light from hitting the receiver detector, but an additional blocking filter coated on the top of the glass slab is designed to prevent laser light above or below a specified wavelength from hitting the detector.

3.11.2 SURFACE EMITTERS AND DETECTORS

Up until now our discussion for the packaging of lasers and detectors has always assumed that the light path was running nearly parallel to the silicon surface. The silicon waferboard approach, including passive alignment, is also suitable for surface emitters and detectors. Figure 3.28 shows a silicon waferboard substrate, sometimes called a bowtie, that has mounted on it a light-emitting diode and a detector. The part is very small because no V-groove structure needs to be included, so two to three times more parts can be fabricated per wafer, as compared to V-groove-based assemblies. Instead of (100) silicon processing, (110) silicon was processed, where the (111) slow etching planes created diamond holes with vertical side walls in the silicon part. These (111) side walls are placed to engage with alignment pins for a connector. In this example, an MT connector having both a transmit and a receive optical path was flanked with two metal pins that

FIGURE 3.28 "Bowtie" mount for surface emitters and detectors.

engaged the side walls of the silicon waferboard bow-tie substrate. This integration capability of the silicon waferboard enabled the mounting of a preamplifier in close proximity to the detector.

3.11.3 High-Speed Packaging Design Using Silicon Waferboard

Silicon waferboard can be used for high-speed circuits, but the electrical paths must be designed correctly to prevent parasitics — capacitances, inductances, and resistances — from destroying the high-speed signal. Usually, for high-speed transmission the design calls for either a microstrip line or a coplanar line. Of these two, the coplanar transmission system is the simplest because it can be made as a monolithic integration form of the design rather than as an add-on part, as done with an assembled microstrip. High-resistance silicon is used as the foundation for the assembly. This type of high resistance silicon wafer is called float zone or FZ silicon and can be purchased with bulk resistances in the megaohm to centimeter range. For low-speed work the less expensive Czochralski (CZ) wafers can be used.

Once the FZ silicon is obtained, the design of the coplanar transmission lines needs to be considered. Normally one would expect to have a signal line flanked by grounds in parallel geometry sitting on top of the silicon, as shown in Figure 3.29A. This does, in fact, work very well, but it is not suitable for detectors used in

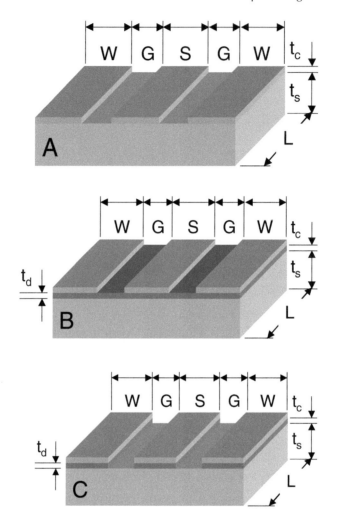

FIGURE 3.29 High speed structure for silicon submounts.

receiver circuits because a very low DC (direct current) leakage will persist through the high-resistance FZ silicon wafer. The logical next step is to add a thin dielectric layer under the signal and ground lines to block this leakage, as shown in Figure 3.29B, but in doing so, it has been observed that the parasitics get much worse. The circuit behaves as if the thin dielectric layer was grown on top of a metallic conductor, rather than on top of silicon, and the parasitic capacitance becomes very high. This phenomenon is caused by interface states between the dielectric layer and the silicon, probably caused by alkalie impurities, which cause conduction at that boundary. Some workers have tried to reduce this problem by forming the thin dielectric using high-temperature oxidation of silicon under extremely clean conditions, such as forming the oxidizing steam directly from hydrogen and oxygen gas. In our case, we found this approach unreliable and discovered that the problem can be eliminated much more easily by using design shown in Figure 3.29C. In this case, the dielectric is mostly only under the conductors, and the space in between is bare of dielectric, which disrupts the interface conduction. In use, the dielectric layer should be slightly larger than the conductor to prevent vertical leakage paths from the conductor to the silicon. Using this approach, a transmission loss of only 1.2 dB/cm was obtained at 10 GHz [119], with a dielectric layer thickness of only 4000 Å.

Another approach that was used to achieve very high speed circuits on silicon was to use the glass metal insulator circuit process (GMIC) [120]. In this case, a very low electrical transmission loss glass, expansion matched to silicon, is melted at high temperature and caused to sag into the surface of a silicon waferboard that has been overcoated with silver where ground planes are desired. This is done either using a vacuum oven and a thin glass wafer or by using glass frit material. The wafer is then lapped back to expose base silicon islands where chips are mounted so they can be heat sunk. Electrical transmission between chips is provided by electrical traces on the glass above the silver ground plane. The silicon material is actually not involved in the electrical signal propagation, and transmission losses are extremely low and transmissions above 40 GHz can be easily handled.

3.11.4 CONNECTORIZED LOW-COST LASER

The versatility of applying silicon waferboard to new package designs is clearly shown in the development of the connectorized low-cost laser [121, 122]. This unit is shown in Figure 3.30. Usually, with the packaging of optical components, the connector ferrule, which is used to align optical cables to optical fiber in a connector, is one of the smallest high-precision parts. In this laser design, however, the silicon waferboard technology is able to be miniaturized to the point that the transmission laser, monitor detector, and passively aligned fiber stub are

FIGURE 3.30 Connectorized low cost laser.

FIGURE 3.31 Silicon waferboard optical subassemblies connected to an MT connector.

all mounted on a tiny silicon waferboard that sits into the side of the connector ferrule itself. Once sealed with a small cylinder, the resulting part looks like a bullet. The part is very versatile because, in effect, it can make a transmission laser out of any connector body that uses that ferrule, and the alignment is included. This also enables the use of one small optical subassembly across a wide product line. In the industry, these types of parts are sometimes known as optical subassemblies and are an important design strategy to leverage large-volume production across a wider product suite having lower volumes of production. Another variation on this is to feed the optical subassemblies into an MT connector body, as shown in Figure 3.31.

REFERENCES

1. W. Hunziker, et al., "Low Cost Packaging of Semiconductor Laser Arrays Using Passive Self-Aligned Flip Chip Technique on Si Motherboard," The Proceedings of 46th Electronic Components & Technology Conference (ECTC), (1996) pp. 8–12.
2. G.S. Oehrlein and J.F. Rembetski, "Plasma-Based Dry Etching Techniques in the Silicon Integrated Circuit Technology," IBM *J. Res. Dev.* Vol. 36, No. 2, (1992) pp. 140–156.
3. H.V. Jansen, Plasma Etching in Microtechnology, Ph.D. thesis, MESA Research Institute, University of Twente, Enschede, the Netherlands (1996).
4. M.C. Wu, "MicroMachining for Optical and Optoelectronic System," Proceedings of the IEEE. Vol. 85, No. 11, (1997) pp. 1833–1856.
5. L.S. Huang, et al., "MEMS Packaging for Micro Mirror Switches," Proceedings of 48th Electronic Components and Technology Conference (1998) pp. 592–597.

6. B. Chapman, *Glow Discharge Processes,* John Wiley & Sons, New York (1980).
7. D.L. Flamm and V.M. Donnelly, *Plasma Chemistry and Plasma Processing,* Vol. 1, No. 4, (1981) p. 317.
8. J.A. Mucha and D.W. Hess, *Plasma Etching.* American Chemical Society Washington. (1983) p. 215.
9. R.A. Morgan, *Plasma Technology Vol. 1: Plasma Etching,* Edited by L. Holland, Elsevier, Amsterdam (1985).
10. H.W. Lehmann, *Thin Film Processes II: Plasma-Assisted Etching,* Edited by John L. Vossen and Werner Kern, Academic Press, San Diego, CA, (1991) p. 673.
11. W.R. Runyan and K.E. Bean, *Semiconductor Integrated Circuit Processing Technology,* Addison-Wesley, Reading, MA, (1990).
12. G.S. Oehrlein, *Reactive Ion Etching: Handbook of Plasma Processing Technology,* Edited by S.M. Rossnagel, Noyes Publications, Park Ridge, NJ, (1990) p. 196.
13. H.F. Winters and J.W. Coburn, "Surface Science Aspects of Etching Reactions," *Surface Sci. Rep.* 14, North-Holland, (1992) p. 161.
14. AEDEPT (Automatic Encyclopedia of Dry Etch Process Technology), Du Pont Electronics, Wilmington, DE, (1989). D11. H.V. Jansen, The Black Silicon Method IV: The Fabrication of the Three-Dimensional Structures in Silicon with High Aspect Ratios for Scanning Probe Microscopy and Other Applications, Proceedings of IEEE Micro Electro Mechanical Systems, Amsterdam, The Netherlands, (1995) pp. 88–93.
15. Y. Kuo, Reactive Ion Etching Technology in Thin-Film-Transistor Processing, IBM *J. Res. Dev.* Vol. 36, No. 1, (1992) pp. 69–75.
16. *Semiconductor Materials and Process Technology Handbook,* Edited by Gary E. McGuire, Noyes Publications, Park Ridge, NJ, (1988).
17. S.W.S. Ruska, *Microelectronic Processing: An Introduction to the Manufacture of Integrated Circuits,* McGraw-Hill, New York, (1987).
18. H.V. Jansen, "A Survey on the Reactive Ion Etching of Silicon in Microtechnology," Proceedings Micro Mechanics Europe, MME'95, Copenhagen, Denmark, (1995) p. 18.
19. E.G. Spencer and P.H. Schmidt, "Ion-Beam Techniques for Device Fabrication," *J. Vac. Sci. Technol.* Vol. 8, No. 5, (1971).
20. J.M.E. Harper, et al., "Low Energy Ion Beam Etching," *J. Electrochem. Soc. Solid-State Sci. Technol.* Vol. 123, No. 5, (19881) pp. 1083.
21. P.G. Gloersen, "Ion-Beam Etching," *J. Vac. Sci. Technol.* Vol. 12, No. 1, (1975), pp. 28.
22. C.M. Melliar-Smith, "Ion Etching for Pattern Delineation," *J. Vac. Sci. Technol.* Vol. 13, No. 5, (1976), pp. 1008.
23. J.W. Coburn, *Plasma Etching and Reactive Ion Etching,* American Vacuum Society, New York, 1982.
24. D.L. Flamm, *Solid State Technol.* Vol. 34 (1991), 47.
25. J. Hopwood, "Review of Inductively Coupled Plasmas for Plasma Processing," *Plasma Source Sci. Technol.* Vol. 1, (1992) pp. 109–116.
26. J. Asmmussen, *J. Vac. Sci. Technol. A.* Vol. 7 (1989), p. 883.
27. Henri Jansen, "The Black Silicon Method IV: The Fabrication of the Three-Dimensional Structures in Silicon with High Aspect Ratios for Scanning Probe Microscopy and Other Applications," Proceedings of IEEE Micro Electro Mechanical Systems, Amsterdam, The Netherlands, (1995), pp. 88–93.
28. R.A. Haring et al., *Appl. Phys. Lett.* Vol. 41, (1982) p. 174.

29. J. Maa, et al., "Effects on Sidewall Profile of Si Etched in BCl3/Cl2 Chemistry," *J. Vac. Sci. Technol.* Vol. B8, No. 4, (1990), pp. 581.

30. G.C. Tyrrell et al., "Ion-Beam Assisted Etching of Silicon with Bromide," *Appl. Surf. Sci.* Vol. 43, (1989) 439.

31. K. Hirobe, et al., *J. Vac. Sci. Technol.* Vol. B5, No. 594, (1987).

32. M. Nakamura, et al., *Jpn. J. Appl. Phys.* Vol. 28, (1989) pp. 2142.

33. M. Nakamura, et al., "Mechanism of High Selectivity and Impurity Effects in HBr RIE: In-Situ Surface Analysis," *Jpn. J. Appl. Phys.* Vol. 31, (1992) pp. 1999–2005.

34. K. Koshino, et al., "Chemical States of Bromine Atoms on SiO_2 Surface after HBr Reactive Ion Etching: Analysis of Thin Oxide," *Jpn. J. Appl. Phys.* Vol. 32, No. 6B, (1993) pp. 3063–3067.

35. G.K. Herb, et al., *Solid State Technol.,* Vol. 30, (1987) p. 109.

36. R.N. Carlile, et al., *J. Electrochem. Soc.* Vol. 135, (1988) p. 2058.

37. M. Sato and Y. Arita, *J. Electrochem. Soc.* Vol. 134, (1987) p. 2856.

38. I.W. Rangelow, A. Fichelscher, "Chlorine or Bromine Chemistry in RIE Si-Trench Etching?" *Advanced Techniques for Integrated Circuit Processing,* SPIE Vol. 1392, (1990) pp. 240–245.

39. Kurt E. Petersen, "Silicon as a Mechanical Material," Proceeding of the IEEE. Vol. 70, (1982) p. 1982.

40. D.B. Lee, "Anisotropic Etching of Silicon," *J. Appl. Phys.* Vol. 40, (1969) p. 4569.

41. R.C. Kragness and H.A. Waggener, U.S. Patent 3506509 (1970).

42. L.P. Bovin, "Thin-Film Laser-to-Fiber Coupler," *Appl. Optics.* Vol. 13, (1974) p. 391.

43. H.P. Hsu and A.F. Milton, "Single Mode Optical Fiber Pickoff Coupler," *Applied Optics,* Vol. 15, (1976) p. 2310.

44. J.T. Boyd and S. Sriram, "Optical Coupling from Fibers to Channel Waveguides Formed on Silicon," *Applied Optics,* Vol. 17, (1978) p. 895.

45. J.W. Faust, "Etching of Metals and Semiconductors," *The Surface Chemistry of Metals and Semiconductors,* edited by Harry Gatos, John Wiley & Sons, Inc., New York, 1960, pp. 151.

46. J.M. Crishal and A.L. Harrington, "A Selective Etch for Elemental Silicon," *J. Electrochem. Soc.,* Vol. 109, (1962) p. 71C.

47. R.M. Finne and D.L. Klein, "A Water-Amine-Complexing Agent System for Etching Silicon," *J. Electrochem. Soc.,* Vol. 114, (1967) p. 965.

48. A.I. Stoller, "The Etching of Deep Vertical-Walled Patterns in Silicon," *RCA Review,* 1970, pp. 271.

49. I.J. Pugacz-Muraszkiewicz, "Detection of Discontinuities in Passivating Layers on Silicon by NaOH Anisotropic Etch," *IBM J. Res. Develop.,* (1972) p. 523.

50. J.B. Price, "Anisotropic Etching of Silicon With KOH-H_2O-Isopropyl Alcohol," *Semiconductor Silicon 1993,* edited by H.R. Huff and R.R. Burgess, The Electrochemical Sociaty Softbound Proceedings Series, Princeton, NJ, (1973) pp. 339.

51. Won-Tien Tsang, Cheng-Chung Tseng, and Shyh Wang, "Optical Waveguides Fabricated by Preferential Etching," *Applied Optical,* Vol. 14, (1975) p. 1200.

52. Donald F. Weirauch, "Correlation of the anisotropic etching of single-crystal silicon spheres and wafers," *J. Applied Physics,* Vol. 46, (1975) p. 1478.

53. Won-Tien Tsang and Shyh Wang, "Preferentially etched diffraction gratings in silicon," *J. of Applied Physics,* Vol. 46, (1975) p. 2163.

54. Michel J. Declercq, Levy Gerzberg, and James D. Meindl, "Optimization of the Hydrazine-Water Solution for Anisotropic Etching of Silicon in Integrated Circuit Technology," *J. Electrochem. Soc.*, Vol. 122, (1975) p. 545.

55. E. Bassous, U.S. Patent 3, 921, 916 (1975).

56. E. Bassous and E.F. Baran, "The Fabrication of High Precision Nozzles by the Anisotropic Etching of (100) Silicon," *J. Electrochem. Soc.*, Vol. 125, (1978) p. 1321.

57. Ernest Bassous, "Fabrication of Novel Three-Dimensional Microstructures by the Anisotropic Etching of (100) and (110) Silicon," *IEEE Transactions on Electron Devices*, Vol. ED-25, (1978) p. 1178.

58. A. Reisman, M. Berkenblit, S.A. Chan, F.B. Kauman, and D.C. Green, "The Controlled Etching of Silicon in Catalyzed Ethylenediamine-Pyrocatechol-Water Solutions," *J. Electrochem. Soc.*, Vol. 126, (1979) p. 1406.

59. W.H. Ko et al., "Development of a miniature pressure transducer for biomedical applications," *IEEE Trans. Electron Devices*, ED-26, (1979) p. 1897.

60. J.M. Borky and K.D. Wise, "Integrated signal conditions for silicon pressure sensors," *IEEE Trans. Electron Devices*, ED-26, (1979) p. 1906.

61. X.P. Wu, M.H. Bao and W.X. Ding, "An integrated pressure transducer for bio-medical applications," *Sensors and Actuators*, 2, (1982) p. 309.

62. S.C. Kim and K.D. Wise, "Temperature sensitivity in silicon piezoresistive pressure transducers," *IEEE Trans. Electron Devices*, ED-30, (1983) p. 802.

63. L. Csepregi, "Micromechanics: A Silicon Microfabrication Technology," *Microelectronic Engineering*, 3, (1985) p. 221.

64. Xian-Ping Wu, Quing-Hai Wu, and Wen H Ko, "A Study of Deep Etching of Silicon Using Ethylenediamine-Pyrocatechol-water," *Sensors and Actuators*, Vol. 9, (1986) p. 333.

65. Lloyd D. Clark, Jr. and David J. Edell, "KOH:H_2O Etching of (110) Si, SiO_2, and Ta: An Experimental Study," Proceeding of IEEE Micro-Robots and Teleoperators Workshop, Hyannis Massachusetts (1987).

66. H. Seidel, L. Csepregi, A. Heuberger, and H. Baumgartel, "Anisotropic Etching of Crystalline Silicon in Alkaline Solutions," *J. Electrochem. Soc.*, Vol. 137, (1990) p. 3612.

67. U. Schnakenberg, W. Benecke, and B. Lochel, "NH_4OH-based Etchants for Silicon Micromachining," *Sensors and Actuators*, A, 21–23, (1990) p. 1031.

68. U. Schnakenberg, W. Benecke, B. Lochel, S. Ullerich, and P. Lange, "NH_4OH-based Etchants for Silicon Micromachining: Influence of Additives and Stability of Passivation Layers," *Sensors and Actuators*, A, 25–27, (1991) p. 1.

69. Osamu Tabata, Ryouji Asahi, Hirofumi Funabashi, and Susumu Sugiyama, "Anisotropic Etching of Silicon in $(CH_3)_4NOH$ Solutions," *Techn. Digest Transducers '91*,' San Fransisco, USA, 1991, pp. 811.

70. U. Schnakenberg, W. Benecke, and P. Lange, "TMAHW etchant for silicon micromachining," in *Tech. Digest Transducers '91*,' San Fransisco, USA, 1991, pp. 815.

71. G. Findler, J. Muchow, M. Koch and H. Munzel, "Temporal Evolution of Silicon Surface Roughness during Anisotropic Etching Processes," IEEE Workshop on Microelectromechanical System '92, Travemunde, Germany, Feb., 1992 (MEMS '92) pp. 62.

72. H. Linde and L. Austin, "Wet Silicon Etching with Aqueous Amine Gallates," *J. Electrochem. Soc.*, Vol. 139, (1992) p. 1170.

73. P. Allongue, V. Costa-Kieling, and H. Gerischer, "Etching of Silicon in NaOH Solutions," *J. Electrochem. Soc.,* Vol. 140, (1993) p. 1009.

74. Yuji Uenishi, Masahiro Tsugai, and Mehran Mehregany, "Micro-Opto-Mechanical Devices Fabricated by Anisotropic Etching of (110) Silicon," IEEE Workshop on Microelectromechanical System, Oiso, Japan, Feb., 1994 (MEMS' 94) pp. 319.

75. Carola Strandman, Lars Rosengren, Hakan G.A. Elderstig, and Ylva Backlund, "Fabrication of 45° Mirrors Together with Well-Defined V-Grooves Using Wet Anisotropic Etching of Silicon," *J. of Microelectromechanical Systems,* Vol. 4, (1995) p. 213.

76. Harold G. Linde and Larry W. Austin, "Catalytic control of anisotropic silicon etching," *Sensors and Actuators,* Vol. A 49, (1995) p. 181.

77. H. Camon, Z. Moktadir, and M. Djafari-Rouhani, "New trends in atomic scale simulation of wet chemical etching of silicon with KOH," *Materials Science and Engineering,* Vol. B37, (1996) p. 142.

78. P.M. Sarro, S. Brida, C.M.A. Ashrut, W.V.D. Vlist, and H.V. Zeijl, "Anisotropic Etching of Silicon in Saturated TMAHW Solutions for IC-Compatible Micromachining," *Sensors and Materials,* Vol. 10, (1998) p. 201.

79. Irena Zubel and Irena Barycka, "Silicon anisotropic etching in alkaline solutions I, The geometric description of figures developed under etching Si (100) in various solutions," *Sensors and Actuators,* A 70, (1998) p. 250.

80. Irena Zubel, "Silicon anisotropic etching in alkaline solutions II, On the influence of anisotropy on the smoothness of etched surfaces," *Sensors and Actuators,* A 70, (1998) p. 260.

81. Y. Nemirovsky and A. El-Bahar, "The nonequilibrium band model of silicon in TMAH and in anisotropic electrochemical alkaline etching solutions," *Sensors and Actuators,* A 75, (1999) p. 205.

82. Donald F. Weirauch, "Correlation of the anisotropic etching of single-crystal silicon spheres and wafers," *J. Applied Physics,* Vol. 46, (1975) p. 1478.

83. Don L. Kendall, "On etching very narrow grooves in silicon," *Applied Physics Letters,* Vol. 26, (1975) p. 195.

84. Kenneth E. Bean, "Anisotropic Etching of Silicon," IEEE Transactions on Electron Devices, Vol. ED-25, (1978) p. 1185.

85. D.L. Kendall, "Vertical Etching of Silicon at Very High Aspect Ratios," In *Annual Review of Materials Science* 1979, edited by R.A. Huggins, Vol. 9, (1979) p. 373.

86. D.M. Allen and I.A. Routledge, "Anisotropic etching of silicon: a model diffusion-controlled reaction," *IEE Proc.,* Vol. 130, (1983) p. 49.

87. J.W. Faust, Jr. and E.D. Palik, "Study of the Orientation Dependent Etching and Initial Anodization of Si in Aqueous KOH," *J. Electrochem. Soc.,* Vol. 130, (1983) p. 1413.

88. Don L. Kendall and G.R. de Guel, "Orientation of The Third Kind: The Coming of Age of (110) Silicon," Micromachining and Micropackaging of Transducers, edited by C.D. Fung, P.W. Cheung, W.H. Ko, and D.G. Fleming, Elsevier Science Publishers B.V., Amsterdam, 1985, pp. 107.

89. T.L. Poteat, "Submicron Accuracies in Anisotropic Etched Silicon Piece Parts – A Case Study," Micromachining and Micropackaging of Transducers, edited by C.D. Fung, P.W. Cheung, W.H. Ko, and D.G. Fleming, Elsevier Science Publishers B.V., Amsterdam, 1985, pp. 151.

90. Egon Herr and Henry Baltes, "KOH Etch Rates on High-Index Planes from Mechanically Prepared Silicon Crystals," *Techn. Digest Transducers 91,* San Fransisco, USA, 1991, pp. 807.

91. Chishein Ju and Peter J. Hesketh, "High index plane selectivity of silicon aniso-tropic etching in aqueous potassium hydroxide and cesium hydroxide," *Thin Solid Films,* 215, (1992) p. 58.

92. Peter J. Hesketh, Chishein Ju, and Sanjay Gowda, "Surface Free Energy Model of Silicon Anisotropic Etching," *J. Electrochem. Soc.,* Vol. 140, (1993) p. 1080.

93. Mattias Vangbo and Yiva Backlund, "Precise mask alignment to the crystallo-graphic orientation of silicon wafers using wet anisotropic etching," *J. Micromech. Microeng.,* Vol. 6, (1996) p. 279.

94. Alexander Holke and H. Thurman Henderson, "Ultra-deep anisotropic etching of (110) silicon," *J. Micromech. Microeng.,* Vol. 9, (1999) p. 51.

95. J.C. Greenwood, "Ethylene Diamine-Catechol-Water Mixture Shows Preferential Etching of p-n Junction," *J. Electrochem. Soc.,* Vol. 116, (1969) p. 1325.

96. A. Bohg, "Ethylene Diamine-Pyrocatechol-Water Mixture Shows Etching Anom-aly in Boron-Doped Silicon," *J. Electrochem. Soc.,* Vol. 118, (1971) p. 401.

97. J.B. Price, "Anisotropic Etching of Silicon With KOH-H_2O-Isopropyl Alcohol," *Semiconductor Silicon 1993,* edited by H.R. Huff and R.R. Burgess, The Electro-chemical Sociaty Softbound Proceedings Series, Princeton, NJ, (1973) pp. 339.

98. Irena Barycka, Helena Teterycz, and Zbigniew Znamirowski, "Sodium Hydroxide Solution Shows Selective Etching of Boron-Doped Silicon," *J. Electrochem. Soc.,* Vol. 126, (1979) p. 345.

99. E.D. Palik, J.W. Faust, Jr., H.F. Gray, and R.F. Greene, "Study of the Etch-Stop Mechanism in Silicon," *J. Electrochem. Soc.,* Vol. 129, (1982) p. 2051.

100. N.F. Raley, Y. Sugiyama, and T. Van Duzer, "(100) Silicon Etch-Rate Dependence on Boron Concentration in Ethylenediamine-Pyrocatechol-Water Solutions," *J. Electrochem. Soc.,* Vol. 131, (1984) p. 161.

101. M.M. Abu-Zeid, "Corner Undercutting in Anisotropically Etched Isolation Con-tours," *J. Electrochem. Soc.,* Vol. 131, (1984) p. 2138.

102. Xian-Ping Wu and Wen H. Ko, "Compensating Corner Undercutting in Anisotro-pic Etching of (100) Silicon," *Sensors and Actuators,* Vol. 18, (1989) p. 207.

103. B. Puers and W. Sansen, "Compensation Structures for Convex Corner Microma-chining in Silicon," *Sensors and Actuators,* A21–A23, (1990) p. 1051.

104. H. Sandmaier, H.L. Offereins, K. Kuhl, and W. Lang, "Corner Compensation Techniques in Anisotropic Etching of (100)-Silicon Using Aqueous KOH," in *Tech. Digest Transducers '91',* San Fransisco, USA, 1991, pp. 456.

105. H.L. Offerins, K. Kuhl, and H. Sandmaier, "Methods for the Fabrication of Convex Corners in Anisotropic Etching of (100) Silicon in Aqueous KOH," *Sensors and Actuators A,* 25–27, (1991) p. 9.

106. B. Block and M. Sierakowsky, "The Use of a Certain Fluorocarbon Surfactant and Fluorocarbon Conformal Coating Improves KOH Silicon Etching Quality," *Microma-chining and Micropackaging of Transducers,* edited by C.D. Fung, P.W. Cheung, W.H. Ko, and D.G. Fleming, Elsevier Science Publishers B.V., Amsterdam, 1985, pp. 125.

107. Y.K. Bhatnagar and A. Nathan, "ON pyramidal protrusions in anisotropic etching of <100> silicon," *Sensors and Actuators A.* 51, (1993) p. 233.

108. S. Tan, M.L. Reed, H. Han, and R. Boudreau, "Morphology of etch hillock defects created during anisotropic etching of silicon," *J. Micromech. Microeng,* 4, (1994) p. 147.

109. S. Tan, M.L. Reed, H. Han, and R. Boudreau, "Mechanisms of Etch Hillock Forma-tion," *J. of Microelectromechanical Systems,* Vol. 5, (1996) p. 66.

110. L.M. Landsberger, S.N. Mojtaba, and M. Paranjape, "On Hillocks Generated During Anisotropic Etching of Si in TMAH," *J. of Microelectromechanical Systems,* Vol. 5, (1996) p. 106.

111. P.M.M.C. Bressers, J.J. Kelly, J.G.E. Gardniers, and M. Elwenspoek, "Surface Morphology of *p*-Type (100) Silicon Etched in Aqueous Alkaline Solution," *J. Electrochem. Soc.,* Vol. 143, (1996) p. 1744.

112. Y.K. Bhatnagar, A. Nathan, and Y. Lu, "New Observations on Pyramidal Hillocks in the Anisotropic Etching of <100> Silicon," *Sensors and Materials,* Vol. 8, (1996) p. 423.

113. A. Goto, et al., "Hybrid WDM Transmitter/Receiver Module Using Alignment Free Assembly Techniques," in Proceeding of 1997 Electronic Components and Technology Conference, San Jose, May 1997, pp. 620–625.

114. B.W. Batterman, "Hillocks, Pits, and Etch Rate in Germanium Crystals," *J. Applied Physics,* Vol. 28, (1957) p. 1251.

115. S. Tan, R. Boudreau, and M. Reed, "Anisotropic Etching of Silicon on {111} and Near {111} Planes," *Sensors and Materials,* Vol. 13, (2001) pp. 303–313.

116. H.C. Gatos and M.C. Lavine, "Chemical Behaviour of Semiconductors: Etching Characteristics," in *Progress in Semiconductors,* Vol. 9, pp. 3 edited by A.F. Gibson and R.E. Burgess.

117. S. Tan, M. Ye, and A.G. Miles, "Diffusion Limited Chemical Etching Effects in Semiconductors," *Solid-State Electronics,* Vol. 38, (1995) p. 17.

118. D.W. Shaw, *J. Electrochem. Soc.,* 113, (1966) p. 958.

119. R. Boudreau, S. Tan, et. al., US Patent 6, 490, 379 December 3, 2002.

120. S. Iezekiel, et al., "High Speed Optoelectronic Multi-chip Modules Packaged in Glass on Silicon," Proceedings of the SPIE, Vol. 2149 (1995).

121. R. Wallace Roff, "Pigtailed Package for an Optoelectronic Device," US Patent 5, 764, 836 June 4, 1998.

122. R. Wallace Roff, "Package for an Optoelectronic Device," US Patent 5, 937, 124 Aug 10, 1999.

4 Silicon Waferboard Mechanical Passive Alignment II

Werner Hunziker and Werner Vogt

CONTENTS

4.1 INTRODUCTION

The major challenge of the packaging of III–V semiconductor optoelectronic devices lies in the optical connection between their waveguides and the single-mode optical fibers (SMFs). A very high alignment precision in the micron or even submicron range is required for efficient optical coupling. Section 4.2 of this chapter is therefore dedicated to the specific optical properties of the III–V semiconductor waveguides to determine the relevant geometrical requirements of the optical interconnection. First, the different optical coupling loss mechanisms for single-mode waveguide connections are discussed, including their sources and possible improvements for reduction. Second, the semiconductor waveguide to single-mode fiber coupling is discussed, regarding the influence of geometrical dimensions on alignment tolerances and the principles of optoelectronic device packaging.

The passive self-alignment technique for fiber-to-chip coupling and packaging described in this chapter has been developed to fulfill the alignment precision requirements not only for single-fiber but also for SMF array connections, including processes for broadband electrical interconnections. Section 4.3 gives the basic principles of the Si V-groove self-alignment technique. First, the optical alignment principle is described with its main features and the resulting advantages for the Si motherboard fabrication and the packaging process. Then, different possible solutions are given for the electrical connections of these flip-chip mounted devices. Process requirements, advantages, and drawbacks of the various techniques are summarized.

Section 4.4 is dedicated to different implementations of optoelectronic modules using the Si V-groove self-alignment technique. The packaging of optical space switch matrices covers the topic of large semiconductor chip packaging. It includes the fiber array connection on the input and output side as well as electrical interconnection and transmission lines for high-speed telecommunication applications. The major challenges in packaging of laser arrays include the handling of a small device size and the need for excellent heat dissipation. Very tricky aspects come into play for the self-aligned packaging of semiconductor optical amplifier arrays, which exhibit waveguides that are tilted relative to the perpendicular direction to the chip facet. The light refraction at the semiconductor-to-air interface complicates the alignment arrangement significantly, together with the fact that the small chips require optical coupling on both sides as well as electrical connections and good heat removal. Optohybridization (i.e., the combination of different optoelectronic, passive optical, and electronic devices on a single mounting platform) is important for realizing cost-effective modules for applications in optical communication systems. The Si V-groove self-alignment techniques offers such possibilities, as discussed in the subsections on direct chip-to-chip interconnection and on module fabrication using similar technologies.

4.2 OPTICAL COUPLING OF III–V SINGLE-MODE DEVICES

The optical coupling interface of optoelectronic devices represents the largest packaging difference and challenge compared to electronic chip packaging. The coupling of light between an optoelectronic integrated circuit (OEIC) and an optical fiber or vice versa has many more topological restrictions and interface problems than the connection between electrical circuits. This is especially the case for the optical interconnection between two single-mode waveguides. To achieve only one transverse optical mode, the light-guiding optics have very small geometrical profiles of a few microns, which result in very difficult alignment problems. The size of the optical mode depends on the refractive index difference between the waveguide core and cladding. Larger modes, preferred for packaging, are usually realized in dielectric-type waveguides such as optical fiber, silica on silicon, or lithium niobate.

Larger modes are much more difficult to achieve in III–V semiconductor materials based on InP and GaAs. Furthermore, large waveguides in the active region strongly affect the desired device properties because optical confinement in small cross sections allows higher efficiencies. In view of the large variety of semiconductor single-mode devices that need to be packaged and their much higher requirements compared to multimode devices, this section will cover the coupling properties of single-mode waveguides with special emphasis on semiconductor waveguide-to-SMF coupling. The resulting geometrical properties, restrictions, and possibilities for easing this interconnection challenge provide a fundamental basis for the development of appropriate alignment and packaging technologies.

4.2.1 Coupling Losses

The main parameter affecting the optical interconnection between two waveguides is the optical loss. Maximizing the amount of light transferred through the optical coupling interface is a major goal for most applications, as optical losses reduce the performance not only of devices but especially of systems. Less fiber-coupled power diminishes the repeater distance or the performance of diode pumped devices. High coupling loss reduces the amplifier gain and the signal-to-noise ratio, or the sensitivity of photodiodes and receivers.

Figure 4.1 summarizes the sources of the different optical coupling losses. They are grouped in three parts. First, the inherent losses at the device interface resulting from the material properties and the interface preparation (Figure 4.1a) are almost independent of the optical mode shapes and, therefore, of the alignment between the coupling partners. These losses are, however, sensitive to environmental influences during the packaging process and operation. It is important to note that the optical reflections play an important role at the semiconductor-to-fiber interface. The semiconductor materials have high refractive indices, close to 3.5, which result in a Fresnel reflection at each chip-to-air transition of about 30%. The optical active devices, such as lasers and especially amplifiers, are very sensitive to back-reflections for values above 10^{-5}. Antireflection coatings therefore have to

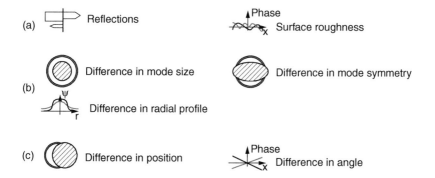

FIGURE 4.1 Summary of the sources of optical coupling losses with (a) device interface inherent losses, (b) mode shape-dependent losses, and (c) alignment-dependent losses.

be applied to the chip facets. To reduce the transmission loss, single-quarter wave dielectric coatings can be used. For especially sensitive devices, multilayer coatings and direct control during coating process are employed [1]. These layers can also improve the lifetime of high-power devices by preventing facet oxidation and corrosion [2]. A further method to drastically reduce the back-reflections from the facets into the device itself is tilting the waveguide away from the perpendicular direction of the facet [3]. Angles of 6 to 10° are widely used to ease the reflection problem. The new packaging problems caused by the light refraction at the facets with angles of 18 to 35° are treated in the corresponding subsections on optical alignment principle and packaging of tilted amplifier arrays.

Losses resulting from surface roughness can be avoided for the III–V semiconductor-to-waveguide coupling geometry by proper cleaving of chip facets and fibers. They become important if on-chip facets are desirable (e.g., for full-wafer processing and testing [4] or for motherboards with integrated waveguide devices). For the latter, mainly silica-on-silicon waveguide components are used. The surface requirements are in this case less crucial than for the semiconductor case because of the smaller refractive index step and the larger optical modes.

Second, the mode shape-dependent losses caused by the different refractive index profiles defining the waveguides of the coupling partners (Figure 4.1b) result just from the different geometrical properties of the modes to be coupled and are present even for optimum alignment. The difference in mode size and mode symmetry between partners to be coupled contributes to a large extent to the overall coupling losses. Semiconductor waveguides are realized in high-index material based on a horizontally layered structure resulting in very small and elliptical mode shapes. The standard SMF, in contrast, is based on a much smaller index difference and is rotationally symmetric. Figure 4.2 shows, on the left side, different

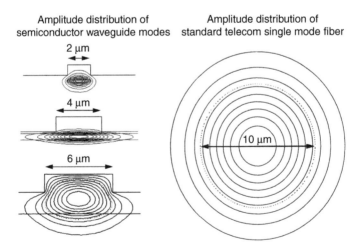

FIGURE 4.2 Various device amplitude mode distributions representing lasers, electrooptic switches, or modulators and diluted or tapered waveguide devices in comparison with the mode size of a standard telecom fiber.

amplitude mode distributions of semiconductor waveguide structures that are also used later on to illustrate coupling and alignment properties. The very small and elliptical shapes represent modes of common semiconductor devices such as lasers, amplifiers, modulators, and switches. The larger mode corresponds approximately to special low-index difference waveguides or, with increasing importance, to devices with integrated mode shape transformers, which are also referred to as waveguide tapers. The small mode of the waveguide in the active section of the component is enlarged toward the chip facet [5]. On the right side of Figure 4.2, the amplitude mode distribution of a standard telecom fiber for the 1.3- and 1.5-μm wavelength range is given for comparison. The resulting coupling losses are very obvious if fiber and waveguide have to be connected. A reduction of these losses can be achieved using optical imaging elements between the modes of each partner. For mode size adaptation, lenses of various shapes are used. This includes small ball lenses, GRIN (graded index) lenses [6], or so called lensed fibers, in which the fiber end itself is processed to act as a lens [7–11]. To correct for the differences in mode symmetry, anamorphotical optics or elliptical tapers are used. The difference in radial profile provides, in most cases, a minor contribution to the coupling losses because the single-mode waveguides guide similar mode profiles that are often approximated by Gaussian beams. Correction would require complicated, mode-specific, aspherical optics.

Third, the alignment-dependent losses caused by the nonoptimal position of the devices relative to each other (Figure 4.1c), represented by alignment tolerance curves, play an important role in the development of a packaging technique because of the high placing precision needed, which also has to be maintained during the lifetime of the device. Alignment tolerances depend strongly on the mode shapes, as illustrated here. The difference in alignment position includes misalignments in both lateral (horizontal and vertical) as well as longitudinal (gap) direction. The lateral alignment tolerances are more restrictive. In all cases, alignment precision is relaxed for larger optical modes. The difference in angular alignment becomes critical for coupling systems with free optics between the waveguides, as the angular misalignment is transferred into a lateral shift. For direct coupling, the angular misalignment is not as critical as the lateral misalignment, but in contrast to the latter, it becomes more sensitive for larger modes.

For a precise quantitative analysis of the optical properties of the interconnection, the mode theory of guided optical modes is used. Kogelnik [12] studied in detail the coupling of modes between different laser resonators with different geometries and misalignments. The modes of light-guiding structures are regarded as transversal electromagnetic waves. They are represented by their transverse amplitude field distributions $\Psi(x,y)$ or $\Psi(r, \varphi)$. Coupling between modes is then described by the overlap of the incident Ψ_i and the coupled field distribution Ψ_c. The coupled power is given by the power transmission coefficient T (sometimes also referred as the overlap integral η):

$$T = \frac{|\iint \Psi_i(x,y)\overline{\Psi_c(x,y)}dxdy|^2}{|\iint |\Psi_i(x,y)|^2 \, dxdy \iint |\Psi_c(x,y)|^2 \, dxdy} \tag{4.1}$$

Optimum coupling, which means no power loss and $T = 1$, is achieved for identical-amplitude and phase-field distributions ψ_i and ψ_c. Remarkably, there is no difference between light coupled from the device to the fiber or vice versa because of the symmetry of the integral. Thus, the same amount of light is coupled from a small to a large mode, as in the opposite direction.

A very useful tool for evaluating coupling properties and interconnection setups with optical imaging elements is the Gaussian beam approximation [13]. The description of the optical beam by only two parameters, beam waist radius and its position, results in analytical solutions for the overlap integral and easy calculation of the beam transformation by optical elements. Modes of step index fibers can be approximated very well by Gaussian beams, which allow us to describe the coupling properties between SMFs appropriately [14]. The mode profiles of semiconductor waveguides are not as close Gaussian as those of the fibers, but good predictions of coupling characteristics are still possible. The coupling of two Gaussian beams is characterized by their beam waist radii w_i and the misalignment between the waist positions. The equations for the power transmission coefficients or coupling efficiencies for beams with different waist sizes and the various misalignment possibilities are summarized in Figure 4.3. The evaluation of the first equation shows that coupling of two modes results in losses greater than 2 dB if one mode is twice the size of the other, and more than 8 dB if one mode is five times larger than the other. The most critical alignment is in the lateral direction,

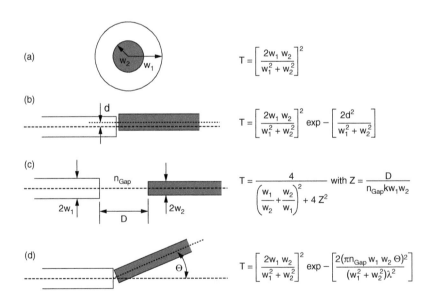

FIGURE 4.3 Power transmission coefficients (or coupling efficiencies) for coupling of two Gaussian beams with (a) different beam waist radii w_i as well as for (b) lateral, (c) longitudinal, and (d) angular misalignment (based on reference [14]). n is the refractive index, λ the vacuum wavelength, and $k = 2\pi/\lambda$ the wave number.

where for identical beam sizes, an approximately 1-dB loss for a shift of a quarter of the mode diameter and a loss of more than 4 dB for a misalignment of half a diameter are obtained. With respect to the very small semiconductor waveguide modes, the high-precision requirements of the packaging technology can be seen. These equations also show that the longitudinal and lateral alignment tolerances become tighter for smaller modes, but larger for the tilt. Because the differences in mode size and lateral alignment tolerance contribute the most to the coupling losses, larger modes are preferred for coupling. However, more attention has to be paid to the angular alignment.

4.2.2 Efficient Packaging of Single-Mode Devices

Efficient packaging of single-mode devices not only means low optical coupling losses but also involves the packaging process itself. Large alignment tolerances and a small number of components to be aligned are essential to reduce packaging costs. This section describes the influence of mode shapes and mode shape transformation on the critical trade-off between efficient coupling and efficient packaging.

The first point is to investigate the properties of the waveguide mode to SMF coupling, including the influence of using a lens for mode size transformation. Figure 4.4 shows the calculated power transmission coefficients between the three different waveguide modes given on the right, and rotationally symmetric Gaussian beams are plotted as a function of the Gaussian beam radius w. A radius of about 5 µm represents the guided fiber mode in a standard telecom fiber without an optical imaging element. The resulting values for butt fiber coupling clearly

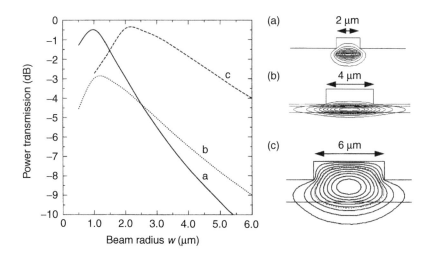

FIGURE 4.4 Calculated power transmission coefficients (or overlap integrals) between different waveguide modes (amplitude mode distribution given on the right) and rotational symmetric Gaussian beams as functions of their beam waist radius w.

demonstrate the large effect of mode size mismatch, with about 10 dB of coupling loss between lasers and cleaved fibers (mode distribution and curve a) and the large improvements that are made possible by larger waveguide structures (curve c). Coupling losses can be decreased significantly by using lenses to reduce the fiber mode and adapt it to the waveguide mode. Coupling losses of below 1 dB can be achieved. The effect of the difference in mode symmetry is illustrated with the elliptical waveguide mode (curve b), where the overlap with the rotational mode never results in very low coupling losses. Anamorphotical optics would be required to adapt the mode shapes. Another important point is the value of the beam radius w, where the maximum coupling efficiency occurs. A laser beam radius of about 1 μm requires an optical imaging system with a very high numerical aperture, hardly achievable with glass lenses, whereas values of 2 μm are reached with most types of microlenses.

The reduction of the fiber mode size to a very small value, ideal for laser coupling, drastically increases the coupling efficiency. However, the packaging process has to align the components in the appropriate position and, more important, maintain this position during fixation and the time of operation. Alignment tolerances are therefore important parameters when evaluating the required alignment precision of the package and its sensitivity to small shifts caused by temperature changes or mechanical vibrations and impact. Figure 4.5 summarizes and compares the alignment tolerances for cleaved and lensed fiber coupled to a laser and an expanded mode device. The calculated overlap integrals between the mode distributions shown on top and Gaussian beams with beam

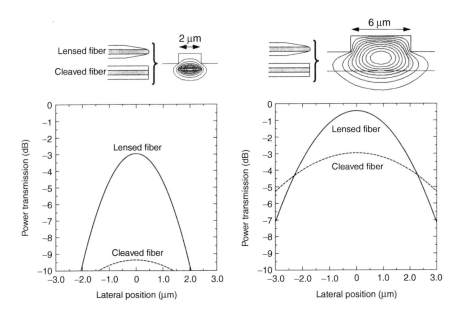

FIGURE 4.5 Calculated coupling efficiencies and lateral alignment tolerances between the waveguide modes given on top and Gaussian beams with beam radii of 2 and 5 μm, representing a lensed and cleaved fiber, respectively, as a function of horizontal displacement.

radii of 2 and 5 μm are shown as a function of the lateral displacement between the two. In the case of the typical small mode on the left, a large improvement in coupling efficiency is achieved by reducing the fiber mode with a 5-μm radius to a mode with a 2-μm radius, using optics (or magnification of the waveguide mode in the opposite direction by a factor of 2.5). However, this results in two negative effects on the alignment tolerances. First, the alignment tolerances for 0.5 and 1 dB excess loss become very small (0.55 and 0.8 μm instead of 1.2 and 1.8 μm), and second, the tolerance curves become much steeper outside the maximum. If, for example, the components are fixed in an imperfectly aligned position, minor shifts will cause a large coupling difference of several decibels for less then half of a micron displacement. This effect influences strongly the stability and reliability properties of a module. Thus, the trade-off between coupling efficiency and alignment precision becomes very important in view of low-cost packaging and stability. The curves on the right side of Figure 4.5 illustrate the coupling to the larger waveguide mode, clearly demonstrating multiple advantages: much better coupling efficiencies, larger alignment tolerances, and flatter tolerance curves. The coupling efficiencies increase from −9.5 to −3 dB and from −3 to −0.5 dB for cleaved and lensed fibers with 5- and 2-μm beam radii, respectively. The alignment tolerances for 0.5- and 1-dB excess loss are relaxed to values of 0.8 and 1.2 μm for lensed and 1.4 and 2.0 μm for cleaved fibers, respectively. Thus, alignment variations cause fewer power fluctuations because of the flatter tolerance curves.

Optical elements introduced between typical III–V waveguides and SMFs enhance coupling efficiency. The use of discrete lenses can increase the tolerances during the packaging process if the fibers are adjusted to the magnified image of the waveguide mode behind the lens in a second step. The stability problem between chip and lens remains the same because a shift between them is also magnified. This approach is very useful if other optical elements, as, for example, isolators, are included in the package or if hermeticity is desired, as a fiber feed through into the package is avoided. However, at least three or more components per coupling have to be adjusted and fixed in place. Furthermore, it is very difficult to realize array connections. The use of lensed fibers is advantageous because lens and fiber are inherently aligned and firmly connected in a very compact form.

The right part of Figure 4.5 indicates that much easier coupling schemes are possible if the mode size of the waveguide itself is better adapted to the fiber mode. If the optical magnification is carried out on the waveguide side of the connection, the fiber alignment requirements are efficiently relieved. The waveguide taper or mode shape adapter is therefore an important contribution to reducing packaging problems. Tapers enlarge the mode on the chip close to the facet through an on-chip mode transformation. Through the monolithic integration of device waveguide and taper, the most critical alignment to the small waveguide mode is inherently achieved in the taper fabrication process. These tapers correspond to self-aligned, magnifying optical elements directly fixed to the waveguide.

There are two major contributions to cost-effective packaging of optoelectronic devices: the waveguide tapers and the self-aligned packaging technologies.

The tapers improve coupling properties and relax alignment tolerances. The self-alignment technology reduces the number of alignment steps and the required precision of the chip placement, while avoiding equipment and time-intensive active device operation during packaging. As both contributions rely on specific features of the OEIC, it is important that the packaging aspects be taken into consideration during the design, development, and processing of optoelectronic components.

4.3 PRINCIPLE OF SI V-GROOVE SELF-ALIGNMENT TECHNIQUE

To use OEICs in optical communication networks, different components, such as lasers, modulators, switches, optical amplifiers, and detectors, have to be connected to optical fibers [either multimode fibers (MMFs), SMFs, or polarization-maintaining fibers (PMF)]. The efficient coupling of light to semiconductor waveguides requires accurate positioning in the submicron range, as described in the previous section. The adjustment of fiber arrays, frequently prebonded in Si V-grooves [15], requires complicated precision alignment stages and very sophisticated techniques for precise and stable fixing of the two parts. Several approaches have been described using flip-chip techniques, as described in detail elsewhere in this book. The optical alignment of the devices is either done by special solder bumps [16] or by perpendicular alignment trenches [17]. In this chapter, a flip-chip mounting technique [18] is described that allows the alignment of both the fibers and the OEIC in the same Si V-groove, using flexible beam-leads or solder for the electrical connections between the OEIC and the silicon motherboard.

The principle of the Si V-groove flip-chip mounting technique is based mainly on two ideas. The first idea is to use a single motherboard. All optically active (e.g., amplifiers, switches, and detectors) and passive (e.g., fibers) components are fixed and aligned on this board. The second idea is to rely on the dimensions defined at surface level only (i.e., on the dimensions given by the mask layout but only slightly influenced by details of the process). The alignment precision should not be related (within certain limits) to the exact depth and the sidewall angle (deviation from verticality) of etched alignment features. The use of photolithographically defined structures allows the realization of high-precision alignment features, both on wafers of III–V materials and on Si. The choice of the III–V material for active components is given by the desired wavelength of operation, whereas the choice of the (100)-oriented Si as the platform material is primarily determined by the thermal and mechanical properties and the processing possibilities. Si has a high thermal conductivity and a low coefficient of thermal expansion. This places its properties between the III–V semiconductor materials and the optical fibers. Self-alignment on a single platform relies on precise geometrical dimensions of the devices and materials involved, so it is important that mechanical and thermal influences are minimized.

As the thickness of the chip substrate is difficult to control within 0.5 µm during all the processing steps, and the vertical alignment is most critical (see previous section), upside-up mounting of optoelectronic devices is difficult. It is therefore best to bring the plane in which the light propagates in the III–V material as close as possible to the plane in which the alignment features of the mounting platform are defined. In other words, it is best to mount the OEIC face upside down onto the mounting platform. The influence of the thermal expansion difference of the III–V material, and that of the optical fibers to the vertical alignment positioning, is thus minimized. The longitudinal movement resulting from temperature changes is a less critical issue for coupling efficiency, as shown earlier. This upside-down mounting technique makes it necessary to bury the fibers in the mounting platform to bring the optical axis of the fiber to the same level as the mode center of the III–V waveguide.

As it is difficult to control the outer dimensions (width and length) of a cleaved OEIC device within an accuracy of better than 0.5 μm, we rely on alignment features that are (almost) independent of thickness, width, and length of the different coupling partners. The alignment principle described in this chapter relies only on geometrical dimensions given at the surface of the mounting platform and the III–V component, but not on height or depth of etched alignment features. In addition, this technique makes use of a self-aligning mechanism for the fiber-to-chip coupling that reduces packaging time and cost. The active driving of the device and its connection to the feedback and control tools can be omitted, and the required precision of the picking and placing tools is also reduced.

4.3.1 Optical Alignment

The Si V-groove self-alignment technique is based on a (100)-oriented Si wafer that acts as a mounting platform. It is processed to hold the optical fibers and the OEIC, providing optical as well as electrical connections. Figure 4.6 shows the

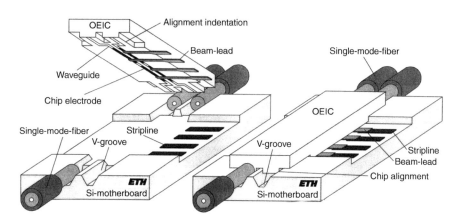

FIGURE 4.6 Principle of switch mounting and packaging technique shown for the example of a electro-optic space switch before (left) and after (right) mounting. The switch is flip-chip mounted onto a Si motherboard containing V-grooves for precise alignment of the single-mode fibers and the switch waveguides. The electrical connections of the switch electrodes to the transmission lines are made with beam-leads (i.e., electroplated electrodes cantilevered beyond the edges of the optoelectronic integrated circuit).

packaging principle for an electro-optic space switch matrix. The motherboard consists of different parts. The front and back areas exhibit anisotropically etched V-grooves to carry the fibers and at the same time provide optical alignment to the waveguides of the flipped-over OEIC. The center section is used for the electrical connection from the chip to the outside world, including the chip-to-board connection and the electrical transmission lines. As a main feature, this optical self-alignment technique uses the etched Si V-groove sidewalls to align the semiconductor waveguide devices to the SMFs. The assembly is sketched in Figure 4.7. Alignment indentations are etched into the III–V device in parallel to the waveguide. The indentations align the chip on the same planes as the fiber, leading to different specific advantages. Only one parameter, the distance a between the center of the waveguide and the edge of the alignment indentation, defines the optical alignment in both the horizontal and vertical direction. The alignment precision is only minimally affected by the effective V-groove width (see following for restrictions), as fiber and chip move up or down in parallel for narrower or wider V-grooves and maintain their alignment. It is not necessary to initially place the chip and fiber with a submicron precision, as the sidewalls guide the components into the final position. The restriction of the important alignment parameter to an on-chip

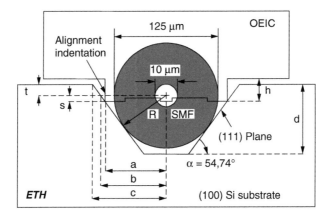

FIGURE 4.7 Cross section of optoelectronic integrated circuit (OEIC) to single-mode optical fiber (SMF) alignment. The OEIC is mounted upside down onto a V-grooved Si substrate. The edges of the alignment indentations glide on the (111) plane of the V-groove to their final position. The SMF is fixed on the (111) plane of the same V-groove. The center of the optical mode in the rib waveguide of the OEIC and the one in the fiber are in the same vertical and horizontal position. R is the radius of the SMF, s the vertical distance between the mode center and the top of the rib waveguide, t the distance between the mode center and the top surface of the Si substrate, h the height of the alignment indentations, d the depth of the V-groove, a the distance between the mode center and the alignment indentation, b the horizontal distance between the mode center and the V-groove side wall in the plane through the fiber center parallel to the Si-surface, c the distance between the center of the fiber and the edge of the V-groove, and the angle between the (100) and the (111) crystal plane of the Si substrate.

distance allows for the best alignment performance, as the waveguide and the alignment indentations can be defined in the same photolithographic mask. Furthermore, the processing precision of the Si board is reduced, which is very advantageous with respect to the mask alignment to the (111) Si plane (see following) and the control of the etch parameters for the 70-μm-deep grooves.

During mounting, the position of the fibers must only be optimized in the longitudinal direction (i.e., along the propagation axis of the light in the fiber). As the lateral fiber position is defined by the sidewalls of the V-groove, any time-consuming manual high-precision alignment procedure in the horizontal and vertical direction is omitted. In addition, no angular adjustment is necessary. Furthermore, arrays of fibers may be placed very closely. The plane that carries the strip lines lies below the original surface of the Si wafer to prevent the switch device from touching the Si except in the alignment region.

For the layout of the V-groove mask, the Si etch process parameters have to be considered. The preferential etching of the (100) Si-crystal plane compared to the (111) plane requires the V-grooves to be adjusted parallel to the (111) plane. Angular misalignment of the V-groove mask results in somewhat wider V-grooves. The angle of the V-groove sidewalls is given by the (100) and (111) crystal plane of the silicon substrate and has a value of $\alpha = 54.74°$ (see Figure 4.7). For a radius of $R = 62.5$ μm of a SMF, the width of the V-groove is $2c$, where $c = b + t/\tan\alpha$ and $b = R/\sin\alpha$ (see Figure 4.7). The V-groove depth d must be larger than $R + t$ to prevent the fiber from touching the bottom of the V-groove. This guarantees that the position of the fiber is given by the position of the V-groove sidewalls only. The exact width of the V-groove is not very critical (±1 μm) because both the fiber and the OEICs fall deeper into the V-groove by the same amount for wider V-grooves. The tolerance of the V-groove width is limited by the V-groove depth (the condition $d > R + t$ must be fulfilled) and the etch depth of the alignment indentation ($h > s + t$). Figure 4.8 shows the realization of a flip-chip-mounted OEIC on a motherboard with waveguides in the center of the V-groove and with the alignment indentations aligned on the Si sidewalls (left), as well as

 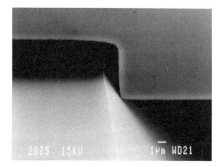

FIGURE 4.8 SEM picture of a flip-chip-mounted OEIC with a waveguide in the center of the Si V-groove and alignment indentations aligned on the sidewalls (left). The latter is magnified on the right.

a magnified detail of an alignment edge (right). These pictures show clearly that the sidewall angle (deviation from vertical) of the dry-etched alignment trench does not influence the alignment accuracy. Furthermore, the surface roughness of the etched bottom is not relevant as long as it does not touch the silicon surface.

The optimum distance a of the alignment indentations on the OEIC from the center of the corresponding waveguide depends on the location of the modal light distribution in the waveguide. The distance a depends on layer thickness, rib geometry, and refractive index profile. The center of the mode usually lies a few μm below the top of the waveguide. If s is the vertical distance between the center of the light distribution and the top level of the alignment indentation, the distance a relevant for the optical alignment becomes $a = b - s/\tan\alpha$. The length of the alignment indentations depends on the device size and ranges typically from 200 to 500 μm. If the waveguides on the III–V device are positioned perpendicular to the cleaved facets, the alignment indentations are positioned parallel to the waveguides. This means that the exact position of the facet is not very critical (± 50 μm) because the light direction remains in the center of the V-groove, and the fiber can be moved along the groove.

For devices using waveguides that are tilted relative to the facet (e.g., for semi-conductor optical amplifiers with reduced back-reflection), this becomes very different. Figure 4.9 illustrates the optical arrangement required to be compatible with the Si V-groove self-alignment technique. It shows that the alignment indentations are no longer parallel to the waveguide. The distance and the angle between the

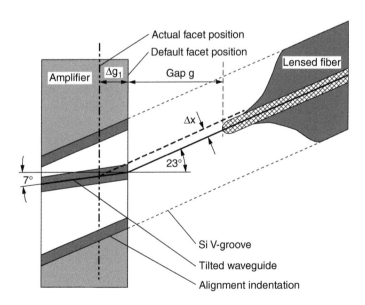

FIGURE 4.9 Tilted amplifier arrangement with lensed fiber for compatibility with the Si V-groove self-alignment technique. A chip length error Δg_1 results in an offset Δx of the optical beam compared to the optical axis of the fiber.

waveguide and the alignment indentations become dependent on the tilt angle of the waveguides, as determined by the refractive index. Hence, precise cleaving of the facet becomes very important because the correct position is given by the mask design. In addition, the distance a is not the same on both sides of the waveguide. For a refractive index of 3.2 in the waveguide section and a tilt angle of $7°$, the light outside the waveguide device (i.e., in the air) is tilted $23°$, according to Snell's law. A difference of $\Delta g_1 = 5\ \mu m$ between the actual facet position and the default position given by the mask design (see Figure 4.9) results in a horizontal beam shift of $\Delta x = 1.4\ \mu m$, which is not acceptable in most cases. In addition, this construction works preferably with lensed fibers, but not efficiently with butt fibers. The outer edge of the butt fiber end face with a $62.5\text{-}\mu m$ radius touches the amplifier facet when a gap, g, of $25\ \mu m$ still remains.

In addition, a very good antireflection coating on the fibers is required to prevent the degradation of the low-reflection properties achieved by waveguide tilting. A possible solution to solve these problems is shown in Figure 4.10, where the lensed fiber is replaced by an angle-polished butt fiber. The angle of the fiber facet is given with the condition that the light leaving the fiber and the light leaving the optical amplifier are parallel. The exact value of this angle is given by the refractive index of the fiber and the refractive index in the gap. By moving the fiber in the V-groove, the center of the fiber is shifted horizontally relative to the center of the light beam direction to compensate for the above-mentioned problems. In the design of the alignment indentations, a default gap is defined, together with a default cleaving position. Again, a deviation of $\Delta g_1 = 5\ \mu m$ from the designed

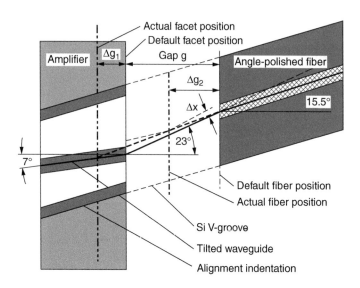

FIGURE 4.10 Tilted amplifier arrangement with angle-polished fiber showing compatibility with Si V-groove self-alignment technique. A chip-length error Δg_1 results in an offset Δx of the optical beam compared to the optical axis of the fiber. This offset can be compensated by changing the device-to-fiber gap g by an amount of Δg_2.

cleaving position (see Figure 4.10) results in a horizontal beam shift of $\Delta x =$ 1.4 μm. This Δx can now be compensated for by changing the default gap width by $\Delta g_2 = 10.2$ μm. The excess loss depends on the sign of the miscleaving and especially on the mode shape of the amplifier. As the fiber facet is tilted relative to the optical beam, an antireflection coating on the fiber facet is not required. If arrays of fibers are prefixed in an auxiliary V-groove substrate before polishing, no rotational alignment is required during flip-chip mounting. The mode shape influences not only the coupling efficiency but also, to a large extent, the tilt angle and the related tolerances. The mode shape adapters are thus highly desirable for tilted amplifier array packaging (see Section 4.4.3).

4.3.2 ELECTRICAL CONNECTION

As optoelectronic devices are used in many applications of broadband information transmission, high-speed electrical connections play an important role. Transmission lines have to provide low-loss, impedance-matched connections (low electrical reflection) from the package connector over the mounting platform to the OEIC. Of the various types of transmission lines [19], micro-strip and coplanar lines are employed most often. The coplanar type of line is mostly used because the ground and signal line are on the same surface, allowing for an easy shunt configuration without through-plated holes. It allows a geometrical adaptation of the line width from connector to chip side and maintains constant impedance. The main advantage of the micro-strip line is the much smaller substrate area required, which allows dense connections of array devices. The silicon substrate has also little electrical influence on these lines if they consist of metal–dielectric–metal sandwiches on top of the motherboard. Micro-strip lines are also preferred for low-impedance lines ($< 50 \, \Omega$) because of the geometrical dimensions.

For the electrical connection from the chip electrodes to the transmission lines on the motherboard, three main technologies are used: wire bonding, beam-leads, and solder bumps. Important parameters characterizing the electrical connection are electrical impedance, quantity and space of connections, electrical crosstalk, and physical properties of the connections. A comparison of the properties of the different connection techniques is given in Table 4.1. For optoelectronic packaging, the influence of the electrical connection process on the optical coupling must be considered. As the OEIC is mounted face-down in the case of the flip-chip mounting technique, it is not possible to use bond-wires to connect the electrodes (unless big holes are made into the substrate, as shown in Figure 4.29). Hence, either beam-leads (i.e., electroplated electrodes cantilevered beyond the edges of the device) or solder areas are used for electrical interconnections. The beam-lead technology was invented for electronic IC packaging [20] but has not been widely used for OEIC modules, as a special technology is needed for their fabrication. Our beam-leads consist of a polyimide carrier and electroplated electrodes. Figure 4.11 shows an SEM picture of the outer part of an optical space switch, with the beam-leads projecting over the chip edge. The leads can be bonded directly to the transmission lines on the motherboard after flip-chip

TABLE 4.1
Comparison of Properties of Different Transmission Line-to-Chip Connections

Property	Wire Bonding	Beam-Leads	Solder Bumps
Chip position	Upside up	Upside down	Upside down
Connection mechanism	Wires of 17–50 μm in diameter (Au, Al, Pd, Cu, and alloys)	Thickened electrodes projecting over chip edge	Solder balls or stripes
Connection process	Thermocompression, ultra/thermosonic bonding	Thermocompression, ultrasonic bonding, or soldering	Soldering
Connection chip features	Electrode pads	Beam-lead process	Solder wettable pads
Connection versatility	High flexibility in horizontal and vertical direction	All connections guided to chip edges	Midchip connections possible
High speed-capability	Limited	Good	Good
Connections per process	Single	Single or multi	Multi
Connection space	Large	Small	None
Use in Si V-groove self-alignment technique	Needs through hole in Si board (see Figure 4.29)	Does not affect optical alignment because of flexibility, needs extra chip fixation	Good heat dissipation and chip fixation; can affect optical alignment

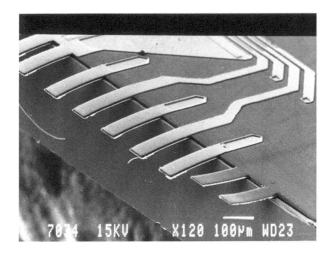

FIGURE 4.11 SEM picture of a part of an electro-optic space switch showing the beam-leads projecting over the chip edge.

mounting. Because beam-leads are flexible, they do not influence the positioning accuracy of the OEIC during bonding. In addition, a gap of typically 10 to 30 μm between the Si motherboard surface and the surface of the optoelectronic device (outside the alignment features) is easily bridged. Thus, the different vertical levels (typically several μm) of ground and signal contacts do not cause any problems. The center part of the Si substrate containing the electrical strip-line connections does not influence the alignment precision because this area is etched to a depth of ~20 μm (see Figure 4.6). For longer OEIC devices, the etch depth of this level might be increased to correct for a bow of the Si substrate or the OEIC.

The situation is different if lasers, optical amplifiers, or other temperature-sensitive or heat-producing devices need to be mounted. In this case, not only an electrical connection is required but also an efficient heat removal or temperature stabilization. Solder must be used instead of beam-leads. The flip-chip solder technology is used increasingly in electronic packaging because of its access to all chip positions on a board, the multiconnection process, the large bandwidth, and because no extra space other than the chip is needed [21]. For OEIC applications, the solder bumps are smaller in diameter, about 20 μm instead of about 100 μm, to achieve higher placement precision for self-alignment and to allow higher bandwidth [22]. If the optical alignment is not provided by the solder bumps, as for the Si V-groove technique, the solder areas can be large and the thickness small to provide optimum heat removal. In contrast to the beam-lead technique, ground and signal contacts must be leveled better to achieve good solder wetting of all contact pads and to allow optimal alignment in the V-grooves of the Si motherboard. The alignment indentations must be inserted (loosely) into the V-grooves before soldering. Precise alignment is achieved and maintained during the soldering process.

4.4 APPLICATIONS OF SI V-GROOVE SELF-ALIGNMENT TECHNIQUE

4.4.1 SPACE SWITCH MATRICES

Optical space switches are key components for optical networking and signal processing applications. Major requirements for the successful introduction of the switches into fiber-optic systems are polarization-independent operation, low optical losses, high on–off ratios, and low switching voltages, allowing for high switching speeds at multiple wavelengths in the 1.55-μm range. The possibility of integrating waveguides with optical amplifiers, detectors, and electronic drivers makes the InP/InGaAsP material system very attractive for this purpose.

Different versions of electro-optic space switches using flip-chip mounting have been realized in our lab [23–25]. The conceptual view and the cross section at the facet of a 4×4 switching matrix is shown in Figure 4.12. The InP/InGaAsP switch matrix is configured as a four-switch unit with four 2×2 Mach-Zehnder Interferometer (MZI) type switches, with two input and four output waveguides. Each basic 2×2 MZI switch consists of two phase-shifting arms, two 3-dB multimode interference (MMI) couplers, and two input and two output waveguides. The refractive index change for phase shifting is produced by applying a reverse bias voltage to the waveguide pin-junction. To achieve polarization-independent operation with respect to switching efficiency, a special orientation of the phase-shifting waveguides was chosen relative to the (011) cleavage plane of

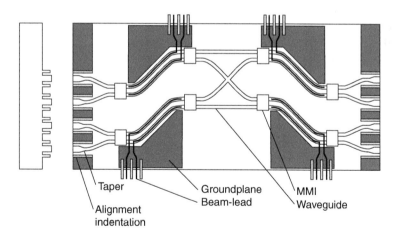

Taper
Alignment indentation
Groundplane
Beam-lead
MMI
Waveguide

FIGURE 4.12 Conceptual view and cross section at the facet of a 4×4 switch matrix consisting of four switch units with four 2×2 Mach-Zehnder Interferometer (MZI) type switches. Each MZI-switch is composed of two multimode interference couplers (MMI, used as 3-dB splitter/combiner) and two waveguides with electrodes. Beam-leads provide the electrical connection. Tapers and alignment indentations are used for efficient optical coupling when flip-chip mounted on the motherboard.

the standard (100) InP wafers. In this direction, the polarization dependences of the Pockels- and Franz-Keldysh effect cancel out [26].

MMI couplers are used as polarization-insensitive and fabrication-tolerant 3-dB power splitters and combiners [27]. The phase-shifting sections are combined with the curved waveguides to align the input and output waveguides perpendicular to the cleavage planes. The guided optical modes in the waveguides of the switches are confined to about 3×1 μm^2. Combined vertical and horizontal tapers (mode shape adapters) [28] are integrated on the chip to reduce the optical losses resulting from mode mismatch between waveguides and single-mode fibers and to increase the alignment tolerances. The waveguide tapers allow for an increase of the mode size by about a factor of four in the vertical and two in the horizontal direction, thereby leading to low waveguide-to-lensed-fiber coupling losses. Lensed fibers are prepared by etching and melting as described in reference [8] to further improve the coupling efficiency. A waveguide spacing of 250 μm for the input and output ports has been chosen for compatibility with fiber ribbons. Waveguides and alignment indentations are structured using H_2/CH_4-reactive ion etching and a SiO_2 mask. During the deep etch of the alignment indentations, the waveguides are protected by an additional photoresist mask, leaving the trenches and a small part of the original SiO_2 mask around the edges of the alignment indentations open. The alignment indentations are thus self-aligned to the waveguides. Microstrip lines of 25 Ω connect the switch electrodes with the beam-leads at the chip edges. These serve as electrical interconnections when the matrix is flip-chip mounted on the silicon motherboard. Figure 4.13 shows a photograph of the realized switch matrix. The taper and the alignment indentations are indicated as bright lines close to the short side of the chips, and the beam-leads extend over the long sides.

The fabrication of the Si motherboard includes two etch levels: the V-grooves (75 μm deep) and the plane containing the strip lines for the electrical connection

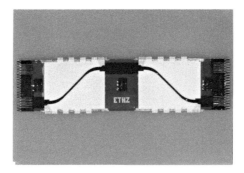

FIGURE 4.13 Photograph of a realized switch matrix showing the taper and alignment indentations as bright lines close to the short chip sides and the beam-leads projecting over the long chip sides.

(20 µm deep). Both are anisotropically wet etched using thermally grown SiO_2 as the etch mask. The underetching of the mask has been considered in the design. After silicon etching, the SiO_2 mask is removed and the wafer is reoxidized to improve the electrical insulation. The ground and signal electrodes of the microstrip lines consist of an evaporated Ti/Au metallization electroplated to a thickness of 2 to 3 µm. A polyamide layer in between is used as dielectric to form the impedance-matched microstrip lines.

The switch matrices are placed face-down onto the V-groove motherboard for packaging. Tweezers mounted on a micropositioner hold the chip for positioning. The beam-leads are used as alignment marks in the longitudinal direction. They have to mate with the strip lines on the Si-substrate. In the horizontal direction, positioning can be controlled by a magnified view of the facet and the V-grooves or by moving the chip slightly on the silicon until the alignment indentations snap into the V-grooves. Arrays of fibers are then inserted into the V-grooves and positioned in front of the facets. The required alignment precision for cleaved fibers is 10 to 20 µm, which is reduced to about 5 µm for lensed fibers. The fibers are clamped together to the motherboard, using a small piece of a inverted V-groove array before cementing the chip and fiber with an optically transparent epoxy. The primary coating of the fibers is fixed at the border of the Si motherboard for strain relief and to prevent the fibers from being bent and broken at the edges of the Si substrate. Next, the beam-leads are bonded to the strip lines on the Si motherboard by means of a wedge-bonder. Figure 4.14 illustrates both interfaces (i.e., the optical and the electrical) of the mounted switch matrix. The left SEM picture shows the optical self-alignment of the device waveguides and the optical fibers in the Si V-grooves. The right picture shows the beam-leads bonded to the transmission lines on the motherboard. Finally, the modules are connected to the InP/InGaAs HBT driver circuit [29, 30].

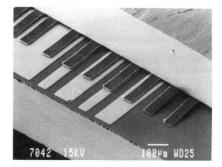

FIGURE 4.14 SEM pictures of a flip-chip-mounted switch matrix illustrating both interconnections; the optical on the left showing the self-alignment of waveguides and fibers in the V-grooves, and the electrical on the right with the beam-leads bonded onto the striplines.

4 Input fibers   InP high-speed driver   Si-motherboard

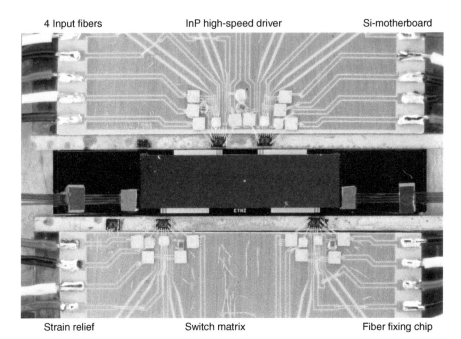

Strain relief   Switch matrix   Fiber fixing chip

FIGURE 4.15 Photograph of the center part of a high-speed switch module. The main parts are the flip-chip mounted switch matrix on the Si board self-aligned to four input and four output fibers with strain relief, and the electrical connections to the high-speed drivers.

Figure 4.15 shows a photograph of the center part of a high-speed switch module. The main parts are the self-aligned flip-chip-mounted switch matrix on the Si board, with four input and four output fibers having strain relief, and the electrical connections to the high-speed drivers. The overall insertion losses for the matrix chip with two switches in series are below 5 dB [23], and they are below 11 dB [24] for a more complex matrix with three switches in a series. The excess loss caused by the Si V-groove self-alignment mounting is in the range of 0.5 dB per coupling. A comparison with alignment tolerance measurements before mounting shows that the achieved alignment precision is in the submicron (0.5 to 0.8 μm) range. The polarization dependence of the on state is around 0.5 dB. On–off ratios for worst-case polarization are better than 15 dB. Large-signal high-speed switching experiments using flip-chip mounted and packaged matrices show optical rise and fall times of the modulation switches of less than 180 ps. This allows switching speeds of up to 3 Gbit/sec. The best results for speed were achieved with modules on which the transmission lines were designed as traveling wave electrodes, driven in a push–pull mode by the InP/InGaAs HBT drivers. With these modules, 12 Gbit/sec:3 Gbit/sec optical time domain demultiplexing [25] was performed, demonstrating the high-speed performance of the beam-lead and transmission line technology.

4.4.2 LASER ARRAYS

After LEDs, semiconductor laser diodes are the most numerously packaged opto-electronic devices, dominated by the CD lasers. Fiber pigtailing has become more important for the increasing fiber-optic communication market [31]. The demand for higher transmission rates in broadband telecommunication networks, data communication systems with high-density interconnections, and high data throughput with low cross talk is increasing. For datacom systems, parallel optical links offer a method to accomplish the data transfer requirements using their high bandwidth, low attenuation, low cross talk, high packaging density, and insensitivity to electromagnetic interference [32, 33]. To improve the performance of long-distance telecom networks, wavelength division multiplexed systems (WDM) are introduced to increase network capacity where switching speed improvements are reaching their limits. Even though the application of vertical cavity surface-emitting laser (VCSEL) arrays for parallel optical links is a hot topic of research [34–37], arrays of edge-emitting lasers are also used for high-speed links [38, 39]. For WDM systems, single-chip laser arrays emitting different specific wavelengths are being developed [40] to reduce the packaging effort, compared to the packaging of many single lasers [41]. Laser array packaging is also used for high-power applications by bundling the fiber-coupled output.

In view of the above-mentioned applications, the Si V-groove self-alignment technique has been adapted for laser array packaging. Although this technology can be used for single-fiber connections, it shows its full strength for array coupling where a high number of coupling partners have to be aligned precisely. Compared to the electro-optic switch packaging shown earlier, different new component properties have to be addressed. First, more power must be dissipated from a much smaller device. The flip-chip mounting with the alignment indentations inserted into the V-grooves and the use of beam-leads for electrical connections, as shown before, provide only a very small direct chip-to-board contact and, therefore, a modest heat transfer. Substantial heat dissipation is obtained if solder covers the laser stripe areas. In this case, the upside-down mounting process removes the heat closest to the origin of the heat source. Very efficient heat dissipation is achieved because Si exhibits a much larger thermal conductivity (150 W/m°K) than GaAs (40 W/m°K) or InP (70 W/m°K). The small separation with a few microns of solder further improves efficiency. In a standard upside-up mounting technology, the heat has to be transferred through a chip substrate of 100 to 200 µm thickness.

The laser array packaging design and realization described below on one hand addresses the problem of heat dissipation by using solder areas instead of beam-leads. On the other hand, the Si V-groove self-alignment process should also be investigated with regard to the following questions: Is the technology applicable if the semiconductor device and Si motherboard are realized in different labs? Do minor postprocess modifications on an existing device fabrication process allow the use of this flip-chip technology? The answers to these questions are important in terms of the diversity of possible applications and the influence

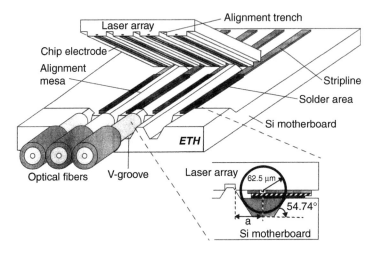

FIGURE 4.16 Principle of optical self-aligned laser array packaging technique on a Si motherboard using V-grooves and alignment mesas for precise alignment of optical fibers and device waveguides, as well as solder areas for electrical connection, heat dissipation, and fixation.

of the packaging process on the device fabrication. Figure 4.16 illustrates the basic principle of the Si V-groove self-alignment technique for laser arrays. There are three main differences between this technique and the mounting of space-switch matrices. Fibers are coupled only to one facet, solder areas between Si and chip replace the beam-leads, and the optical alignment relies on both V-grooves and separate alignment mesas. The fiber connection to a single facet is very helpful for the layout of the electrical interconnection lines. This is especially important for larger arrays, because the "backside" can be used to connect the solder areas for the small chips. Crossings of transmission lines underneath the laser array can be avoided. The solder pad areas are matched in size to the chip electrode pads. If not already defined, the pads are chosen to cover the laser stripe over the full length and to be as large as possible to achieve optimum heat dissipation as well as low electrical resistance and good die bond. The use of the V-groove sidewalls alone for optical passive alignment is not possible in most cases for devices that are postprocessed because precise etching relatively close to the device waveguide is required. Special mesa structures on the Si board are then used to provide the self-alignment. The alignment in both the horizontal and the vertical direction is again defined by the distance a between waveguide center and the alignment edge, as shown in the inset of Figure 4.16. However, the alignment is not carried out on the identical etched surfaces. If the sidewalls of these mesas, relevant for the optical alignment, are etched in the same step as the V-grooves, the same alignment precision is achieved as that in the case of V-groove alignment only. Low demands regarding sidewall slope and roughness of the chip trench etching as well as relaxed placing precision for the exact final alignment hold for this packaging design too.

The developments presented for laser array packaging were carried out in cooperation with IBM Zurich Research Laboratory. The first modules were realized with laser array chips provided by IBM that included monitor diodes [4]. Alignment trenches at the chip borders were first used to investigate self-aligned flip-chip bonding to silica on silicon waveguides, using intentionally misaligned solder bumps to pull the chip on the standoffs to the microstops [42]. With a Si motherboard layout adapted to these existing alignment trenches, the feasibility of the V-groove self-alignment technique using solder pads for electrical connection and die bonding has been shown. Four-channel laser array modules with integrated monitor diodes coupled to optical fibers have been realized [43]. The more recent modules are based on 980-nm lasers. This type of laser has been developed as a pump for Erbium-doped fiber amplifiers [2]. These lasers can be produced with high yield and show high efficiency, high output power, and high reliability [44]. So far, 4-, 8-, and 12-channel laser array modules have been realized using these lasers and the V-groove self-alignment technology described below [45], [46].

For the self-aligned packaging of laser arrays on the Si motherboard, the goal was to achieve the highest possible alignment accuracy with minor modifications of the existing laser fabrication process, while maintaining the good laser characteristics. After laser fabrication, the wafers are postprocessed to realize the precise alignment trenches on the chips. Self-alignment between laser waveguides and alignment trenches is achieved by using the same photoresist mask for the etching of the laser rib as for the evaporation of a Ni layer. This Ni layer acts afterward as an etch mask for the trenches. This process requires only one additional evaporation and one additional etching step without an additional mask level for the definition of the alignment trenches to achieve the highest precision. The opening width and distance from the laser stripe could be reproduced with an accuracy of ± 0.3 μm. The anisotropic etching of the alignment grooves was carried out with reactive ion etching in a chlorine plasma. At the point where the material is normally cleaved into single devices, 30-μm-wide and 10-μm-deep alignment trenches with nearly vertical sidewall into the GaAlAs were realized. Figure 4.17 shows SEM pictures of a processed laser bar and a magnified detail of one individual laser. The alignment trenches, the laser rib, and the electroplated pads on both sides can be clearly seen. The alignment of the laser chip on the silicon motherboard relies only on the accuracy of the trench edges. Therefore, the roughness and the angle of the sidewalls of the etched grooves are not critical. Also, the roughness of the trench bottom causes no problem as long as it does not touch the Si mesa surface. The finished lasers show threshold currents of 14 mA at room temperature and can be operated CW linearly up to an optical power of 120 mW. The device yield was between 90 and 95%, which allowed cleaving out arrays of different size (4-, 8-, and 12-channel lasers). All these lasers work within the same specifications as the lasers without alignment trenches from the standard process.

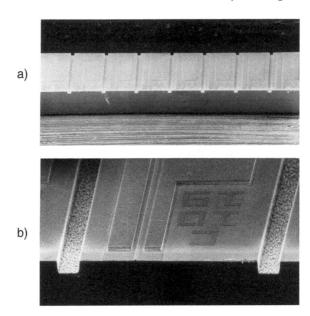

FIGURE 4.17 SEM pictures of laser bar showing (a) an overview and (b) a detail magnification with waveguide rib, plated electrode pad, and the 30-μm-wide and 10-μm-deep alignment trenches used for self-alignment on the Si motherboard.

The layout of the Si motherboard for the flip-chip packaging has been designed on the basis of the geometrical dimensions and positions of the alignment trenches and laser waveguides. The design of the Si board, along with its profile, is shown in Figure 4.18. The board includes the V-grooves for the fiber array in the front part. A second Si etch step is used in the center and the back part of the silicon board. This level exposes the alignment mesas, prevents the laser chip from unwanted contact with the board, and supports the electrical connections to the solder areas underneath the chip. For suspension and alignment of the laser chip, every second alignment trench has a Si mesa counterpart. This is a trade-off between the definition of precise position, the correction for chip or board bending, and the suspension of the chip with respect to solder forces during the solidification of the solder, especially for large array mounting. The solder areas are adapted to the electroplated chip contact pad area and include the laser rib part to achieve optimum heat removal. The geometrical dimensions for the optical alignment have been designed such that the laser chip penetrates 6 μm into the silicon motherboard. This results in chip-placing tolerances of ±4 μm and allows for large tolerances in the Si etching.

To fabricate the motherboards, the Si is anisotropically etched to a depth of 18 μm for the connecting plane and to about 80 μm for the fiber grooves. The wafers are then thermally oxidized to a thickness of 0.5 μm before metallization to avoid a direct metal to semiconductor contact. The transmission lines are Au electroplated to a thickness of approximately 3 μm for lower resistivity. The solder

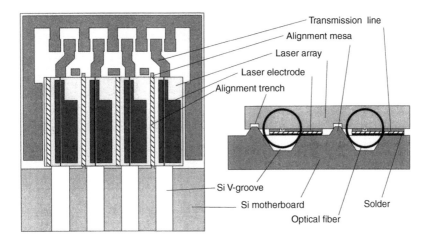

FIGURE 4.18 Layout of the Si motherboard with profile for self-aligned flip-chip mounting of the 980-nm laser arrays.

areas are formed with electroplated In, having a thickness of about 6 μm. This thickness is not very critical because the liquid In either bumps up to wet the laser pads or is pushed down by the laser chip during soldering. For the packaging process, the Si motherboard is placed on a heating chuck with a small amount of flux applied to the solder areas. The turned-over laser chip, carried by vacuum tweezers, is then positioned over the alignment mesas and set down. The insertion of the alignment trenches over the mesas, as well as the self-positioning effect, can be observed with a stereo microscope or a camera system. Subsequently, the chip can no longer be moved laterally to the trenches; however, it can still slide longitudinally along the mesas. This axial position can be defined with microstops on the Si. For simple alignment structures such as the straight trenches, the chip facet itself is used for this alignment. For rectangular alignment trenches, the corresponding Si mesas can directly provide the positioning in both directions [43]. The design takes into account that the lateral alignment is much more critical than the longitudinal one. The soldering of the chip is then carried out by heating the chuck up to 180°C for about 30 seconds. Afterward, ribbons of 4-, 8-, or 12–channel multimode fibers (MMFs), which have been cleaved together to the same length, are placed in the V-grooves and positioned close to the chip facet. The fiber to the chip distance can be adjusted by looking from the top, as the axial alignment tolerances are about 20 μm for cleaved and 10 μm for lensed fibers or using microstops. The SEM photographs in Figure 4.19 show an overall view and an alignment detail of such a flip-chip mounted laser array with eight-channel lasers. The etched alignment trenches and the silicon mesas that provide the optical alignment to the fibers are clearly shown in the right picture. The packaging is completed by fixing the fiber array close to the chip and the primary coating close to the Si motherboard edge. The additional fixing of the primary coating to the motherboard acts as a strain relief and prevents fiber breaking at the Si edge.

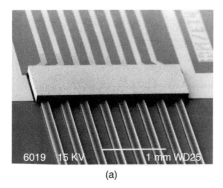

(a)   (b)

FIGURE 4.19 SEM pictures of packaged eight-channel laser array with (a) an overview and (b) a detail magnification showing the chip-to-fiber alignment with chip trench inserted over Si mesa and fibers in V-grooves.

The 4×9-mm size of the board allows easy packaging of the pigtailed laser arrays in butterfly packages, as illustrated in Figure 4.20. The low temperature sensitivity of these lasers creates the possibility of operating them without active temperature control, thereby lowering the costs for packaging itself and for the modules.

Measurements of the laser characteristics have been carried out before and after mounting. Coupling characteristics have been investigated by first measuring with a large-area photodiode in front of the lasers and by then comparing that measurement to the light coupled to optimally positioned 50 μm core/125 μm diameter MMFs and 5 μm core/125 μm diameter SMFs at 980 nm wavelength. Best coupling efficiencies of from −3.3 to −3.5 dB with cleaved and −1.1 to −1.3 dB with lensed MMFs are measured, as well as −4.6 to −4.8 dB for SMF. The packaged

FIGURE 4.20 Photograph of a four-channel laser array mounted in a butterfly package.

4-, 8-, and 12-channel laser arrays with MMF ribbons show coupling efficiencies of -3.4 ± 0.2 dB for all channels with a mounting excess loss of less than 0.3 dB. A four-channel array has been coupled to lensed MMF, achieving -1.3 ± 0.2 dB. Coupling to 980-nm SMF results in a 5.5 ± 0.8-dB coupling loss, including a less than 1.5-dB mounting excess loss. This loss includes differential loss between the channels resulting from different rotational position of the fibers. By rotating the fibers into appropriate position, the excess loss is reduced to less than 0.5 dB, clearly indicating that for techniques relying on fiber surface alignment, core eccentricities and tolerances in outer fiber diameter play an important role. These problems are less critical when standard telecom fibers for the 1.3 to 1.5-μm wavelength range are used because of their much higher precision resulting from high-volume production and tighter process control. Regarding alignment precision, the Si V-groove self-alignment technique is feasible for SMF array coupling and provides simple passive alignment of MMF arrays. Unchanged power vs. current curves both before and after mounting for currents up to 150 mA also indicate that the heat removal through the solder areas to the silicon is sufficient. Furthermore, this work shows that the passive self-alignment can be achieved with only minor modifications to existing device processes. The motherboard design and fabrication can also be carried out separately. This mounting technology is therefore very promising for low-cost laser array module packaging.

4.4.3 TILTED AMPLIFIER ARRAYS

Semiconductor optical amplifiers (SOAs) are of great interest to all-optical network systems because of their signal amplification, fast switching, and wavelength conversion capabilities. The introduction of these components into fiber-optical systems forces specific component requirements such as polarization independence, low gain ripple, and low coupling losses to SMFs.

Because of the rotational symmetry of the standard telecom fibers, the state of polarization is not maintained along the transmission path, making polarization insensitive devices necessary. The SOA has therefore to amplify all the different states of polarization by the same amount [47, 48]. The low gain ripple (i.e., smooth gain vs. wavelength behavior) is important to achieve equal amplification for different wavelengths. This parameter is critically dependent on light reflections into the amplifier. Back-reflections larger than 1×10^{-4} begin to deteriorate the performance. The packaging of the SOA is therefore strongly affected by the required high-quality antireflection (AR) coatings on the chip facets and the optically coupled fibers. An elegant way to achieve low facet reflectivities is the use of tilted waveguides. As the light intersects the chip facet off the perpendicular, direct reflection into the waveguide is strongly diminished [3, 49]. This effect depends on the angle and the waveguide structure. For smaller waveguide modes and therefore larger light divergence, a larger deviation from the perpendicular to the facet is required to achieve the same reduction in back-reflection. This angle is very important for the packaging, because the large refractive index of the semiconductor material causes an angle increase at the facet by a factor of 3 to 3.5 from light refraction. Tilt angles of 10° result in angles of up to 35° outside the chip, which complicates the packaging process significantly.

Mode converters or waveguide tapers can not only provide better coupling and alignment properties but also reduce the required tilt angles [50, 51]. The small tilt angles are important, especially for coupling array devices for which tolerances in angle strongly affect the fiber array alignment (see subsection on optical alignment and following text). Devices with multiple SOAs on one chip are becoming more and more important, not only as compact, multichannel devices for amplification and gating [52] but also for monolithic integrated devices. These components can be used, for example, to compensate for splitter losses [53] or to build space switch matrices [54], wavelength-selective components [55], or high-speed all-optical devices [56]. All these devices require optical array connections and exhibit the sensitivity to back-reflections. Their connection to fibers or silica-on-silicon wave-guides is complicated either by the requirements of the AR coatings [57] or by the tolerances and alignment problems of tilted waveguide devices [58].

The application of the self-aligned Si V-groove technology to SOA array device packaging is now described. The alignment principle remains the same, and the arrangement is similar to the one used for the laser arrays, in that the fibers and chips are not aligned using the same etched Si sidewalls. This is caused by the special geometry and the device fabrication procedure. The self-aligned coupling difficulties resulting from the tilted waveguides and the resulting problems of light refraction at the facet have already been discussed in Section 4.3.1. The main problems are, first, the propagation direction of the light outside of the amplifier chip and the distance between the amplifiers for a fixed fiber spacing of 250 µm depend on the magnitude of the mode effective index. Second, the alignment indentations become tilted relative to the waveguides, and cleaving of the chip beside the predetermined position causes horizontal misalignment (see Figure 4.9). An incorrect chip length Δg of 5 µm on a side results in a horizontal shift Δx of the outside beam axis of 1.4 µm, for a 7° tilted waveguide. Our solution to solve this problem is illustrated in Figure 4.10. Angle-polished fibers are used for the optical coupling. The light is refracted back at the 15.5° tilted surface. The horizontal shift Δx can be compensated for by a difference in fiber-to-amplifier distance Δg_2, which allows more alignment tolerances than the horizontal misalignment. The use of the angle-polished fiber surface parallel to the chip facet and not perpendicular to the optical beam also makes an AR coating on the fiber unnecessary. For a successful application of this principle, however, amplifiers with tapered waveguides are needed. Their advantages are manifold. A small tilt angle can be used to achieve low back-reflection and reduce the sensitivity of all the angle-dependent tolerances. Furthermore, the larger modes allow better coupling efficiency to the "butt"-ended fibers and relax the alignment tolerances.

The amplifier array module described here [59] has been developed in coop-eration with Alcatel Alsthom Recherche in France [60]. The amplifier array provided by Alcatel is based on the polarization-independent bulk InGaAsP/InP

FIGURE 4.21 Photograph of amplifier array chip with the 7° tilted waveguides in the center and the 15.5° tilted alignment indentations closer to the chip corners.

devices developed for high fiber-to-fiber gain and high-saturation output power at a 1.55-µm wavelength [61]. The nearly square active-stripe waveguide structure ensures low polarization sensitivity. The short windows and lateral tapers of ~100 µm in length on both sides lower the far-field divergence and enhance the coupling efficiency. The process and geometry have been adapted to realize the four-channel self-aligned SOA module. The resulting array of four amplifiers has 7° tilted waveguides to provide low facet reflectivity. Their spacing was chosen to fit with the 250-µm separated SMFs. The 15.5° tilted alignment indentations for flip-chip alignment are defined in the same step as the waveguides to achieve maximum precision. This precision is maintained as a selective regrowth process for the upper amplifier cladding is used, leaving the alignment regions open. After thinning down to 120 µm, the chips are cleaved with a precision of within 5 µm on both sides. The facets are AR coated with a TiO_2/SiO_2 coating. Figure 4.21 shows a picture of an amplifier array chip with the four 7° tilted waveguides underneath the wide electrode pads in the center of the chip, while the four alignment indentation regions are closer to the edges. The indentations each consist of two alignment mesa stripes that fit into the special alignment V-grooves on the silicon board. Figure 4.22 shows both the layout (a) and a photograph (b) of the corresponding motherboard. The board consists of two arrays 70 µm deep and 250-µm-spaced, anisotropically etched V-grooves at both sides for the fiber arrays, with simultaneously etched grooves for the chip alignment. The 20-µm-deep etched center part is used for the electrical connections. The connections include electroplated Au lines to the edge of the mounting board and electroplated In solder stripes to contact and fix the amplifier array. The fibers are fixed in auxiliary Si V-groove arrays and polished together to the desired angle of 15.5°. No AR coating is applied to the polished fiber ends. The array with four tilted amplifiers is flip-chip soldered onto the Si motherboard, as described previously for the laser array, with the alignment trenches inserted in the alignment grooves. Two angle-polished fiber arrays are then placed in the V-grooves, held down with short pieces of turned-over V-grooves, and fixed with epoxy. Only the distance between the fiber and the amplifier facet has to be adjusted to optimize the coupling efficiency to compensate for the misplaced cleaving of the SOA length. No other

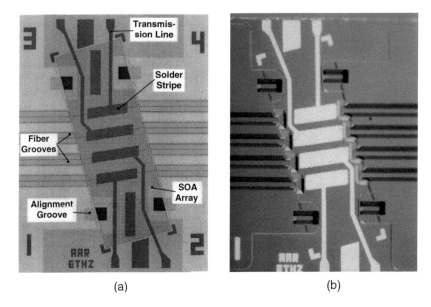

(a)   (b)

FIGURE 4.22 Design of the Si motherboard (a) and photograph of its realization (b) used for flip-chip mounting of the tilted amplifier array with V-groove arrays for the fibers and special alignment grooves for the alignment features of the SOA chip, including transmission lines and solder stripes for electrical connections, fixation, and heat dissipation.

adjustments, neither horizontal, vertical, nor rotational have to be made. Figure 4.23 shows a photograph of a flip-chip-mounted SOA array coupled to an angle-polished fiber array. Compared with an optimized coupling of an individual amplifier, the collective coupling of four channels with a fiber array results in a coupling penalty of less than 0.5-dB.

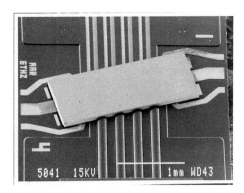

FIGURE 4.23 Scanning electron microscopy picture of a tilted amplifier array mounted on a Si motherboard and coupled to an angle-polished fiber array.

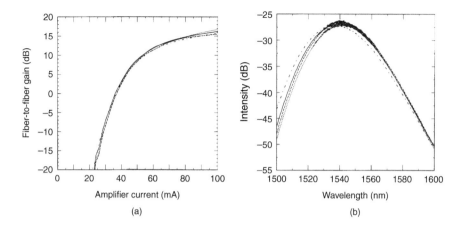

(a)   (b)

FIGURE 4.24 Fiber-to-fiber module gain of the four amplifiers measured at 1554 nm and 20°C with a 1-nm optical filter (a) and the amplified spontaneous emission spectrum at 100-mA amplifier current, illustrating the low ripple without antireflection coating on the fibers (b).

The characteristics of a packaged device are summarized in Figure 4.24. The module shows a fiber-to-fiber gain of 16 ± 1 dB for all four amplifiers and a low ripple of less than ±0.2 dB, all without AR coating on the fibers. The use of "butt" fibers in the arrangement reduces the coupling efficiency by about 4 dB compared to lensed fibers but allows the use of passive self-alignment with small differential loss between channels. With further improvements in waveguide tapering to achieve larger modes, fiber-to-fiber gains of greater than 20 dB are possible, using the described self-aligned mounting technology. The low polarization dependence and the high switching capability of the packaged amplifier array have been used to realize a high-speed gate array module, as shown in Figure 4.25. As shown, the

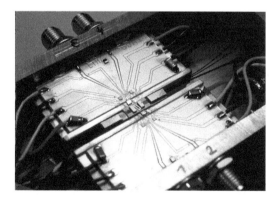

FIGURE 4.25 Photograph of the center part of a four-channel gate array module for 2.5 Gb/s realized with an optically self-aligned, tilted amplifier array connected to high-speed electronic drivers.

amplifier itself is the tiny chip slightly tilted in the center, mounted upside down on the Si motherboard, which also carries the fiber arrays and the fiber-fixing chips beside the SOA. The fiber arrays are fixed beside the mounting board between the Si parts that have been already used for polishing. The boards also hold the fiber coating that provides the primary fiber strain relief to prevent breakage. The electrical lines on the Si board are connected with short bonding wires directly to high-speed electronic drivers, which are soldered directly onto heat sinks that are thermally separated from the amplifier array. A thermoelectric cooler stabilizes the temperature of the amplifier array. Impedance-matched electrical transmission lines connect the drivers to standard high-speed connectors. The module can thus be used in optical networks for routing and switching applications with switching speeds of up to 2.5 Gb/s [62, 63].

4.4.4 CHIP-TO-CHIP COUPLING

Different optoelectronic devices are combined to achieve the functionality of a port or access node of a fiber-optical network. Many electrical and optical interfaces between ports or nodes degrade performance and reliability. One way to reduce optical interfaces between components is the monolithic integration of several optic devices on a single substrate [64]. A lower number of interfaces as well as alignment and fastening steps require, however, a larger number of—and more complex—processing steps. The complex regrowth with different material composition on recessed substrates, for example, influences yield and reliability [65]. Yield may be low, as each device cannot be optimized separately, and new parameters, such as optical and electrical cross-talk on the single chip, are involved. In contrast, hybridization techniques offer advantages for the implementation of very different waveguide materials—even nonsemiconductor ones. Hybridization approaches that use the same technique for waveguide device interconnections as for fiber-to-waveguide coupling are preferable. The pigtailing of a single chip can then be extended to optoelectronic multichip modules. The use of Si boards as hybridization platforms, including silica-on-silicon waveguide devices, is described in other chapters of this book.

The Si V-groove self-alignment technique described in this chapter has also been investigated for its suitability as a hybridization technology. Figure 4.26 shows a sketch of a hybridization arrangement. Different independently optimized waveguide devices are directly coupled optically. These chips can be based on III–V semiconductor materials with different structures and functionalities but also can be made of other materials such as silica on silicon. Furthermore, the mounting board can hold flip-chip-mounted driving electronics and provide the electrical transmission lines to the OEICs or the drivers. The electrical substrate-to-chip connections can be carried out using beam-lead or solder bump technology. The optical alignment, including the connection to the optical fibers, is carried out using the principle described before. Etched trenches on the chips are inserted in the fiber V-grooves or in special alignment grooves. For multichip applications, a common level for all the optical mode centers relative to the Si surface has to

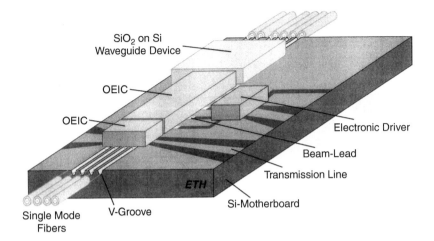

FIGURE 4.26 Silicon motherboard for hybridization of different integrated optic devices and electronic chips using the V-groove self-alignment technique for optical coupling between the waveguide components and fiber arrays. Electrical chip-to-board connections are provided either by solder bumps or by beam-leads.

be defined to determine the proper distance a between waveguide center and alignment edge (see Figure 4.7). The design of board and alignment structures on the chips has to consider the possible etch dimensions to achieve proper alignment without contact with the Si surface. Nevertheless, high alignment precision, even with relatively large Si etch tolerances, remains if the Si sidewalls for the alignment are defined and etched in the same step.

The feasibility of the Si V-groove self-alignment technique for hybridization has been investigated with two set-ups. First, the technology was set up for coupling the silica-on-silicon waveguide array to the fiber array. Second, direct chip-to-chip coupling experiments with semiconductor waveguide arrays were performed.

Single-mode silica-on-silicon waveguide devices were made using a fabrication process, which allows formation of the high-precision alignment features [66]. Chips with arrays of four waveguides were then self-aligned to SMFs by inserting the alignment ribs of the turned-over silica chip and the fibers into the same V-grooves of a Si board [67]. Figure 4.27 shows a SEM picture of such a flip-chip-mounted silica-on-silicon device. Submicron alignment precision was achieved. Mounting excess losses of 0.2 dB for butt-coupled fiber and of about 0.6 dB for lensed-coupled fiber were achieved using silica waveguide modes of about 5 μm in diameter. This shows the feasibility of the mounting technology for silica waveguide devices and the possible optohybridization of these devices with active components and fibers on a single mounting board.

The chip-to-chip coupling experiments were carried out with two types of compound semiconductor waveguide structures [68]. The devices include arrays of four

FIGURE 4.27 SEM picture of flip-chip-mounted silica waveguide chip with alignment ribs inserted into the V-grooves of the Si mounting board.

waveguides each, showing elliptical mode diameters ($1/e^2$ intensity) of about 3×5 and 1×5 μm^2, respectively. The larger mode corresponds to devices with integrated waveguide tapers, whereas the other, very thin, mode is used in high-performance electro-optic devices. Coupling loss measurements to butt-ended and lensed fibers show values of ~4 and ~1.5 dB for the large and ~9 and ~4.5 dB for the small waveguide modes, respectively. Direct chip-to-chip coupling of two devices with the same flip-chip aligned waveguide structure in the same Si V-grooves results in ~1.5-dB coupling loss for waveguide arrays with the large modes, and a ~5-dB loss with the small ones. Alignment tolerance calculations and measurements indicate clearly the problem of direct interchip coupling with small modes. The alignment in the vertical direction becomes extremely critical and is influenced strongly by the chip-to-chip gap. A gap of 5 μm already introduces a coupling loss of more than 3 dB. Nevertheless, the achieved interchip coupling efficiencies are similar to those achieved for only one waveguide-to-lensed fiber connection (i.e., direct chip-to-chip coupling shows about half the loss compared to an interconnection with an optical fiber lensed on both ends). However, angle tolerances between waveguides, the respective direction of the alignment features, and the cleaved facet are important. If two 2-mm-wide chips are coupled, exhibiting a total angle error of $1°$, the gap at one end of the facet is 35 μm when the chips touch each other.

The performed experiments show that the Si V-groove self-alignment technique can be used for hybridization of different waveguide components and fibers on a single board. A successful design has to take into account many parameters such as the waveguide structures, device processing regarding alignment features, many geometrical tolerances, electrical connections, heat dissipation, and other packaging relevant aspects.

4.4.5   DEVICE PACKAGING USING A SIMILAR
ALIGNMENT PRINCIPLE

This subsection will shortly review two self-aligned packaging techniques using similar alignment principles to those of the Si V-groove technology already described in this chapter. The first approach is sketched in Figure 4.28 and is intended for the application of direct laser to silica-on-silicon waveguide coupling [69]. The Si board is used for the silica-on-silicon device and acts at the same time as the mounting board with alignment features for the laser chip. The Si board shows a recessed area in front of the silica waveguide and an alignment groove in parallel. For optimum precision, the waveguide core and the opening for the V-groove etch process are defined in the same step. This procedure is also chosen to provide an alignment ridge in parallel to the laser ridge on the InP. This ridge fits into the Si alignment groove when flip-chip mounted and provides the horizontal alignment. The vertical position for placement is defined by the thickness of the silica pad, including contact metallization and solder. A specific gap between the laser and the silica waveguide is achieved by moving the laser until it abuts the etched microstops on each side of the waveguide. The large coupling loss of ~12 dB is attributed to the mode mismatch of ~4 dB and to coupling losses of ~8 dB resulting from the gap and a lateral misalignment of ~1.5 μm. With the development of a tapered laser with a much larger mode size [70], the passive alignment is eased (see Chapter 10). A newer process for laser pigtailing

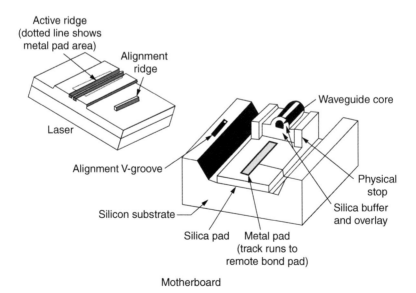

FIGURE 4.28 Schematic diagram showing the mechanisms for passive alignment of a laser diode to a silica waveguide on a silicon motherboard (from reference [69]).

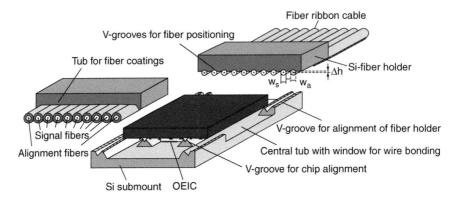

FIGURE 4.29 A self-aligned switch packaging technique on a silicon submount. It contains V-grooves to take up the alignment ribs defined on the InP chip, and alignment fibers that are exclusively used for the positioning of the whole fiber array (from [73]).

is based on precision-cleaved laser facets that are moved toward the silica microstops on the Si board, carrying the fiber in a V-groove for horizontal alignment. The vertical alignment is again defined by metallization thickness [71]. This second packaging principle is now similar to those described elsewhere in this book, such as in Chapter 3.

The second approach is used for optical space switch matrices and is based on the same alignment principle as described in this chapter [72, 73]. Figure 4.29 illustrates the principle of coupling eight-channel fiber arrays self-aligned to an InP-based space switch matrix. The fiber arrays are premounted on Si V-groove boards to carry the signal fibers from a ribbon in the center while having two alignment fibers at the outside. The InP base switch matrix exhibits alignment ribs that are simultaneously processed with the waveguide ribs. The Si submount includes three major features for the self-aligned packaging. There are V-grooves in which the alignment ribs of the turned-over switch are inserted, V-grooves that receive the alignment fibers of the fiber carriers, and a central tub underneath the chip center. The first V-grooves and the alignment ribs of the chip provide a defined horizontal and vertical position for the optical modes relative to the Si board. With the fiber alignment grooves, the fiber carrier is then appropriately positioned in front of the switch interface. Fiber array carriers with a different V-groove width, w_s relative to w_a allow for the correction of vertical differences Δh in mode height caused by process variations in the alignment rib and waveguide processing from wafer to wafer. The opening in the Si board provides direct access to the chip electrode pads and allows wire bonding to a printed circuit board. Excellent butt-ended fiber-to-chip coupling losses of 0.5 dB are achieved using waveguide tapers, the adjustment of the fiber array in vertical direction, and index matching-glue between the fibers and the chip.

4.5 SUMMARY

The optical coupling between single-mode waveguide devices and SMFs sets severe demands on the packaging process. Alignment precision in the micron range or even better is required for efficient light transfer. After alignment, the position must be maintained during the fixation process and the lifetime of the module. The Si V-groove self-alignment technique described in this chapter provides passive self-alignment between single-mode waveguide devices and SMFs, or SMF arrays on a Si motherboard. The main feature of this technology is the use of the etched silicon V-groove sidewalls together with the alignment indentations on the waveguide chip for high-precision alignment. In performing an alignment, the alignment features on the turned-over chips are inserted into the Si V-grooves, and the chips self-align themselves by sliding on the sidewalls. The position in both the horizontal and the vertical direction is thereby given only by the precision of the chip alignment features and is decoupled from the Si board processing precision. The following very distinct advantages of this technology are especially useful for array coupling and packaging: the alignment precision in the horizontal and vertical direction is defined by a single parameter — the distance between the waveguide center and the alignment indentation on the optoelectronic chip; the alignment relevant distance can be defined in the same photolithographic mask, providing maximum precision; the achieved alignment precision for the optical coupling is, to a certain degree, independent of the Si V-groove width and this opening size is more difficult to control to submicron precision because of the required deep etch — on the order of 70 µm and higher; a passive self-alignment is achieved as the devices and fibers slide on the Si sidewalls from the initial placement position into the final alignment position, which reduces the precision requirements of the placing tools; and a stable fixation is achieved as all the components are "pushed" into a mechanically defined position and fixed on a single mounting board.

The achieved submicron alignment precision allows efficient waveguide-to-SMF array coupling for most of the waveguide structures, as butt-ended or lensed-fibers can be used. The geometrical properties of the fibers play an important role because the alignment is carried out on the outer glass fiber surface, using self-alignment techniques. For nonstandard telecom SMF or special fibers such as polarization-maintaining fibers, this can lead to severe excess loss from mounting and especially differential loss between channels in array coupling. Tapered waveguide devices therefore simplify packaging in many aspects. They enlarge the mode on the chip close to the facet through an on-chip mode transformation; through the monolithic integration of device waveguide and taper, the most critical alignment to the small waveguide mode is inherently achieved; and the tapers not only provide much better coupling efficiencies and larger alignment tolerances but also flatter tolerance curves, which results in less differential loss and fewer power fluctuations caused by alignment variations during the packaging and over the device lifetime.

For the electrical connections to the flip-chip-mounted devices, either beam-leads or solder is used. The former are mechanically decoupled from the critical optical alignment because of their flexibility, whereas the thermal stress of the soldering process is avoided. Disadvantages are the special fabrication process required for their implementation, extra mechanical fixation of the chip, and modest heat dissipation. The use of solder opposite the chip electrode pads eliminates those drawbacks. The critical points are the solder thickness definition and the soldering process itself. This includes wetting and flux problems, as well as thermal and mechanical stress during solder solidification. Common to both connection principles are the good electrical high-speed properties.

The self-alignment principle has been used to realize optoelectronic modules, including very different devices. Packaging of large-area electro-optic space switch matrices with input and output fiber arrays, mounting of much smaller laser arrays with up to 12 channel fibers, and even the self-aligned connection of fiber arrays to amplifier arrays with tilted waveguides have been shown. Along with the fiber connection of silica-on-silicon waveguide array devices and the direct optical chip-to-chip coupling, this technology enables optohybridization on a Si motherboard. Furthermore, this development has shown that passive self-alignment can be achieved with minor modifications on existing device processes. The motherboard design and fabrication can even be carried out separately. The Si V-groove self-alignment technique offers great versatility and a wide range of application for the packaging of optoelectronic devices.

REFERENCES

1. M. Serenyi and H.-U. Habermeier, "Directly controlled deposition of antireflection coatings for semiconductor lasers," *Appl. Optics*, **26**(5), 1987, pp. 845–849.
2. H. Meier, "Recent developments of 980 nm pump lasers for optical fiber amplifiers," *Proc. Eur. Conf. Opt. Comm. ECOC'94,* 1994, pp. 947–954.
3. P. A. Besse, J. S. Gu, and H. Melchior, "Reflectivity minimization of semiconductor lasers with coated and angled facets considering two dimensional beam profiles," *J. Quant. Electron.,* **27**, 1991, pp. 1830–1836.
4. P. Vettiger, M. K. Benedict, G. Bona, P. Buchmann, E. C. Cahoon, K. Däetwyler, H. P. Dietrich, A. Moser, H. K. Seitz, O. Voegeli, D. J. Webb, and P. Wolf, "Full wafer technology—A new approach to large-scale laser fabrication and integration," *IEEE J. Quant. Electron.,* **27**(6), 1991, pp. 1319–1331.
5. I. Moerman, G. Vermeire, M. D'Hondt, W. Vanderbauwhede, J. Blondelle, C. Coudenys, P. Van Daele, and P. Demeester, "III-V semiconductor waveguide devices using adiabatic tapers," *Microelectron. J.,* **25**, 1994, pp. 675–690.
6. SELFOC® Product Guide, Nippon Sheet Glass Co., Tokyo 1995.
7. M. C. Farries and W. J. Stewart, "Fibre Fresnel phaseplates with efficient coupling to semiconductor lasers and low reflective feedback," *Proc. Eur. Conf. Opt. Comm. ECOC'90,* 1990, pp. 291–294.
8. W. Hunziker, E. Bolz, and H. Melchior, "Elliptically lensed polarisation maintaining fibres," *Electron. Lett.,* **28**(17), 1992, pp. 1654–1656.
9. H. M. Presby and C. A. Edwards, "Near 100% efficient fibre microlenses," *Electron. Lett.,* **28**(6), 1992, pp. 582–584.

10. C. A. Edwards, H. M. Presby, and C. Dragone, "Ideal microlenses for laser to fiber coupling," *J. Lightwave Technol.,* **11**(2), 1993, pp. 252–257.
11. K. Shiraishi, N. Oyama, K. Matsumura, I. Ohishi, and S. Suga, "A fiber lens with a long working distance for integrated coupling between laser diodes and single-mode fibers," *J. Lightwave Technol.,* **13**(8), 1995, pp. 1736–1744.
12. H. Kogelnik, "Coupling and conversion coefficients for optical modes," *Micro-wave Res. Inst. Symp. Ser.* **14**, Polytechnic Press, New York, 1964, pp. 333–347.
13. R. D. Guenther, *Modern Optics,* Wiley, New York, 1990.
14. D. Marcuse, "Loss analysis of single-mode fiber splices," *Bell Syst. Tech. J.,* **56**(5), 1977, pp. 703–718.
15. A. Greil, H. Haltenorth, and F. Taumberger, "Optical 4×4 InP switch module with fibre-lens-arrays for coupling," *Proc. ECOC'92,* Berlin, 1992, pp. 529–532.
16. M. J. Wale, C. Edge, F. A. Randle, and D. J. Pedder, "A new self-aligned technique for the assembly of integrated optical devices with optical fibre and electrical interfaces," *Proc. ECOC'89,* Gothenburg, Sweden, 1989, pp. 368–371.
17. C. A. Armiento, M. Tabasky, C. Jagannath, T. W. Fitzgerald, C. L. Shieh, V. Barry, M. Rothman, A. Negri, P. O. Haugsjaa, and H. F. Lockwood, "Passive coupling of InGaAsP/InP laser array and singlemode fibres using silicon waferboard," *Electron. Lett.,* **27**(12), 1991, pp. 1109–1110.
18. H. Kaufmann, P. Buchmann, R. Hirter, H. Melchior, and G. Guekos, "Self-adjusted permanent attachment of fibres to GaAs waveguide components," *Electron. Lett.,* **22**(12), 1986, pp. 642–644.
19. K. C. Gupta, R. Garg, I. Bahl, and P. Bhartia, *Microstrip Lines and Slotlines,* 2nd Edition, Artech House, Norwood, MA, 1996.
20. M. P. Lepselter, "Beam-lead technology," *Bell Syst. Technol. J.,* **45**(2), 1966, pp. 233–253.
21. J. H. Lau, *Flip Chip Technologies,* McGraw-Hill, New York, 1966.
22. H. Tsunetsugu, K. Katsura, T. Hayashi, F. Ishitsuka, and S. Hata, "A new pack-aging technology for high-speed photoreceivers using micro solder bumps," *Proc. 41st Electron. Components Technol. Conf. ECTC'91,* 1991, pp. 479–482.
23. R. Krähenbühl, "Low-Loss Polarization-Insensitive InP/InGaAsP Optical Space Switch Matrix for Optical Communication Systems," *IEEE Photonics Technol. Lett.,* **8**(5), 1996, pp. 632–635.
24. R. Kyburz, W. Vogt, R. Krähenbühl, M. Bachmann, T. Brenner, and H. Melchior, "Polarization-insensitive high-speed InP/InGaAsP access node space switch matrix for bidirectional optical ATM networks," *Proc. 8th Int. Conf. InP Related Mater. (IPRM '96),* Schwaebisch Gmuend, Germany, pp. 447–449, 1996.
25. R. Krähenbühl, "Electro-optic space switches in InGaAsP/InP for optical communi-cation," *Ser. Quant. Electron.,* **4**, Hartung-Gorre, Germany, 1998.
26. M. Bachmann, M. K. Smit, P. A. Besse, E. Gini, H. Melchior, and L. B. Soldano, "Polarization-insensitive low-voltage optical waveguide switch using InGaAsP/InP four-port Mach-Zehnder interferometer," *Proc. OFC/IOOC,* San Jose, 1993, pp. 32–33.
27. P. A. Besse, M. Bachmann, H. Melchior, L. B. Soldano, and M. K. Smit, "Optical bandwidth and fabrication tolerances of multimode interference couplers," *J. Light-wave Technol.,* **12**, 1994, pp. 1004–1009.
28. T. Brenner, M. Bachmann, and H. Melchior, "Vertically tapered InGaAsP/InP waveguides for highly efficient coupling to flat-ended single-mode fibres," *Appl. Phys. Lett.,* **65**(7), 1994, pp. 798–800.

29. R. Bauknecht, H. P. Schneibel, J. Schmid, and H. Melchior, "A 12 Gb/s laser and optical modulator driver circuit with InGaAs/InP double heterostructure bipolar transistors," *Proc. 8th Int. Conf. InP Related Mater. (IPRM '96),* Schwaebisch Gmuend, Germany, pp. 61–63, 1996.

30. R. Bauknecht, "InP double heterojunction bipolar transistors for driver circuits in fiber communication systems," *Ser. Microelectron., 72,* Hartung-Gorre, Germany, 1998.

31. S. Forrest, L. A. Coldren, S. Esener, D. Keck, F. Leonberger, G. R. Saxonhouse, and P. W. Shumate, *JTEC (Jpn. Technol. Evaluation Center) Panel on Optoelectronics in Japan and the United States,* Loyola College, 1996.

32. R. A. Nordin, W. R. Holland, and M. A. Shahid, "Advanced optical interconnection technology in switching equipment," *J. Lightwave Technol.,* **13**(6), 1995, pp. 987–994.

33. H. Karstensen, C. Hanke, M. Honsberg, J.-R. Kropp, J. Wieland, M. Blaser, P. Weger, and J. Popp, "Parallel optical interconnection for uncoded data transmission with 1Gb/s-per-channel capacity, high dynamic range and low power consumption," *J. Lightwave Technol.,* **13**(6), 1995, pp. 1017–1030.

34. L. J. Norton, F. Carney, N. Choi, C. K. Y. Chun, R. K. Denton Jr., D. Diaz, J. Knapp, M. Meyering, C. Ngo, S. Planer, G. Raskin, E. Reyes, J. Sauvageau, D. B. Schwartz, S. G. Shook, J. Yoder, and Y. Wen, "OPTOBUS I: A production parallel fiber optical interconnect," *Proc. 47th Electron. Components Technol. Conf. ECTC'97,* San Jose, CA, 1997, pp. 204–209.

35. H. Karstensen, L. Melchior, V. Plickert, K. Drögemüller, J. Blank, T. Wipiejewski, H.-D. Wolf, J. Wieland, G. Jeiter, R. Dall'Ara, and M. Blaser, "Parallel optical link (PAROLI) for multichannel gigabit rate interconnections," *Proc. 48th Electron. Components Technol. Conference ECTC'98,* Seattle, WA, 1998, pp. 747–754.

36. K.H. Hahn, "POLO — Parallel optical links for gigabyte data communication," *Proc. 45th Electron. Components Technol. Conf. ECTC'95,* Las Vegas, NV, 1995, pp. 368–375.

37. M. Usui, N. Matsuura, N. Sato, M. Nakamura, N. Tanaka, A. Ohki, M. Hikita, R. Yoshimura, K. Tateno, K. Katsura, and Y. Ando, "700 Mb/s × 40 channel parallel optical interconnection module using VCSEL arrays and bare fiber connectors," *Proc. IEEE Lasers Electro-Optics Soc. 1997 Ann. Meeting (LEOS '97),* San Francisco, CA, **1**, 1997, pp. 51–52.

38. N. Tanaka, Y. Arai, H. Takahara, Y. Ando, N. Ishihara, and S. Hino, "3.5 Gb/s × 4 ch optical interconnection module for ATM switching system," *Proc. 47th Electron. Components Technol. Conf. ECTC'97,* San Jose, CA, 1997, pp. 210–216.

39. R. Nagarajan, W. J. Sha, B. Li, P. Braid, R. Furmanak, J. Marchegiano, and B. Booth, "Gigabit parallel fiber optic link based on edge emitting lasers," *Proc. 47th Electron. Components Technol. Conf. ECTC'97,* San Jose, CA, 1997, pp. 231–233.

40. M. G. Young, T. L. Koch, U. Koren, D. M. Tennant, B. I. Miller, M. Chien, and K. Feder, "Wavelength uniformity in λ/4 shifted DFB laser array WDM transmitters," *Electron. Lett.,* **31**(20), 1995, pp. 1750–1752.

41. N. Suzuki, Y. Muroya, J. Sasaki, H. Yamada, and T. Torikai, "Multiwavelength DFB-LD array module using self-aligned solder bump bonding," *Proc. Eur. Conf. Opt. Comm. ECOC'97,* Edinburgh, **2**, 1997, pp. 212–215.

42. K. P. Jackson, E. B. Flint, M. F. Cina, D. Lacey, J. M. Trewhella, T. Caulfield, and S. Sibley, "A compact multichannel transceiver module using planar-processed optical waveguides and flip-chip optoelectronic components," *Proc. 42nd Electron. Components Technol. Conf. ECTC'92,* 1992, pp. 93–97.

43. W. Hunziker, W. Vogt, H. Melchior, P. Buchmann, and P. Vettiger, "Passive self-aligned, low-cost packaging of semiconductor laser arrays on Si motherboard," *Photonics Technol. Lett.,* **7**(11), 1995, pp. 1324–1326.

44. A. Oosenbrug, C. S. Harder, and P. Roentgen, "Threshold current as acceleration parameter for degradation of 980 nm pump lasers," *Proc. IEEE Lasers Electro-Optics Soc. 1995 Ann. Meeting (LEOS '95),* San Francisco, CA, **1**, 1995, pp. 252–254.

45. W. Hunziker, W. Vogt, H. Melchior, R. Germann, and C. Harder, "Low-cost packaging of semiconductor laser arrays using passive self-aligned flip-chip technique on Si motherboard," *Proc. 46th Electron. Components Technol. Conf. ECTC'96,* Orlando, FL, 1996, pp. 8–12.

46. W. Hunziker, "Low-cost packaging of semiconductor laser arrays," *IEEE Circuits Devices,* **13**(1), 1997, pp. 19–25.

47. Ch. Holtmann, P. A. Besse, T. Brenner, and H. Melchior, "Polarization independent bulk active region semiconductor optical amplifiers for 1.3 μm wavelengths," *Photonics. Technol. Lett.,* **8**(3), 1996, pp. 343–345.

48. L. F. Tiemeijer, P. J. A. Thijs, T. v. Dongen, J. J. M. Binsma, E. J. Jansen, and A. J. M. Verboven, "27 dB gain unidirectional 1300 nm polarization-insensitive multiple quantum well laser amplifier module," *Photonics. Technol. Lett.,* **6**(12), 1994, pp. 1430–1432.

49. C. E. Zah, J. S. Osinski, C. Caneau, S.G. Menocal, L.A. Reith, J. Salzman, F. K. Shokoohi, and T. P. Lee, "Fabrication and performance of 1.5 μm GaInAsP travelling-wave laser amplifiers with angled facets," *Electron. Lett.,* **23**(19), 1987, pp. 990–992.

50. T. Brenner, E. Gini, and H. Melchior, "Low coupling losses between InP/InGaAsP optical amplifiers and monolithically integrated waveguides," *Photonics. Technol. Lett.,* **5**, 1993, pp. 212–214.

51. A. E. Kelly, I. F. Lealman, L. J. Rivers, S. D. Perrin, and M. Silver, "Polarisation insensitive, 25 dB gain semiconductor laser amplifier without antireflection coatings," *Electron. Lett.,* **32**(19), 1996, pp. 1835–1836.

52. S. Kitamura, K. Komatsu, and M. Kitamura, "Polarization-insensitive semiconductor optical amplifier array grown by selective MOVPE," *IEEE Photon. Technol. Lett.,* **6**(2), 1994, pp. 173–175.

53. U. Koren, M. G. Young, B. I. Miller, M.A. Newkirk, M. Chien, M. Zirngibl, C. Dragone, B. Glance, T. L. Koch, B. Tell, K. Brown-Goebeler, and G. Raybon, "1 × 16 photonic switch operating at 1.55 μm wavelength based on optical amplifiers and a passive optical splitter," *Appl. Phys. Lett.,* **61**(14), 1992, pp. 1613–1615.

54. M. Gustavsson, B. Lagerström, L. Thylen, M. Janson, L. Lundgren, A.-C. Mörner, M. Rask, and B. Stoltz, "Monolithically integrated 4 ×4 InGaAsP/InP laser amplifier gate switch arrays," *Electron. Lett.,* **28**(24), 1992, pp. 2223–2225.

55. C. Joergensen, S. L. Danielsen, T. Durhuus, B. Mikkelsen, K. E. Stubkjear, N. Vod jidani, F. Ratovelomanana, A. Enard, G. Glastre, D. Rondi, and R. Blondeau, "Wavelength conversion by optimized monolithic integrated Mach-Zehnder interferometer," *Photonics Technol. Lett.,* **8**(4), 1996, pp. 521–523.

56. R. Hess, M. Caraccia-Gross, W. Vogt, E. Gamper, P. A. Besse, M. Duelk, E. Gini, H. Melchior, B. Mikkelsen, M. Vaa, K. S. Jepsen, K. E. Stubkjear, and S. Bouchoule, "All-optical demultiplexing of 80 to 10 Gb/s signals with monolithic integrated high-performance Mach-Zehnder interferometer," *Photonics Technol. Lett.,* **10**(1), 1998, pp. 165–167.

57. Y. Yamada, H. Terui, Y. Ohmori, M. Yamada, A. Himeno, and M. Kobayashi, "Hybrid-integrated 4 × 4 optical gate matrix switch using silica-based optical waveguides and LD array chips," *J. Lightwave Technol.,* **10**(3), 1992, pp. 383–389.

58. I. Ogawa, F. Ebisawa, F. Hanawa, T. Hashimoto, M. Yanagisawa, K. Shuto, T. Ohyama, Y. Yamada, Y. Akahori, A. Himeno, K. Kato, N. Yoshimoto, and Y. Tohmori, "Hybrid integrated four-channel SS-SOA array module using planar lightwave circuit platform," *Electron. Lett.,* **34**(4), 1998, pp. 361–363.

59. W. Hunziker, W. Vogt, H. Melchior, D. Leclerc, P. Brosson, F. Pommereau, R. Ngo, P. Doussiere, F. Mallecot, T. Fillion, I. Wamsler, and G. Laube, "Self-aligned flip-chip packaging of tilted semiconductor optical amplifier arrays on Si motherboard," *Electron. Lett.,* **31**(6), 1995, pp. 488–490.

60. D. Leclerc, P. Brosson, F. Pommereau, R. Ngo, P. Doussiere, F. Mallecot, P. Gavignet, I. Wamsler, G. Laube, W. Hunziker, W. Vogt, and H. Melchior, "High-performance semiconductor optical amplifier array for self aligned packaging using Si V-groove flip-chip technique," *Photonics Technol. Lett.,* **7**(5), 1995, pp. 476–478.

61. P. Doussiere, P. Garabedian, C. Graver, D. Bonnevie, T. Fillion, E. Derouin, M. Monnot, J. G. Provost, D. Leclerc, and M. Klenk, "1.55 μm polarisation independent semiconductor optical amplifier with 25 dB fiber to fiber gain," *Photonics. Technol. Lett.,* **6**(2), 1994, pp. 170–172.

62. P. Gavignet, M. Sotom, J. C. Jacquinot, P. Brosson, D. Leclerc, W. Hunziker, and H. Duran, "Penalty-free 2.5 Gbit/s photonic switching using a semiconductor four-gate-array module," *Electron. Lett.,* **31**(6), 1995, pp. 487–488.

63. F. Masetti, J. Benoit, H. Melchior et al., "High speed, high capacity ATM optical switches for future telecommunication transport networks," *IEEE J. Selected Areas Commun.,* **14**(5), 1996, pp. 979–998.

64. P. J. Williams, P. M. Charles, I. Griffiths, N. Carr, D. J. Reid, N. Forbes, and E. Thom, "WDM transceiver OEICs for local access networks," *Electron. Lett.,* **30**(18), 1994, pp. 1529–1530.

65. M. Erman, "Monolithic vs. hybrid approach for photonic circuits," *Proc. 5th Eur. Conf. Integrated Optics ECIO'93,* 1993, pp. 2–1 to 2–3.

66. Q. Lai, P. Pliska, J. Schmid, W. Hunziker, and H. Melchior, "Formation of optical slab waveguides using thermal oxidation of SiO_x," *Electron. Lett.,* **29**(8), 1993, pp. 714–716.

67. Q. Lai, W. Hunziker, and H. Melchior, "Silica on Si waveguides for self-aligned fibre array coupling using flip-chip Si V-groove technique," *Electron. Lett.,* **32**(20), 1996, pp. 1916–1917.

68. W. Hunziker, W. Vogt, R. Hess, and H. Melchior, "Low-loss self-aligned flip-chip technique for interchip and fiber array to waveguide OEIC packaging," *Proc. IEEE Laser Electrooptics Soc. 7th Ann. Meeting LEOS'94,* Boston, MA, **2**, November 1994, pp. 269–270.

69. C. A. Jones, K. Cooper, M. W. Nield, J. D. Rush, R. G. Waller, J. V. Collins, and P. J. Fiddyment, "Hybrid integration of a laser diode with a planar silica waveguide," *Electron. Lett.,* **30**(3), February 1994, pp. 215–216.

70. I. F. Lealman, L. J. Rivers, M. J. Harlow, and S. D. Perrin, "InGaAsP/InP tapered active layer multiquantum well laser with 1.8 dB coupling loss to cleaved singlemode fibre," *Electron. Lett.,* **30**(20), September 1994, pp. 1685–1687.

71. J. V. Collins, I. F. Lealman, P. J. Fiddyment, A. R. Thurlow, C. W. Ford, D. C. Rogers, and C. A. Jones, "The packaging of large spot-size optoelectronic devices," *IEEE Trans. Components, Packaging, Manufacturing Technol. Part B,* **20**(4), November 1997, pp. 403–408.

72. B. Acklin, J. Bellermann, M. Schienle, L. Stoll, M. Honsberg, and G. Müller, "Self-aligned packaging of an optical switch array with integrated tapers," *Photonics Technol. Lett.,* **7**(4), April 1995, pp. 406–408.

73. G. Wenger, M. Schienle, J. Bellermann, B. Acklin, J. Müller, S. Eichinger, and G. Müller, "Self-aligned packaging of an 8×8 InGaAsP-InP space switch," *IEEE J. Selected Topics Quant. Electron.,* **3**(6), December 1997, pp. 1445–1456.

5 Soldering Technology for Optoelectronic Packaging

Qing Tan, Yung-Cheng Lee, and Masataka Itoh

CONTENTS

5.1 INTRODUCTION

Integration of high-speed Internet access, telephone, and cable is a major driving force for fiber-optical networking and related technologies. Cost reduction is critical to accelerating the market growth of optoelectronic modules and systems, and it is estimated that packaging contributes 60 to 90% of the overall cost of an optoelectronic module (Iezekie et al., 1997), and alignment can contribute up to 90% of the packaging cost. As a result, it is necessary to connect and maintain hundreds of optical precision alignments through a batch assembly process that is compatible with the existing manufacturing infrastructure. Soldering is the technology of choice for such a cost-effective assembly process. In addition to providing electrical connections, solder is useful in the formation of passive, precision alignments for optoelectronic packaging. It can be used to couple optical fibers or waveguides to devices such as lasers, light-emitting diodes, or photodetectors. The alignments can vary from submicrometer to micrometer levels for single- or multimode fiber applications. Different designs have demonstrated precision alignments, and the aligned structures are becoming more and more complex (Lee and Basavanhally, 1994).

Solder has been used widely as a die-attach material for optoelectronic packaging, and its application range is being further expanded as a result of the development of the flip-chip assembly. Flip-chip soldering was introduced by IBM as a controlled-collapse chip connection soldering technology (Miller, 1969). Since then, many advantages of the technology have been realized: superior electrical performance, high reliability, reduced footprint, high I/O (input/output) density, low cost, efficient heat conduction, batch assembly, and self-alignment during the chip jointing. In particular, batch assembly capability and the self-alignment mechanism are critical to precision optical alignments.

Figure 5.1 illustrates the mechanism of solder self-alignment. As the temperature of the solder is raised above its melting point, the molten solder starts wetting the metal pad and moving the chip. The chip movement is driven by a surface tension force in an effort to minimize the surface area and to reach the lowest total energy of the assembly. At the final position, the system will achieve its lowest total energy, and the position of the chip is locked by cooling the solder joint. Using solder, hundreds or thousands of such alignments can be accomplished with a single batch reflow process, and the cost per alignment can be reduced by orders of magnitude.

Different soldering technologies have already been developed for various optoelectronic modules. The packaging strategies have been focused on precision alignment. This chapter presents representatives of the modules in different

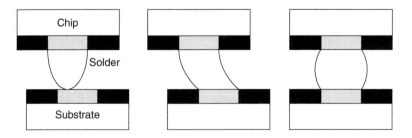

FIGURE 5.1 Self-alignment of soldering technology.

categories, as defined by the roles of the solder. For a detailed understanding of the technology, we review a case study of NEC's Au/Sn-based optoelectronic assembly technology. In addition, critical issues are discussed, with an emphasis on solder materials, fluxless reflow, design, and reliability.

5.2 SOLDER-ASSEMBLED OPTOELECTRONIC MODULES

Table 5.1 lists 18 cases taken from 68 solder-assembled modules reported between 1990 and 1997. The number of publications in this area has increased significantly recently, and therefore this review could not cover all of them. The 18 studies discussed in the table are listed in the reference section; other publications reviewed are listed in Tan and Lee (1996). Solder has been used to align photonic devices to substrate, chip carrier, or fiber; fiber to photonic device; and microlens to photonic device. The alignment methods varied from active to passive ones, both with and without solder self-alignment, and the alignment accuracy ranged from ±5 to 0.5 μm. The metallization used Ti or Cr as the adhesion layer; Cu, Ni, or Pt as the barrier metal; and Au coating to prevent surface oxidation. Length scales (diameter or width) of the solder bumps varied from 26 to 330 μm, and heights ranged from 5 to 50 μm. The solder materials were eutectic 63Sn/37Pb, In, 50In/50Pb, or eutectic 80Au/20Sn. The solder reflow used liquid flux, formic acid vapor, or forming gas. Couplings were efficient for single- or multimode applications.

Representatives of the modules are further reviewed, with the following four assembly categories: solder assembly with no precision self-alignments, self-aligned solder assembly with no mechanical stops, self-aligned solder assembly with one mechanical stop, and self-aligned solder assembly with two mechanical stops.

5.2.1 SOLDER ASSEMBLY WITH NO PRECISION SELF-ALIGNMENTS

In this assembly category, solder is used to provide electrical connections and maintain mechanical connections for optical alignments. Solder self-alignment is not an important concern. Figure 5.2 shows an optical head for CD-ROM drives (Nagano et al., 1993). The photodetector was used as a substrate to carry a preamplifier through flip-chip soldering and a laser diode through die-bonding. Before the bonding, the laser diode's position was adjusted according to an alignment mark on the photodetector. With the preamplifier and the laser diode, the photodetector assembly was flip-chip soldered to a ceramic case to form a very compact module. This solder assembly approach was better than the wire-bonding approach because of its reduced size, electromagnetic noise, capacitance, and inductance. In addition, the design increased the alignment tolerance between the optical module and a hologram element because the distance between the photodetector assembly and a glass window was well controlled. The module was 6 × 6 × 1.64 mm and 0.25 g, which was much smaller and lighter than the old module (7.5 × 8 × 3.05 mm and 0.6 g).

TABLE 5.1
Examples of Optoelectronic Modules Using Solder

Package	Alignment Method and Accuracy (µm)	Solder Bumping and Size (µm)	Solder Material	Metallization	Reflow Atmosphere/ Temperature	Main Feature	Reference
Laser diode on silicon motherboard	±20 and 5° tilt Self-aligned	Height: 8.5	Pb/Sn Eutectic	Cr/Cu/Au	Flux	Hybrid assembly	GEC-Marconi; Edge et al., 1991
Light-emitting diode/photodiode on AlN submount	Self-aligned, 3	Electrical plating, 150 × 400 × 50	Pb/Sn	Cr/Ni 0.1 µm /0.2 µm	180°C		NEC; Itoh et al., 1991
Si chip on glass	Self-aligned, 0.8	Evaporate, diameter: 130 µm	50%In/ 50%Pb	Ti/Cu/Au		Long-term alignment stability	NTT; Hayashi, 1992
High-speed photoreceiver	Self-aligned, 0.5	Lift-off photoresistor, diameter: 25	Pb/Sn or In			Microsolder bump	NTT; Tsunetsugu et al., 1992,
Light-emitting diode print head	Self-aligned					Self-align a row of light-emitting diodes	HP; Haitz, 1992
Laser on wafer board	Mechanical stop ±1		In	Ti/Ni/Au	Flux, 200°C	Mech stop only	GTE; Armiento et al. 1992
Laser diode on Si substrate	Self-aligned, ±3 ±1	Mechanically formed, diameter: 50	80Au/ 20Sn eutectic			Au/Sn solder alignment	NEC; Sasaki, 1992
Optical fiber alignment	Self-alignment for fiber					Fiber ⊥ substrate	IBM; Blacha et al., 1993

Application	Alignment	Deposition	Solder	Metallization	Flux	Notes	Reference
Photodetector on Si	Self-aligned, ±60 ±2	E-beam evaporation, height: 5	Au/Sn Eutectic	Ti/Pt/Au	Gaseous formic acid	Fluxless	AT&T; Deshmukh et al., 1993a and 1993b
VLSI/FLC SLM	Self-aligned ±2, self-pulling gap uniformity: 0.3	Diameter: 105	63Sn /37Pb		Forming gas	Mixed joint design	University of Colorado; Lin et al., 1993; Ju et al., 1993
Optical head for CD-ROM	Mechanically aligned					Flip-chip used for save space	NEC; Nagano et al., 1993
OEIC transceiver	Self-aligned, X: ±2 Y: ±0.75, vertical standoff	Diameter: 125 μm	Pb/Sn	Ti/Ni/Au Cr/Cu/Au	Flux	Self-alignment and mech. stop	IBM; Jackson et al., 1994
Optical subsriber system	Self-aligned, ±3, standoff	Thermal evaporate, diameter: 50		Ti/Pt/Au	Flux	Self-aligned/standoff	ETRI; Lee et al., 1995a and 1995b
Light-emitting diode on Silicon	Self-aligned, ±5	330 × 330 × 25	Au78/ Sn22	Ti/Pt/Au	Formic acid 300 to 320°C		AT&T; Dautartas et al., 1995
Multiplex data link	Active aligned		Au/Sn	Au/AuZn		Solder for fiber fixing	Fujitsu; Yano et al., 1995
LASER-PAC	±50 Self-aligned	Evaporated thickness 6	Au80/ Sn20	Ti/Pt/Au	Fluxless	Batch aseembly	Bell Labs; Gate et al., 1996
Free space module	Self-aligned, <2	Electroplated height 30	Eutectic Pb/Sn	Cr/Cu/Au	Flux	Align microlens to vertical-cavity surface-emitting laser	University of Maryland; Pusarla and Christou, 1996
Waveguide module	Lateral alignment, <2		Pb/Sn	Ti/Cu	Formic acid	Controlled self-alignment	University of Colorado; Morozova et al., 1997

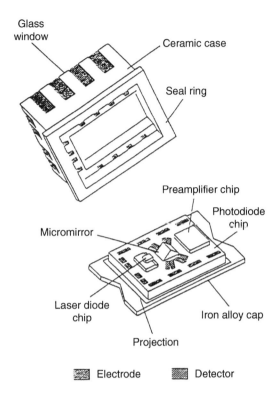

FIGURE 5.2 Optical head for a CD-ROM drive.

Figure 5.3 shows another solder application for optoelectronic packaging. A hybrid transmitter module was developed, using silicon waferboard technology and solder (Armiento et al., 1992). As shown in the figure, pedestals and standoffs were used to define the chip position during robotic placement. They were precisely micromachined, permitting the placement of the laser array to an accuracy of ±1 μm. The height of the standoff was larger than that of the deposited solder thickness, and a gram-level force was applied to the laser array during the solder reflow process. The solder was raised to wet the solder pad on the laser, and the laser was held in contact with the standoffs and pedestals. The assembly accomplished precision alignments with coupling efficiencies comparable to those achieved with active alignments. Solder joints were used to provide electrical contacts and maintain the alignments.

5.2.2 Self-Aligned Solder Assembly with No Mechanical Stops

As shown in Figure 5.1, solder self-alignment can align a component to a substrate automatically during a reflow process. Such a mechanism was used by

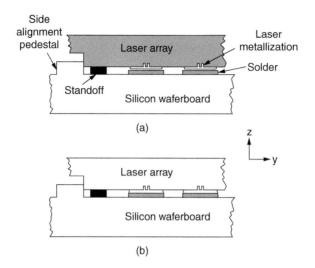

FIGURE 5.3 Precision assembly using standoffs, pedestals, and solder: (a) before reflow, (b) after reflow.

Deshmukh et al. (1993) and Nordin et al. (1992), which were reviewed by Lee and Basavanhally (1994) and another case was demonstrated by Hayashi and Tsunetsugu (1996). The module was similar to a conventional pigtail module. It aimed at providing a package solution for photodiode (PD), light emitting diode, and vertical-cavity surface-emitting laser. A zirconia ferrule was integrated with an alumina chip carrier as the packaging platform. The photonic device was self-aligned to the chip carrier, using solder. Experiments were carried out to achieve the optimum alignment result. These experimental works showed that alignment accuracy improved as the bump diameter decreased or the number of bumps increased. Submicron (0.5 µm) alignment accuracy was achieved by using 32 microsolder bumps with a diameter of 25 µm.

Figure 5.4 is another self-aligned design, disclosed in U.S. patent 5,247,597 by Blacha et al. (1993). An alignment chip was bonded to the silicon substrate. There was a via hole in the alignment chip that defined the fiber position. The coupling loss was less than 1 dB for one of the modules measured. A similar approach for an optical link perpendicular to a substrate was also reported by Benzoni and Dautartas (1994) and Gates et al. (1996). These self-aligned designs are ideal for the alignment between fibers and surface-emitting or surface-receiving devices, which are important to high-density integrated optoelectronic modules.

Figure 5.5 shows a free space optical interconnect module developed by Pusarla and Christou (1996). Self-aligning soldering was used to align photoresist refractive optics by the melting techniques (PROM) microlens array and PD array. The fabrication process is performed as follows: The PD chip was first bonded to a silicon substrate, and the photoresist was then spin coated on the PD and then

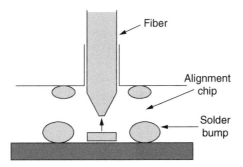

FIGURE 5.4 Optical fiber in an alignment chip soldered onto a substrate.

soft baked, exposed, and developed to define the pattern of solder wettable pads. Cr(300A)/Cu(3000A)/Au(300A) was then thermally evaporated, and the photoresist was lifted off. Microlens array was fabricated on the top side of the glass substrate, using AZ4903 photoresist. Then Eutectic Pb/Sn solder was deposited on the bottom side of the glass substrate. Again, Cr/Cu/Au was thermally evaporated first, and then AZ4904 photoresist was spin coated, soft baked, and exposed. A 30-μm solder layer was then electroplated, and the photoresist was washed away using acetone. The microlens module was then fluxed, reflowed, fluxed, and bonded to the PD chip. However, the self-alignment accuracy was less than 2 μm.

The self-aligning feature of soldering makes it very suitable for batch assembly. An edge-emitting laser module was developed by Gates et al. (1996). The optical subassembly (OSA) for this package was assembled at wafer level, using solder. The OSA contained a silicon substrate, a backface monitor chip, and a laser chip.

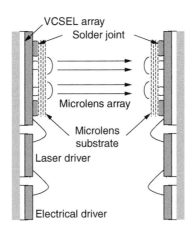

FIGURE 5.5 Free space VCSEL-to-microlens-to-photodiode optical connect structure.

It also had a site for fiber bonding. Ti/Pt/Au (1kÅ/2kÅ/5kÅ) were sputtered, and 6 µm of Au/Sn (80/20) was evaporated on the laser and detector chips. The chips were first attached to the wafer, using thermocompression bonding at a pick-and-place accuracy of ±50 µm. Then the assemblies were reflowed at wafer level in a controlled atmosphere. Solder bumps were used only to self-align the chip, and electrical connections were provided by wire bonding. These OSAs could be electrically probed and linked to identify poor devices. After testing, the Si wafer was diced for OSA separation. More than 2000 OSAs can be assembled at one reflow step.

5.2.3 SELF-ALIGNED SOLDER ASSEMBLY WITH ONE MECHANICAL STOP

The solder self-alignment force can be separated into two components: the restoring force that moves the component laterally and the pulling or pushing force that moves the component vertically. Unfortunately, these two force components are proportional to the misalignment and become very small when the component is close to the well-aligned position. If a strong force is desirable for precision alignments, it is preferable to use misaligned solder joints and mechanical stops. Figure 5.6 shows a mechanical stop to controlling the lateral motion for a vertical alignment or a standoff for the gap (height) control. The misaligned molten solder could push the stops against each other to achieve the precision alignments, and because the force is proportional to the misalignment or spacer height, it can be orders of magnitude larger than that with respect to a nearly well-aligned solder joint.

Figure 5.7 is an optical subscriber interface system developed by Lee et al. (1995). The laser submodule had two V-grooves in a row—one was for the single-mode fiber, and the other was for guiding light from the rear mirror plane of the laser to a monitor photodetector. Both laser diode and photodetector were flip-chip soldered to a silicon substrate. In addition to using solder self-alignment for the lateral alignment, an Au standoff was used to control the height of the laser. The lateral alignment accuracy reached ±2 µm, and the height control was within ±0.7 µm.

Figure 5.8 shows another case using solder self-alignment and a mechanical spacer for a spatial light modulator (Ju et al., 1993a; Lin et al., 1993). During reflow, the large solder joints pulled the cover glass with mechanical spacers

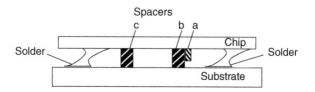

FIGURE 5.6 Self-aligned solder assembly with one mechanical stop using spacers a and b for the horizontal stop or b and c for the vertical standoff.

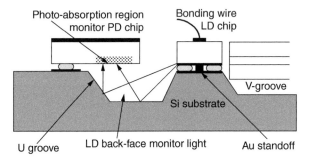

FIGURE 5.7 Silicon mounts for an optical transmitter module using a standoff.

toward the VLSI (very large scale integration); these two components formed a uniform gap to be filled with ferroelectric liquid crystal. The compliant surface tension force ensured a good gap control, and the soldering process was compatible with the existing solder-based manufacturing infrastructure. A micrometer-level gap with a submicrometer-level uniformity was accomplished.

5.2.4 SELF-ALIGNED SOLDER ASSEMBLY WITH TWO MECHANICAL STOPS

The lateral and the vertical mechanical stops shown in Figure 5.6 can be integrated using all three spacers (a, b, and c) in a single-module design. Figure 5.9 is a photo of such a module developed by Jackson et al. (1994) for a four-channel transceiver. The mechanical stops were used to define the position and the height of the laser array that was connected to the planar waveguide for routing optical signals. The use of mechanical stops reduced the effect of solder volume variations on the alignment accuracy. The final misalignment reported was ±2.0 μm for the lateral and ±0.75 μm for the height control. The coupling loss was between 3.9 and 4.2 dB.

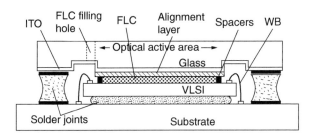

FIGURE 5.8 Solder-assembled liquid crystal-on-VLSI.

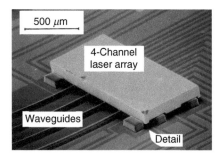
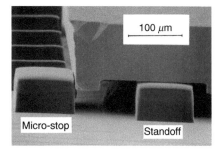

FIGURE 5.9 Four-channel laser array using two mechanical stops and misaligned solder joints.

5.3 CASE STUDY: AuSn SOLDER FLIP-CHIP BONDING AND ITS MODULE APPLICATIONS

In addition to the above overview, we present a case study to discuss some packaging details. In particular, a Au/Sn soldering technology and its applications will be presented. To maintain reliable precision solder alignments, hard solder such as AuSn was used to avoid solder creep (Sasaki et al., 1992). This case study shows a three-dimensional passive alignment technology of optoelectronic devices using AuSn solder bump flip-chip bonding. The study will address three issues: bumping technique, stripe-type bump bonding, and applications.

5.3.1 BUMPING TECHNIQUE

In solder bump flip-chip bonding, soft solder such as Pb/Sn or Pb/In is used for bump material because it makes bump fabrication easy. However, a laser diode (LD)/PD chip mounted with such soft material tends to move out of alignment as a result of solder creep as time passes, and this may cause additional coupling loss. Furthermore, the use of flux in the optical chip mounting reduces chip reliability. To solve these problems, hard solder such as AuSn should be used. However, fabricating AuSn solder bumps poses a specific difficulty. Conventional electroplating technique is not feasible because of the difficulty of precisely controlling both the ratio of Au/Sn and the solder volume. Similarly, vacuum deposition is impractical because of the high cost incurred as a result of the many unavoidable processes involved in vacuum deposition of films with thicknesses of tens of micrometers.

To solve this problem, an AuSn solder bump fabrication technology has been developed in which AuSn bumps are mechanically fabricated using a combination of micropunch and die (Itoh et al. 1996a, 1996b), as shown in Figure 5.10. First, a AuSn alloy ribbon was punched into a small piece with a micropunch and a die onto the submount. The submount was heated and placed under the punch

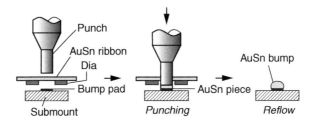

FIGURE 5.10 AuSn bump mechanical fabrication technique.

and the die. It was then positioned so that the electrode on which the piece to be provided was just under the punch.

The submount and the punch were moved synchronously with the punching operation. The punched piece was bonded onto the electrode pad of the submount in a successive manner. Because the submount was heated in advance, the bonding metal piece of AuSn alloy was alloyed with the Au film of the electrode pad to form an alloy layer by the thermocompression bonding, so that the piece could be bonded to the electrode pad. After all the pieces were produced, the submount was heated to the melting point of 280°C to form the solder bumps.

Figure 5.11a and 5.11b are scanning electron microscopy photographs of punched AuSn pieces and AuSn solder bumps after reflow, respectively. Figure 5.12 is a histogram of solder bump height for the case with a punch diameter of 90 μm and a die diameter of 100 μm. The AuSn ribbon was 80% Au–20% Sn eutectic solder, and its thickness was 30 μm. Bumps with a height accuracy of micrometers were successfully fabricated. A high aspect ratio between the bump height and the pad diameter was required for the bumps to achieve the surface tension level needed for a good self-alignment in the horizontal direction. Figure 5.13 is a histogram of the accuracy of the horizontal self-alignment of the optical array chips in the case that had a bump diameter of 50 μm and a bump height of 50 μm. An accuracy of

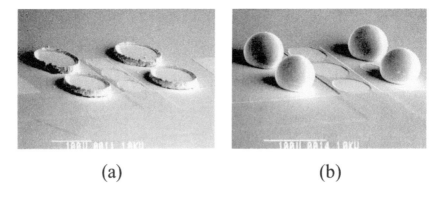

FIGURE 5.11 (a) Punched AuSn pieces, (b)AuSn bumps after reflow.

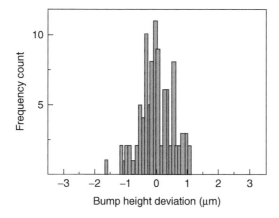

FIGURE 5.12 Bump height deviation

about 1 µm was obtained. The new bumping technique is capable of forming the AuSn solder bumps easily and accurately, even when a submount could not accommodate electroplating or vacuum deposition. The technique is extremely high in productivity and low in cost.

5.3.2 STRIPE-TYPE BUMP BONDING

In the case of LD–single mode fiber coupling, very accurate three-dimensional chip alignment (within 1 µm) is required. In spherical bump bonding, bonding height deviation, which is mainly affected by the unavoidable solder volume deviation, is 1.5 to 2.0 µm for the 50-µm bump diameter usually used for LD

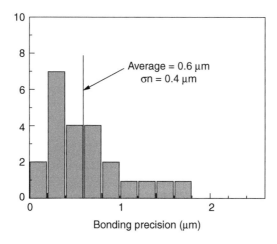

FIGURE 5.13 Horizontal self-alignment accuracy of optical chips.

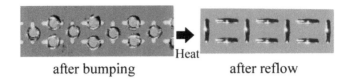

after bumping   after reflow

FIGURE 5.14 Stripe-type AuSn bumps.

chip mounting. Lowering the spherical bump height itself may reduce the bump height deviation. However, it results in a decrease in the aspect ratio, which then decreases the surface tension to a level below that required for the effective self-alignment. Reducing the spherical bump diameter may also reduce the height deviation. However, many smaller bumps are required to achieve the surface tension sufficient to move the LD chip, and this increases bump fabrication cost. Furthermore, reducing the bump size creates a need for precision placement of LD before the reflow process.

An alternative stripe-type bump flip-chip-bonding technique has been developed (Itoh et al., 1996b; Sasaki et al., 1995) to improve the bonding height precision. Figure 5.14 shows scanning electron microscopy photographs of the punched AuSn pieces and of the stripe-type AuSn solder after reflow. Figure 5.15 shows the self-alignment achieved by these stripe-type solder bumps. The left photograph is before reflow, and the right one is after reflow. The precise self-alignment resulting from the stripe-type AuSn solder bumps is clearly demonstrated. In this technique, the ratio of bump height/stripe width is important to good self-alignment in the horizontal direction. The ratio plays a role different from that of the aspect ratio for the spherical bump bonding. Spherical solder bump height could not be reduced without reducing solder volume under a fixed aspect ratio. In contrast, with stripe-type bumps, bump height could be reduced while both solder volume and aspect ratio remain unchanged—the bump width and length could be selected individually. Figure 5.16 shows histograms of bonding height precision for stripe-type bumps (upper) and spherical bumps (lower). In the stripe-type bump bonding, the stripe width was 20 µm, the length was 140 µm, and the bonding height was 20 µm. In the spherical bump bonding, the bump diameter was 50 µm, and the bonding height was 40 µm. This result shows that

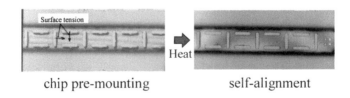

chip pre-mounting   self-alignment

FIGURE 5.15 Self-alignment effect for stripe-type AuSn bumps.

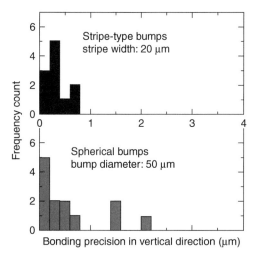

FIGURE 5.16 Bonding height precision for stripe-type AuSn bumps.

the use of stripe-type bumps significantly improved vertical precision. Figure 5.17 is a histogram of horizontal alignment accuracy of the optical array chips. High accuracy of less than 1 µm is obtained.

5.3.3 APPLICATIONS FOR OPTICAL MODULE SI BENCHES

5.3.3.1 PD Array Module

Figure 5.18 shows the schematic diagram of a PD array module optical platform. The output light beams from the optical fiber array are reflected upward from metallized Si V-groove ends and coupled to the PD array's light-receiving regions.

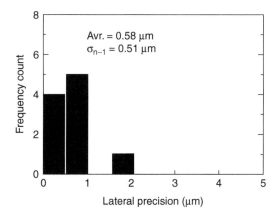

FIGURE 5.17 Horizontal bonding precision for stripe-type AuSn bumps.

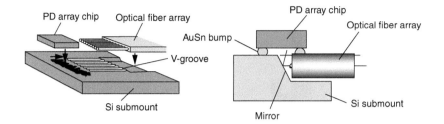

FIGURE 5.18 PD array module optical platform schematic diagram.

The PD array chip was mounted on the Si substrate by AuSn solder bumps. Spherical solder bumps were used because PD chip alignment tolerance was more than 5 μm. The micropunch and die technique was used to deposit AuSn solder alloy pieces onto ten solder pads 70 μm in diameter on the submount. The diameters of the punch and the die were 90 and 100 μm, respectively. The AuSn alloy ribbon was 80%Au–20%Sn eutectic solder with a thickness of 40 μm. The submount was heated to melt the AuSn solder, which balled up to a height of 70 μm. The PD array was a monolithically integrated 1 × 8 linear array, employing InGaAs/InP PIN PDs, each with an 80-μm light-receiving diameter, spaced at a 250-μm pitch. The PD array chip was placed face down on the spherical AuSn bumps, and the submount was heated in N_2 atmosphere to melt the solder. No flux was used. The surface tension of the molten solder aligned the PD array chip automatically (see Figure 5.19; Itoh, et al. 1996a). Three-dimensional alignment accuracy was within ±2 μm. Figure 5.20 shows the PD array module optical platform, composed of the PD array and the fiber array. The fiber array contained eight graded-index, multimode fibers (MMFs) spaced at a 250-μm pitch. Each fiber had a core diameter of 62.5 μm and was tipped with a microlens produced by chemical etching (Honmou and Itoh, 1995) to reduce the coupling loss to the PD array. The lens radius was set to approximately 50 μm to keep the optical output

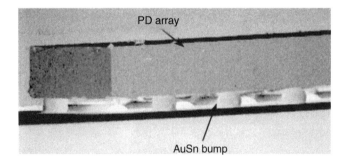

FIGURE 5.19 PD array chip bonded AuSn solder bumps.

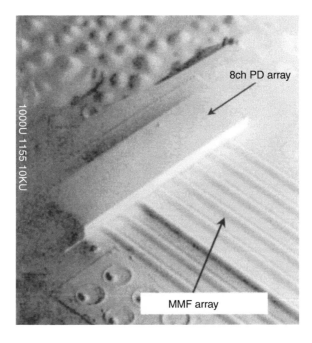

FIGURE 5.20 PD array module optical platform.

light spot size to be within 80 μm of the PD light-receiving diameter. The fiber array was placed into the V-grooves on the Si submount. These V-grooves had been made by an anisotropically wet-etching process and had dimensions controlled to within ±1 μm.

5.3.3.2 LD Array Module

The LD output light beams are coupled directly to a flat-end eight-channel optical fiber array. A slit was fabricated on the Si submount by a dicing saw so that the fiber end could be brought close to the LD facet. The fiber was embedded in the Si V-grooves in the same manner as in the PD submount. Figure 5.21 shows the LD array module optical platform (Itoh et al., 1997), composed of the LD array and the flat-end fiber array. The LD array was a monolithically integrated 1 × 8 linear array, employing 1.3-μm-wavelength InP/InGaAlP LDs with a 250-μm pitch. The LD array chip was bonded on the stripe-type AuSn bumps in the same manner as in the PD chip bonding without flux in a N_2 atmosphere. Figure 5.22 shows the stripe-type bumps on the Si optical bench before LD chip bonding. The fiber array contained eight single-mode fibers spaced at a 250-μm pitch. Each fiber had a core diameter of 10 μm and 10 μm spacing from the LD facet. Average excess coupling loss between an eight-channel LD array and a flat-facet single-mode optical fiber array is 1.0 dB, corresponding to an alignment error of 1.8 μm.

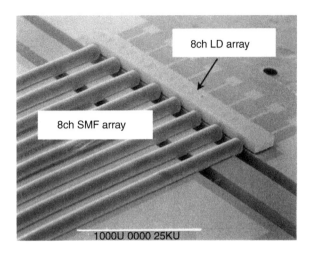

FIGURE 5.21 LD array module optical platform.

The coupling loss was 9.2 ± 1.0 dB. The loss was nearly as good as those achieved using active alignment.

5.3.3.3 4 × 4 Optical Matrix Switch Module

Figures 5.23 and 5.24 show a schematic diagram and a photograph of the hybrid integrated 4 × 4 optical matrix module (Sasaki et al., 1998; Kato et al., 1998). Four sets of silica optical waveguide 1:4 splitter, 4:1 combiner, and some electrical signal lines were formed on a Si platform. Four four-channel spot-size converter-integrated semiconductor optical amplifier (SOA) gate (Hatakeyama et al., 1997) array chips were mounted on the platform by using stripe-type AuSn solder bump flip-chip bonding. The stripe-type bump is the same as one shown in Figure 5.22. The SOA gates were first put on solder bumps formed on the platform. As the solder bump reflowed, they were simultaneously self-aligned to their correct position. The positioning accuracy was less than 1 μm in both lateral and vertical

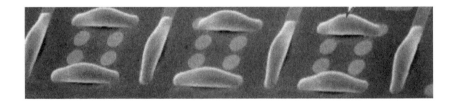

FIGURE 5.22 Stripe-type bumps.

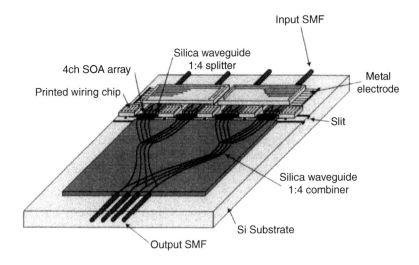

FIGURE 5.23 Diagram of a hybrid integrated 4×4 optical matrix module.

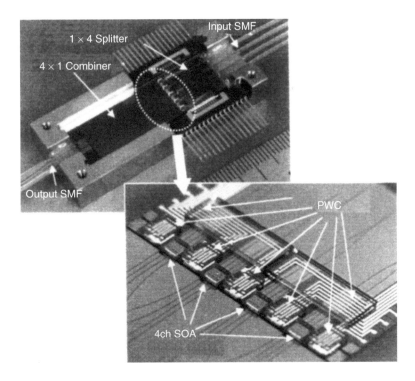

FIGURE 5.24 Hybrid integrated 4×4 optical matrix module.

directions. Two sets of single-mode optical fiber (SMF) arrays were assembled in V-grooves fabricated at both ends of the platform. Here, facets of the silica waveguides were formed by using a mechanical sawing technique with a dicing blade because it is low cost and easy to use. This process, however, also cut the electrical wiring patterns connected to optical chips. To overcome this problem, printed wiring chips (PWCs), which bridged electrical connection across over such a blade-cut region, were employed as shown in Figure 5.25. The PWCs and SOAs were simultaneously mounted by the solder bump reflow. The printed wiring chip could be charged by some electronics such as driver ICs, and the insertion losses at an injection current of 50 mA were 7.4 dB. The coupling loss between an SOA and a silica waveguide was estimated to be about 4.5 dB.

5.4 CRITICAL ISSUES

Solder has been widely used for die attachments in existing optoelectronic products. In addition, most of the aforementioned new modules using solder for precision alignments have been demonstrated in research and development laboratories. As shown in the case study, it is not easy to transfer these new solder technologies from laboratories to manufacturing facilities. Without using mechanical stops, a precise process control of solder volumes is necessary, and external loadings such as gravitational, gas flow, vibration, and other forces should be eliminated. The nearly well-aligned solder joints do not have larger self-alignment forces to overcome the effects of these loadings. Using the mechanical stops, however, the interfacial friction forces among the stops have to be overcome. Misaligned solder joints have large restoring

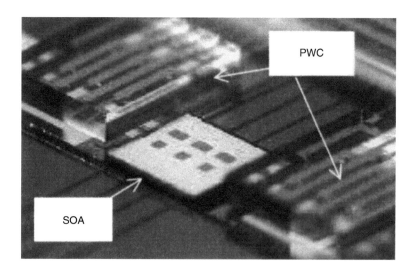

FIGURE 5.25 Use of Printed Wired Chips (PWCs) in the optical matrix module.

or vertical reaction forces, but they may not be large enough to move the component once the stops make the contacts. Using one mechanical stop for the height control (Lee et al., 1995), the lateral self-alignment should be accomplished before the component touches the standoff. Such a dynamic sequence is a challenge. Using two mechanical stops, the solder self-alignment forces must overcome the friction from either the lateral or the vertical interfacial contact.

As a result, repeatability of the alignments is a problem. Large variations are generally reported by Deshmukh et al. (1993), Dautartas et al. (1995a and b), and McGroarty et al. (1993). An example of complete self-aligned optical subassembly shows that the best self-alignment result was 2 μm; however, the average was 5 μm, and the worst was 10 μm (Dautartas et al., 1995a and b). More studies are needed to improve the design and manufacturing processes of these modules for a wide application of low-cost, high-performance, high-reliability modules. In particular, the four critical issues of solder material, fluxless reflow, design, and reliability are discussed in the following sections.

5.4.1 SOLDER MATERIAL AND DEPOSITION

Table 5.2 lists major solder materials used for optoelectronic packaging. Important solder properties are melting temperature, Young's modulus, coefficient of thermal expansion, Poisson ratio, fatigue behavior, and creep rate. Indium-based solder has been used for prototyping, but it is not a good material for high-volume manufacturing because of its high cost, large creep rate, poor corrosion resistance, and high thermomigration rate (Koopman et al., 1988; Shimizu et al., 1994).

TABLE 5.2
Solder Materials Used for Optoelectronic Packaging

Solder Alloy	Melt Temperature (°C)	E	α	γ	Relative Fatigue Life	Relative Creep Rate
63Sn37Pb Eutectic	183	30	21	0.4	10	2
90Pb10Sn	268	20	29	0.4	1	1
95Pb5Sn	310	24	26	0.37	1	0.5
80Au20Sn Eutectic	281	68	14	0.4	poor	N/A
50In50Pb	180		27		3	0.5
100In	157	7	31	0.46	20	Soft
97Sn3Cu	227	41	19	0.33	5	0.01
77.2Sn 20In2.8Ag	175	39	28	0.4	N/A	<0.01

Note: E: Young's Modules, Unit: GPa; α: CTE (coefficient of thermal expansion), Unit: ppm/°C; γ: Poisson ratio.

In addition, indium is a very efficient dopant and demands careful design and manufacturing control to avoid problems caused by indium diffusion. Pb/Sn-based solder is commonly used in electronic packaging. Its fatigue behavior is well accepted, but its creep may result in micrometer-level misalignment and degrade the coupling performance. In addition, eutectic solder might form inter-metallics with Au. The stoichiometric binary compound is generally very brittle, and early failure may result. Therefore, it may not be appropriate if the Au layer is more than 0.25 μm thick (Hinch, 1988). Au/Sn solder is not subject to fatigue or creep rupture during thermal cycling because of its high strength; however, it may transfer the stress to the device and cause cracking in the die.

There is a need for a high-fatigue, low-creep solder alloy that is compatible with the electronic manufacturing infrastructure. In addition, the toxicological and legislative issues might result in future restriction in use of Pb/Sn-based solder. As a result, many new lead-free solder alloys are being developed, and some of them may be better than the existing materials for optoelectronic packaging. For example, as shown Table 5.2, a new solder alloy, 97Sn/3Cu, has been developed for lead-free manufacturing for automotive electronics. The creep rate of this material was shown to be 100 times less than that of 90Pb/10Sn (See Figure 5.26, taken from Pao et al., 1993, and Pao et al., 1995; Lee N. C. et al., 1994), and temperature cycles to failure for the configuration tested were five times longer than that of 90Pb/10Sn. Another new solder alloy, 77.2Sn20In2.8Ag, has been developed as a "drop-in" replacement for eutectic solder. Its creep rate is 2.5 orders of magnitude lower than that of the eutectic solder, which has similar physical characteristics (Lee, N. C., et al., 1994; Lee, N. C., 1998). The relative creep rate and fatigue life of other solder alloys are compared in Table 5.2.

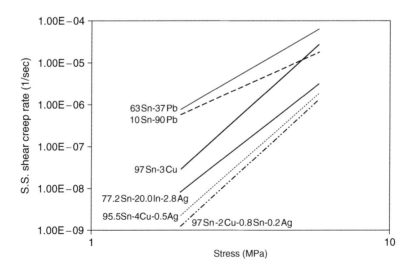

FIGURE 5.26 Creep rates of 97Sn/3Cu and other solder alloys.

The ratios were estimated on the basis of the experimental data in Shimizu et al. (1994), Pao et al. (1995), Darveaux et al. (1995), and N. C. Lee (1998). Creep rates and fatigue lives are strongly dependent on not only materials but also microstructures and configurations. These ratios are listed here for qualitative guidance. Nevertheless, it is clearly indicated in the comparison that new solder alloys (e.g., 97Sn/3Cu and 77.2Sn20In2.8Ag) may enhance solder joint performance substantially for optoelectronic packaging. A recent study by Mavoori and Jin (1998) developed new materials with enhanced creep resistance. Further studies should be conducted to explore different enhancement methods and investigate packaging/manufacturing issues associated with the new materials. For example, the 77.2Sn20In2.8Ag alloy has two drawbacks: first, the wetting time is about twice as long as that of the eutectic solder at 200°C, and second, the low-temperature eutectic point at 118°C was detected for this alloy. It may even define the maximum service temperature of this alloy (Artaki, 1994).

Solder deposition has been a research focus since IBM first introduced the controlled-collapse chip connection technology because the evaporation of solder is expensive. Different bumping technologies were developed to reduce cost in the microelectronic industry. These technologies are listed in Table 5.3. It is important to understand that the requirements of bumping for optoelectronic and microelectronic packaging are different in some aspects. In optoelectronic packaging, an accurate

TABLE 5.3
Solder Bumping Methods (Patterson et al., 1997; Pittroff et al., 1997)

Method	Advantage	Disadvantage
Evaporation	Clean vacuum process Accurate control of chemical composition Accurate bump volume	Expensive Not suitable for eutectic Pb/Sn
Electroplating	Relatively low cost Relatively accurate bump volume	Less accurate control of chemical composition Possible formation of void
Stencil printing	Very low cost Compatible with all solder alloy	Less accurate control of solder bump volume
Solder ball bumping	Simple, quick, low cost Suitable for prototyping	Not suitable for production due to slow sequential process
Micropunch	Low cost Deposition of large solder volume	Sequential process
Solder jet	Capable of using Al as metalization Possible very low cost	Still under development

control of solder volume is critical so that an accurate alignment can be ensured. Therefore, stencil printing of solder is unlikely to be used because of its high bump volume variation, unless such a problem is solved. However, the need for high solder volume, which is one of the major challenges for most of the microelectronic packaging, may not be as critical for optical applications because of their small sizes. For the solder jet process, liquid solder droplets are ejected from a capillary and impinge the bond pads, just as with an ink-jet printer. Deposition of aluminum pad is also demonstrated, which is usually considered as a nonsolderable metalization (Hayes and Wallace, 1996). The solder-jet process might be a low-cost alternative to the evaporation, electroplating, and micropunch processes.

5.4.2 Solder Reflow

Liquid flux is widely used to remove solder oxides for efficient wetting and self-alignment. The flux residue removal is mandatory and constitutes environmental threats (Frear et al., 1995; Lin, 1995; Potier et al., 1995). More important, the use of liquid flux may contaminate the optically active surface by organic residue (Miller, 1969), and a conventional cleaning method may not be effective for the optoelectronic assembly (Lee and Basavanhally, 1994; Lin, 1995). As a result, different fluxless soldering processes have been developed for soldering opto-electronics modules.

There are two different approaches for the fluxless soldering. The first one involves a precleaning process to remove surface oxide and a reflow process protected from oxidation. For example, PADS (plasma-assisted dry soldering), developed by MCNC, uses a microwave reactor (Koopman et al., 1993; Koopman et al., 1996). The following equations explain the PADS process.

PADS:

$$SF_6 \Rightarrow SF_2 + 4\ F$$

$$SnO_x + yF \Rightarrow SnO_xF_y \qquad (5.1)$$

Hydrogen reflow:

$$PbO/SnO_x + SnO_xF_y + H_2 \Rightarrow Pb/Sn + H_2O \qquad (5.2)$$

In the microwave chamber, an inert, nonflammable, fluorine-containing gas such as SF_6 is disassociated to form atomic fluorine, which is active. The fluorine atoms then react with solder oxide and convert oxide to oxyfluoride. This film passivates the solder and has the unique property of breaking up when solder melts, exposing a free solder surface that allows reflow and joining to occur in the absence of flux or reducing gas. The assembly of a light-emitting diode module (Bonda et al., 1997) and flat-panel display module (Nangalia et al., 1997) were demonstrated.

A similar approach at ambient pressure was also reported by Potier et al. (1995). In addition, Lee and Wang (1992) reported another method using a high-vacuum deposition of a gold or copper layer to protect the inner layer from oxidation.

The second approach uses reactive gas to remove solder oxides. By definition, the reactive gas is flux because its function is to remove surface oxides to enhance solderability. However, it is quite different from the liquid flux. With proper control, the gas leaves little residue without a cleaning process and contamination. Two kinds of reactive gases are commonly used in optoelectronic packaging: forming gas (4 to 10% H_2 in nitrogen or argon) by Lin et al. (1993), Boudreau et al. (1993), and Brady and Deshmukh (1993), and formic acid vapor (1.5 to 7% HCOOH in nitrogen) by Deshmukh et al. (1993), Dautartas et al. (1995a and b), Lin (1995), and Brady and Deshmukh (1993).

Figure 5.27 records self-alignment processes with respect to the use of forming gas (10% H_2 and 90% N_2) and formic acid vapor at different concentrations (Lin, 1995). The case with the forming gas was conducted at a reflow temperature of 280°C and was assisted by the outgas from a piece of FR-4 printed wiring board. The solder did reflow, but the self-alignment process was very slow. After 200 seconds, the component had not reached the steady-state, well-aligned position. Similarly, low-concentration formic acid vapor was not effective. However, when the concentration reached 1.6% or higher, efficient self-alignment was completed within a few seconds. The reflow temperature for the formic acid vapor was only 220°C. Formic acid vapor resulted in efficient solder self-alignment.

Several effects should be further considered when a fluxlesss soldering process is to be applied for optoelectronic packaging. For example, in order to avoid

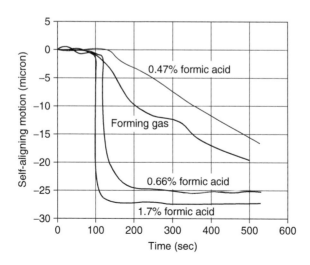

FIGURE 5.27 Dynamic self-alignment processes with different reactive gases or concentrations.

condensation of the vapor, the assembly temperature should be kept higher than 150°C during reflow (Lin, 1995). By introducing the gas into the reflow chamber, the flow may move the component and affect the aligning process (Gershovish and Lee, 1994). This gas flow effect is strongly dependent on the flow rate and the component size. For a chip larger than 5×5 mm, such a displacement caused by gas flow can be significant (Su et al., 1997).

5.4.3 SOLDER DESIGN

Experimental results show that the final alignment accuracy for self-aligned soldering is strongly related to the solder bump parameters such as solder joint diameter and height and the number of joints (Hayashi, 1992). For solder self-alignment, these parameters can be designed for optimum performance. Since the controlled-collapse chip connection was introduced by IBM (Goldmann, 1969; Patra and Lee, 1991a), modeling studies have been conducted to understand the self-alignment mechanism. Patra and Lee (1991a, 1991b) conducted a comprehensive study of solder joint design. The surface profile was calculated on the basis of the energy minimization principle; the profile was affected by solder pad geometry and size, component weight (or spacer height), surface tension coefficient, solder volume, and misalignment level. The quantitative characterizations could correlate these effects with the height variations that are critical to alignments, and the profile and effects can also be modeled using Surface Evolver public domain software (Ju et al., 1993b; Martino et al., 1994; Lin et al., 1995). Figure 5.28 shows the solder profiles calculated and measured for solder with rectangular pads (Ju et al., 1993b). The agreement is very encouraging. More studies are needed to verify the modeling accuracy for micrometer-level solder joints.

Using Surface Evolver, Lin et al. (1995) studied the restoring and normal reaction forces as functions of solder joint height, size, volume, and misalignment level. The case was for a self-aligned module using one mechanical stop for the

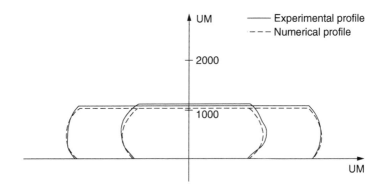

FIGURE 5.28 Solder joint profile prediction and measured profile.

height control (see Figures 5.8 and 5.10). As shown in Figure 5.29, the vertical reaction force may change from a pulling to a pushing force when the solder volume increases under a given spacer height. There also exists an optimum solder joint volume for the maximum pulling force. Using this volume, the spatial light modulator shown in Figure 5.10 was successfully developed by Lin et al. (1993), Ju et al. (1993a), and Lin (1995). However, it should be noted that the shrinkage of the structural elements during cooling also plays a critical role in gap control. A proper mechanical design considering stresses and strains is needed to ensure a good-quality assembly (Hareb, 1998).

Figure 5.30 is a laser-waveguide-fiber module developed by Morozova et al. (1996). A controlled self-alignment motion was achieved by design for self-alignment. The module contained a laser chip, a tapered polymer waveguide chip, and an optical fiber. The laser from the LD was coupled to the fiber through the polymer waveguide. The position of the optical fiber was defined by the V-groove. Both laser chip and waveguide were aligned to the glass substrate using self-aligning soldering. The alignment requirements for this module were that the chips should be precisely aligned laterally and vertically and that the axial gap between the LD and waveguide should be precise. As discussed in Section 5.2, such requirements could be achieved using mechanical stop. Unfortunately, the fabrication of mechanical stop was so expensive that the axial gap was chosen to be close to zero. It means that the LD and the waveguide should make a solid contact. For any self-alignment using one mechanical stop, there is a two-axis alignment problem: the chip could get contact with one of the mechanical stop first, and then the friction force could hinder the alignment motion in the other direction. To achieve an accurate lateral alignment and a solid contact, the alignment motion of the laser chip was controlled. The lateral alignment happened first, followed by the slow axial movement to make the contact. Such a controlled

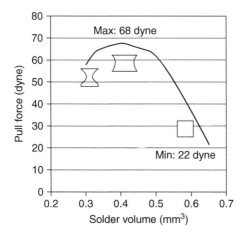

FIGURE 5.29 Solder normal reaction force as a function of solder volume.

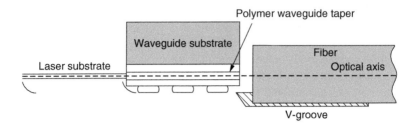

FIGURE 5.30 Laser-waveguide-fiber module: controlled self-alignment motion.

self-alignment sequence was accomplished by using rectangular solder bond pads. As shown in Figure 5.31a, the solder pads were designed to be 300×90 µm in size. With such a pad design, the calculated restoring force was much larger in lateral than that in axial alignment for the same misalignment (Figure 5.31b). As a result, the lateral alignment motion was much faster than the axial motion, and the controlled sequence was achieved with an alignment accuracy of 2 µm at lateral alignment and a solid contact between the laser chip and the waveguide.

The modeling of profiles and self-alignment movement is critical to designing solder parameters to reach desirable performance. The number of trial-and-error runs can be reduced, and more important, the solder self-alignment can be considered in the trade-off analysis for optoelectronic packaging. Such an analysis considers electrical parasitic capacitance and inductance, thermal resistance, stress/strain, and solder joint reliability. For example, a small solder joint height is desirable for a strong restoring force (Lin et al., 1995), and it is also preferred for lowering thermal resistance and inductance across the solder. However, the small height may result in early fatigue failure. Quantitative analysis of these constraints is necessary to derive an optimum solder design. Among these constraints, solder joint reliability has been rarely considered in optoelectronic packaging, although it is a critical factor commonly evaluated for electronics packages.

5.4.4 SOLDER RELIABILITY

Most of the studies on solder-assembled optoelectronics modules have focused on how to obtain high alignment accuracy rather than on how to maintain the alignment obtained. The long-term stability of the alignment accuracy depends heavily on the mechanical integrity of solder joints and their attaching layers.

Solder failure mechanisms are driven by temperature, moisture, and voltage. The thermal expansion mismatch between a chip and a substrate could result in stress/strain/fatigue damage of solder connections, optical devices, and layers of metallization or dielectrics. Moisture and voltage combined could result in corrosion or metal migration (Lowe and Lyn, 1995). Although a variety of mechanisms may result in cracked solder joints, the major cause of such damage is thermal stress and low-cycle, strain-controlled fatigue (Lau and Rice, 1985). Fortunately, for most of optoelectronic modules, the assembly sizes are small (e.g, 300 µm),

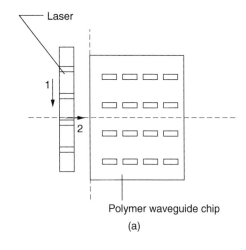

(a)

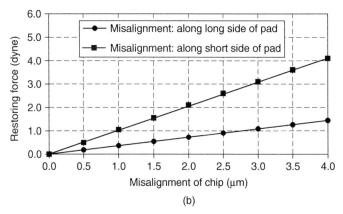

(b)

FIGURE 5.31 (a) design for control self-alignment motion, (b) calculated restoring force.

and the thermal mismatch is much lower than that for microelectronics. Therefore, the reliability issue has not been a major issue to date. However, the chip size is getting larger and larger, with higher integration levels, and reliability assessment should be conducted rigorously.

Table 5.4 lists solder-related reliability tests for laser modules (Bellcore Specifications, 1993). Pass/fail criteria for most of these tests are 10% or 0.5 dB maximum change (or less) in coupling efficiency and no obvious change (beyond measurement error) in the threshold current. In these tests, solder may affect the reliability through the following mechanical behaviors:

- **Stresses/strains:** Soldered assembly may experience large temperature variations during solder reflow processes, on/off operations, and

TABLE 5.4
Solder-Related Reliability Tests for Laser Modules

Test	Conditions
Mechanical shock	Five times/axis; 500 G, 1.0 msec
Vibration	20 G, 20–2000 Hz; 4 minutes/cycle, four cycles/axis
Thermal shock	$\Delta T = 100°C$
Solderability	MIL-STD-883 Method 2003
Temperature cycling	−40°C to +85°C
	400 cycles; pass/fail in room temperature environment
	500 cycles for information
	500 cycles; pass/fail in uncontrolled environment
	1000 cycles for information

environmental changes. In particular, rapid cooling after reflow may result in large thermal stresses/strains in devices, solder joints, substrates, and solder-attached metallization and dielectric thin-film layers. The stresses/strains in these elements and at their interfaces are strongly dependent on the temperature variations, material properties, and physical dimensions (Suhir, 1987).

• **Creep:** Most of solder materials are soft and may creep under stresses. The stresses may be residual thermal stresses resulting from nonuniform shrinkage of different structural elements, or they may be applied by external loadings during manufacturing or operations. Creep properties of some solder materials have been reported with very different behaviors. With the same material, the creep properties can be very different when the solder microstructures are changed by a process change (Guo and Conrad, 1993; McDowell et al., 1994; Pao et al., 1994; Schroeder et al., 1994). Unfortunately, precise control of solder creep properties is necessary because submicrometer creep movement can degrade the coupling efficiency more than 10%. With quantitative understanding and control of solder creep, the best approach is to use no-creep Au/Sn solder, such as in the case study presented in Section 5.3. However, for hard solder, the stresses induced should be controlled.

• **Fatigue:** During temperature cycling, cracks are initiated and propagated across the solder joints. The joints' mechanical strength can be substantially weakened and may not be able to maintain the precision alignments obtained. Even worse, the joints may be broken entirely and lose their electrical and mechanical connections. The following two equations can be used to understand the fatigue behavior of solder joints. These equations are widely used to estimate the fatigue life (number of temperature cycles to failure) for a solder joint away from the center of the assembly (Jeannotte et al., 1988). The fatigue life is correlated with the shear strain resulting from different expansion movements resulting from temperature changes.

$$\text{Shear strain} = \Delta\gamma = L(\alpha_1 - \alpha_2)\Delta T/h \tag{5.3}$$

$$\text{Fatigue life} = N_f = 0.5(\Delta\gamma/0.65)^{-2.26}$$
$$= 0.5(0.65\ h/L(\alpha_1 - \alpha_2)\Delta T)^{2.26} \tag{5.4}$$

where L is the distance between the solder joint and the neutral point of the assembly, α_1 and α_2 are coefficients of the thermal expansion of the component and the substrate, ΔT is the range of the temperature cycles, and h is the height of the solder joint.

Typical length scale (L) for electronic components is in the order of 10 mm, and that for most of optoelectronic components is in the order of 1 mm. The shear strain in an optoelectronic module is small, and the fatigue life is high. In addition, the required "fatigue life" is in the range between 500 and 1000 cycles (see Table 5.4). As a result, this fatigue behavior is not a major concern for the assemblies.

However, the applications of solder-assembled optoelectronic modules expand fast, and their working environment is more and more severe. AT&T is already testing some of its package for telecommunication applications at 0 to 100°C for 5000 cycles (Viera et al., 1993). For nontelecommunication applications, there is a need to increase the number of temperature cycles and the temperature ranges required for the reliability tests. The length scale (L) may also be increased to the 10-mm level because of a large integration of optoelectronic devices.

To deal with these issues, it is necessary to evaluate material properties discussed in Table 5.3. In addition, the solder joint should be designed for reliability. The ongoing development of different fatigue modeling approaches for microelectronic packages should be modified for optoelectronic packages, as solder reliability is a critical issue that needs to be addressed in the near future.

5.5 CONCLUSIONS

Soldering technology for optoelectronic packaging has been reviewed, and representative modules have been presented in four categories: solder assembly with no precision self-alignments and self-aligned solder assembly with no, one, or two mechanical stops. The trend is very clear. In addition to die attachments, soldering technology has been successfully developed for precision alignments with or without using its self-alignment properties.

A case study on the development of a Au/Sn soldering technology has been presented in detail. The study considered bumping technique, stripe-type bump bonding, and three applications. In addition, four critical packaging issues have been reviewed: solder materials, fluxless reflow, design, and reliability. New solder materials may posses high-fatigue, low-creep desirable properties, and PADS and formic acid vapor can be used for an effective "fluxless" reflow process. Profile and force modeling tools are important for engineering solder joints for precision alignments. Solder reliability is becoming more and more important, but it has been rarely studied. All these issues require further studies to support the advancement of opto-electronic packaging for low-cost, high-performance, and high-reliability modules.

ACKNOWLEDGMENTS

We acknowledge support from the National Science Foundation (MIP-9058409, EEC-9015128, MIP-9400655). We also express our appreciation to Mr. Jackson of IBM, Mr. Haugsjaa of GTE, Mr. Tsunetsugu of NTT, and Mr. Nagano of NEC for providing the referenced sketches and photographs.

REFERENCES

Armiento C.A., Nefri A.J., Tabasky M.J., Boudreau R.A., Rothman M.A., Fizgerald T.W., and Haugsjaa P.O., 1992, "Gigabit Transmitter Array Modules on Silicon Wafer-board," *IEEE Trans. Components, Hybrids, and Manufacturing Technology,* Vol. 15, No. 6, pp. 1072–1079.

Artaki I., Jackson A., and Vianco P., 1994, "Fine Pitch Surface Mount Assembly with Lead-Free, Low Residue Solder Paste," Proceedings of Surface Mount Technology, pp. 449–459.

Bellcore Specifications, 1993, "Reliability Assurance for Loop Optoelectronics," TA-NWT-000983, Issue 2, pp. 4–15 to 4–17.

Benzoni A.M. and Dautartas M.F., 1994, "Single In-Line Optical Package," U.S. Patent 5,337,398.

Blacha A., Gfeller F., and Vettiger P., 1993, "Optical Fiber Alignment," U.S. Patent 5,247,597.

Bonda N.R., Fang T., Kaskoun K., Lytle W.H., Marlin B., Swan G., Stafford J.W., and Tam G., 1997, "Physical Design and Assembly Process Development of a Multi-chip Package Containing a Light Emitting Diode (LED) Array Die," IEEE Trans. Components Packaging and Manufacturing Technology, Part B, Vol. 20, No. 4, pp. 389–395.

Boudreau R., Tabasky M., Armiento C., Bellows A., Cataldo V., Morrison R., Urban M., Sargent R., Negri A., and Haugsjaa P., 1993, "Fluxless Die Bonding for Opto-electronics," Proceedings of the 43rd Electronic Components and Technology Conference, Orlando, FL, June 1993, pp. 485–490.

Brady M.F. and Deshmukh R.D., 1993, "Solder Self-Alignment Methods," U.S. Patent 5,249,733.

Darveaux R., Benerji K., Mawer A., and Dody G., 1995, "Reliability of Ball Grid Array Assembly," Chapter 13, *Ball Grid Array Technology,* edited by Lau J. H. McGraw-Hill.

Dautartas M.F., Benzoni A.M., Broutin S.L., Coucoulas A., Moser D.T., Wong Y.-H., and Wong Y.M., 1995a, "Optical Performance of Low-Cost Self-Aligned MCM-d Based Optical Data Links" Proceedings of the 45th Electronic Components and Technology Conference, Las Vegas, May 19996, pp. 1254–1262.

Dautartas M.F., Blonder G.E., Wong Y.H., and Chen Y.C., 1995b, "A Self-Aligned Optical Subassembly for Multi-Mode Devices," IEEE Trans. Components, Hybrids, and Manufacturing Technology, Vol. 18, No. 3, pp. 552–557.

Deshmukh R.D., Brady M.F., Roll R.A., King L.A., Shmulovich J., and Zolnowski D.R., 1993, "Active Atmosphere Solder Self-Alignment and Bonding of Optical Components," *Int. J. Microcircuits Electron. Packaging,* Vol. 16, No. 2, pp. 97–107.

Edge C., Ash R.M., Jones C.G., and Goodwin M.J., 1991, "Flip-Chip Solder Bond Mounting of Laser Diodes," *Electron. Lett.,* Vol. 27, No. 6, pp. 499–501.

Frear D.R., Hosking F.M., Keicher D.M., and Peebles H.C., 1995, "Fluxless Soldering for Microelectronic Applications" *Material for Electronic Packaging,* edited by Chung, Deborah D.L., Butterworth Heinemann.

Gates J.V., Henein G., Shmulovich J., Muehlner D.J., MacDonald W.M., and Scotti R.E., 1996, "Uncooled Laser Package Based on Silicon Optical Bench Technology," Proc. SPIE, Vol. 2610, pp. 127–137.

Gershovish M. and Lee Y.C., 1994, "Gas Flow Effects on Self-Aligning Soldering for Optoelectronics," 8th International Microelectronics Conference, April 1994.

Goldmann L.S., 1969, "Geometric Optimization of Controlled Collapse Interconnections," *IBM J. Res. Dev.,* pp. 251–265.

Goodwin M.J., Moseley A.J., Kearley M.Q., Morris R.C., and Kirkby C.J.G., Thompson J. and Goodfellow R.C., "Optoelectronic Component Arrays for Optical Interconnection of Circuits and Subsystems," *J. Lightwave Technol.,* Vol. 9, No. 12, pp. 1639–1645.

Guo Z. and Conrad H., 1993, "Computer Modeling of Isothermal Low-Cycle Fatigue of Pb-Sn Solder Joints," IEEE 43rd Electronic Components and Technology Conference, Orlando, FL, June 1993, pp. 831–838.

Haitz R.H., 1992, "Light-Emitting Diode Printhead," U.S. Patent 5,134,340.

Hareb S., 1998, "Mechanical Behavior of Solder- and Epoxy-Attached Optoelectronic Assemblies," Ph.D. Thesis, University of Colorado, Boulder.

Hatakeyama H., Kitamura S., Hamamoto K., Yamaguchi M., and Komatsu K., 1997, "Spot Size Converter Integrated Semiconductor Optical Amplifier Gates with Low Power Consumption," Technical Digest of Optoelectronics Communication Conference '97, 9C3-2, pp. 180–181.

Hayashi T., 1992, "An Innovative Bonding Techniques for Optical Chips Using Solder Bumps That Eliminate Chip Position Adjustments," IEEE Trans. Components, Hybrids, and Manufacturing Technology, April 1992, Vol. 15, No. 2, pp. 225–230.

Hayashi T. and Tsunetsugu H., 1996, "Optical Module with MU Connector Interface Using Self-Alignment Technique by Solder-Bump Chip Bonding," Proceedings of 46th Electronic Components and Technology Conference, Orlando FL, May 1996, pp. 13–19.

Hayes D.J. and Wallace D.B., 1996, "Solder Jet Printing for Low Cost Wafer Bumping," Proceedings of SPIE, Vol. 2920, pp. 296–301.

Hinch S.W., 1988, *Handbook of Surface Mount Technology,* Longman Scientific & Technical, Harlow, England.

Honmou H. and Itoh M., 1995, "Optical Coupling of Laser Diode Array to Singlemode-Fiber Array with Heat-Treated Hemispherical Microlens," *Electron. Lett.,* Vol. 31, No. 10, pp. 793–794.

Iezeki el., S., Soshea E.A., O'Keefe M.F., and Snowden C.M., 1997, "Application of Silicon-Glass Technology to Microwave Photonic Multichip Modules," Proceedings of Interpack, Hawaii, pp. 759–764.

Itoh M., Sasaki J., Uda A., Yoneda I., Honmou H., and Fukushima K., 1996a, "Use of AuSn Solder Bumps in Three-Dimensional Passive Aligned Packaging of LD/PD Arrays on Si Optical Benches," Proceedings of 46th Electronic Components and Technology Conference, Orlando, FL, June 1996, pp. 1–7.

Itoh M., Sasaki J., and Uda A., 1997, "Passive Alignment Packaging on Si Optical Bench Using AuSn Solder Bumps," Proceedings of LEOS'97, pp. 126–127.

Itoh M., Yoneda I., Sasaki J., Honmou H., Fukushima K., and Nagahori T., 1996b, "Self-Aligned Packaging of Multi-Channel Photodiode Array Module Using AuSn Solder Bump Flip-Chip Bonding," Proceedings of EuPac 96, pp. 75–77.

Itoh M., Naganori T., Kohashi H., Kaneko H., Honmou H., Watanable I., Uji T., and FujiWara M., 1991, "Compact Multi-Channel LED/PD Array Modules Using New Assembly Techniques for Hundred Mbit/sec/ch Parallel Optical Transmission," Proceedings of 41st Electronic Components and Technology Conference, Atlanta, CA, May 1991, pp. 475–478.

Jackson K.P., Flint E.B., Cina M.F., Lacet D., Kwark Y., Trewhella M., Caufiled T., Buchmann P., Harder C., and Vettiger P., 1994, "A High-Density, Four-Channel, OEIC Transceiver Module Utilizing Planar-Processed Optical Waveguides and Flip-Chip, Solder-Bump Technology," *J. Lightwave Technol.*, Vol. 12, No. 7, pp. 1185–1191.

Jeannotte D.A., Goldmann L.S., and Howard R.T., 1988, "Package Reliability," Chapter 5, *Microelectronic Packaging Handbook,* edited by Tummala R.R. and Rymaszewski E.J., Van Nostrand Reinhold, New York.

Ju T.H., Lin W., Lee Y.C., McKnight D.J., and Johnson K.M., 1993a, "Packaging of a 128 by 128 Liquid-Crystal-On-Silicon Spatial Light Modulator Using Solder," IEEE *Photonics Technol. Lett.,* Vol. 7(9) pp. 1010–1012, Sept. 1995.

Ju T.H., Chan Y.W., Lin W., and Lee Y.C., 1993b, "Experimental Verification of a General Purpose Solder Profile Model," EEP-Vol. 4-2 Advances in Electronic Packaging. American Society of Mechanical Engineers, pp. 1163–1168.

Kato T., et al., 1998, "10 Gbit/sec Photonic Cell Switching with Hybrid 4×4 Optical Matrix Switch Module on Silica-Based Planar Waveguide Platform," Optical Fiber Communication Conference, San Diego, 1998, PD3.

Koopman N., Bobbio S., Nangalia S., Bousaba J., and Piekarski B., 1993, "Fluxless Soldering in Air and Nitrogen," Proceedings of 43rd Electronic Components and Technology Conference, Orlando, FL, June 1993, pp. 595–605.

Koopman N., Nangalia S., Rogers V., Peterson J., Brinkley B., Yow E., Bobbio S., and Pennington M., 1996, "Fluxless, No Clean Solder Processing of Components, Printed Wiring Boards, and Packages in Air and Nitrogen," Proceedings of Surface Movement International, pp. 437–444.

Koopman N.G. Reiley T.C., and Totta P.A., 1988, "Chip-to Package Interconnections," Chapter 6, *Microelectronic Packaging Handbook,* edited by Tummala R.R and Rymaszewski E.J., Van Nostrand Reinhold, New York.

Lau J. and Rice D.W., 1985, "Solder Joint Fatigue in Surface Mount Technology: State of the Art," *Solid State Technol.,* pp. 91–103.

Lee C.C. and Wang C.Y., 1992, "A Low Temperature Bonding Process Using Deposited Gold-Tin Composites," *Thin Solid Films,* Vol. 208, pp. 202–209.

Lee S.H., Joo G.C., Park K.S., Kim H.M., Kim D.G., and Park H.M., 1995, "Optical Device Module Packages for Subscriber Incorporating Passive Alignment Techniques," Proceedings of 45th Electronic Components and Technology Conference, Las Vegas, May 1995, pp. 841–844.

Lee N.-C., Slattery J.A., Sovinsky J.R., Artaki I., and Vianco P.T., 1994, "A Novel Lead-Free Solder Replacement," Proceedings of Surface Mount International Conference, pp. 463–472.

Lee N.-C., 1998, "A Thorough Look at Lead-Free Solder Alternatives," Circuit Assembly, April, p. 64–71.

Lee Y.C. and Basavanhally N.R., 1994, "Overview: Solder Engineering for Optoelectronic Packaging," *Journal of Metals*, pp. 46–50.

Lin W., 1995, "Study of Low Cost Soldering Technology for Liquid-Crystal-On-Silicon (LCOS) Modules Assembly," Ph.D. Thesis, University of Colorado at Boulder.

Lin W., Patra S.K., and Lee Y.C., 1995, "Design of Solder Joint For Self-Aligned Optoelectronic Assembly," IEEE Trans. Components, Hybrids, and Manufacturing Technology, Vol. 18, No. 3, pp. 543–551.

Lin W., Lee Y.C., and Johnson K.M., 1993, "Study of Soldering for VLSI/FLC Spatial Light Modulators," Proceedings of 43rd Electronic Components and Technology Conference, Orlando, FL, June 1993, p. 491–497.

Lowe H. and Lyn R., 1995, "Real World Flip-Chip Assembly: A Manufacture's Experience," Proceedings of Surface Movement International, pp. 80–87.

Martino P.M., Freedman G.M., Racz L.M., and Szekely J., 1994, "Predicting Solder Joint Shape By Computer Modeling," Proceedings of 44th Electronic Components and Technology Conference, Washington D.C., May 1994, pp. 1071–1078.

Mavoori H. and Jin S., 1998, "Significantly Enhanced Creep Resistance in Low Melting Point Solders through Nanoscale Oxide Dispersion," *Appl. Phys. Lett.* Vol. 73, No. 16.

McDowell D.L., Miller M.P., and Brooks D.C., 1994, "A Unified Creep-Plasticity Theory for Solder Alloys," Fatigue of Electronic Materials, ASTM STP 1153, American Society for Testing and Materials, Philadelphia, pp. 42–59.

McGroarty J., Borgensen P., Yost B., and Li C.Y., 1993, "Statistics of Solder Joint Alignment for Optoelectronic Components," IEEE Trans. Components, Hybrids, and Manufacturing Technology, Vol. 16, No. 5, pp. 527–529.

Miller L.F., 1969, "Controlled Collapse Reflow Chip Joining," *IBM J. Res. Dev.* pp. 239–250.

Morozova N., Liew L.A., Zhang W., Irwin R., Su B., and Lee Y.C., 1996, "Controlled Solder Self-Alignment Sequence for an Optoelectronic Module without Mechanical Stop," Proceedings of 46th Electronic Components and Technology Conference, Orlando, FL, May 1996. pp. 1188–1193.

Nagano T., Ueda E., Onayama S., Katayama R., Hamada H., and Ono Y., 1993, "Thin Optical Head with Flip-Chip Bonded Module for Compact Disc Read Only Memory Drives," *Jpn. J. Appl. Phys.,* Vol. 32, Pt. 1, No.11B, pp. 5263–5268.

Nangalia S., Koopman N., Rogers V., Beranek M.W., Hager H.E., Ledbury E.A., Loebs V.A., Miao E.C., Tang C.H., Poco C.A., Swenson E.J., Hatzis D., Li P., and Luck C., 1997, "Fluxless, No Clean Assembly of Optoelectronic Devices with PADS," Proceedings of 47th Electronic Components and Technology Conference, San Jose, CA, May 1997, pp. 755–762.

Nordin R.A., Buchholz D.B., Huisman R.F., Basavanhlly N.R., and Levi A.F.J., 1993, "High Performance Optical Data Link Array Technology," Proceedings of 43rd Electronic Components and Technology Conference, Orlando, FL, May 1996, pp. 795–801.

Pao Y.H., Badgley S., Govila R., and Jih E.. 1994, "Thermomechanical and Fatigue Behavior of High-Temperature Lead and Lead-Free Solder Joints," *Fatigue of*

Electric Materials, ASTM STP 1153, American Society for Testing and Materials, Philadelphia, pp. 60–81.

Pao Y.H., Badgley S., Jih E., Govila R., and Browning J., 1993, "Constitutive Behavior and Low Cycle Thermal Fatigue of 97Sn-3Cu Solder Joint," *J. Electron. Packaging,* Vol. 115, pp. 147–152.

Patra S.K. and Lee Y.C., 1991a, "Quasi-Static Modeling of the Self-Alignment Mechanism in Flip-Chip Soldering. Part I: Single Solder Joint," *J. Electron. Packaging,* Vol. 113, p. 337–342.

Patra S.K. and Lee Y.C., 1991b, "Modeling of the Self-Alignment Mechanism in Flip-Chip Soldering. Part II: Multiple Solder Joints," Proceedings of 41st Electronic Components and Technology Conference, Atlanta, GA, May 1991, pp. 783-788.

Patterson D.S., Elenius P., and Leal J.A., 1997, "Wafer Bumping Technologies-A Comparative Analysis of Solder Deposition Processes and Assembly Consideration," Proceedings of InterPack. EEP-Vol. 19-1, *Advances in Electronic Packaging,* pp. 337–351.

Pittroff W., Barnikow J., Klein A., Kurpas P., Merkel U., Vogel K., Wufil J., and Kuhmann J., 1997, "Flip Chip Mounting of Laser Diode with Au/Sn Solder Bumps: Bumping, Self-Alignment, and Laser Behavior," Proceedings of 47th Electronic Components and Technology Conference, San Jose, CA, May 1997, pp. 1235–1241.

Potier N., Sindzingre T., and Rabia S., 1995, "Fluxless Soldering under Activated Atmosphere at Ambient Pressure," Proceedings of SMI, pp. 453–458.

Pusarla P. and Christou A., 1996, "Solder Bonding Alignment of Microlens in Hybrid Receiver for Free Space Optical Interconnections," Proceedings of 46th Electronic Components and Technology Conference, Orlando, FL, May 1996, pp. 42–47.

Sasaki J., Honmou H., Itoh M., Uda A., and Torikai T., 1995 "Self-Aligned Assembly Technology for Laser Diode Modules Using Stripe-Type AuSn Solder Bump Flip-Chip Bonding," Proceedings of IEEE Laser and Electro-Optics Society, 1995, pp. 234.

Sasaki J., et al., 1998, "Hybrid 4 × 4 Optical Matrix Switch Module on Silica Based Planar Lightwave Circuit by Self-Align Multiple Chip Bonding Technique," Integrated Photonics Research '98, ITuC3.

Sasaki J., Kaneyama Y., Honmou H., Itoh M., and Uji T., 1992, "Self-Aligned Assembly Technology for Optical Devices Using AuSn Solder Bumps Flip-Chip Bonding" Proceedings of IEEE Laser and Electro-Optics Society Annual Meeting, Nov. 1992, pp. 260–261.

Schroeder S.A., Morris W.L., Mitchell M.R., and James M.R., 1994, "A Model for Primary Creep of 63Sn-37Pb Solder," *Fatigue of Electronic Materials,* ASTM STP 1153, American Society for Testing and Materials, Philadelphia, pp. 82–94.

Shimizu K., Akamatsu T., Nakanishi T., Karasawa K., Hahimoto K., and Niwa K., 1994, Solder Joint Reliability of Flip-Chip Interconnection, Proceedings of International Society of Hybrid Microelectronics, 1994, pp. 272–277.

Su B., Gershovish M., and Lee Y.C., 1997, "Gas Flow Effects on Precision Solder Self-Alignment," Proceedings of 47th Electronic Components and Technology Conference, San Jose, CA, May 1997, pp. 797–803.

Suhir E., 1987, "Die Attachment Design and Its Influence on Thermal Stresses in the Die and the Attachment," Proceedings of 37th Electronic Components and Technology Conference, pp. 508–517.

Tan Q. and Lee Y.C., 1996, "Soldering for Optoelectronic Packaging," IEEE Electronic Components and Technology Conference, Orlando, FL, May 28–30.

Tsunetsugu H., Katsura K., Hayashi T., Ishitsuka F., and Hata S., 1992, "A New Packaging Technology Using Microsolder Bumps for High-Speed Photoreceivers," IEEE Trans. Components, Hybrids, and Manufacturing Technology, Vol. 15, No. 4, pp. 578–582.

Viera AP., Boysan P., Tolwaker S., and Foehringer D., 1993, "Attachment Reliability Evaluation and Failure Analysis of Thin Small Outline Packages," Proceedings of 43rd Electronic Components and Technology Conference, Orlando, FL, June 1993, pp. 54–61.

Yano M., Nakagawa G., and Fujimoto N., 1995, "Skew-Free Parallel Optical Links and their Array Technology," Proceedings of the 45th Electronic Components and Technology Conference, Las Vegas, NV, May 1995, pp. 552–564.

6 Plastic-Based Passive Optical Alignment: The Jitney Parallel Optical Interconnect

Sharon M. Boudreau

CONTENTS

6.1   INTRODUCTION

Passive optical alignment can be applied to both single-mode fiber-based products and multimode-fiber-based products. The single-mode fiber-based products provide the best bandwidth, but the alignment tolerances are the most extreme — on the order of ± 0.5 μm — and there is an associated cost. In contrast, multimode-fiber-based products, with alignment tolerances of about ± 5 μm, provide less — but adequate — bandwidth for many datacom applications and are often configured with many parallel channels, allowing parallel paths to achieve a higher combined bandwidth. For multimode fiber-based products, plastic mechanical parts and optics often provide adequate passive alignment if they are incorporated in the design. This chapter reviews the Jitney optical interconnect as an example of a passively aligned multimode interconnect for datacom. The growing data communication industry is looking for an inexpensive data bus that can provide high speed (0.2 to 1.0 Gbit/sec) over relatively short distances of tens to hundreds of meters. The cost factor is critical because the standard copper interconnect technology has undergone recent technical advances [1]. To be cost competitive with the popular data buses made from parallel copper wire, a parallel optical fiber interconnect was developed by OETC (Optoelectronics Technology Consortium). The DARPA-funded industry designers included IBM's Watson Research Center, 3M's fiber optics labs, and Lexmark's Plastics Technology Center. In an aim to cut costs, but not overall performance, three key areas of exploration were included: standard integrated circuit (IC) manufacturing/packaging using overmolded lead-frame technology that could adapt to the addition of optical hardware, passive optical alignment to relax alignment tolerances, and inexpensive plastic molded parts [2]. The result was a small, flexible, low-cost system made from plastic components that may establish an international standard with its optical lens array. In 1996, IBM completed preliminary testing of the 32-channel OETC fiber-optic transmitter/receiver array [3]. This chapter describes the design, testing, and recent advances in the commercial "Jitney" optical bus. It is one example of a plastic-based, passive, optical alignment. This alignment incorporates multimode optical fiber, but at the expense of a lower bandwidth distance product per channel.

6.2 COMPONENTS

6.2.1 LEADFRAME MODULE PACKAGE

The Jitney optical bus designers took advantage of standard leadframe technology to be cost competitive. The starting point was a production line at IBM used for the plastic quad flat pack (PQFP). This IC industrial standard was a 28×28 mm, 208 lead PQFP that was later modified to house the optical bus. The resulting rectangular 20×28 mm design size had 36 leads on each short side. The cable side had no inner leads, but outer leads were placed on the edges for stability during attachment, as shown in Figure 6.1 [2]. This design saved costs and was premolded with a silicon-filled epoxy cavity to house the IC and optoelectronic (OE) chips plus the optical components. The plastic leadframes were made of palladium-plated copper, so immediate outer lead attachments could be made at card level. A palladium-plated copper heat sink was attached to the leadframe using double-sided epoxy tape. One side of the heat sink served as a mounting side for the chips, and the other side was reserved for any necessary heat spreader. The heat slug was the package ground. Fabrication of the plastic lead frame was either single or multiple, with five parts per strip.

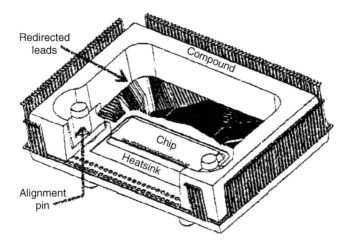

FIGURE 6.1 Premolded Jitney package.

A unique feature of the design was the overmolded alignment pins, which were part of the passive alignment scheme. Reference pins were needed for placement and to align the optical coupler with the chips. Quality control of the molding conditions resulted in center-to-center spacing of the pins of 19.230 mm, with a standard deviation less than 8 µm. The pin lengths of 3.4 mm had diameters of 2.04 mm, with standard deviations less than 3 µm. Outside pins were added for more coarse alignment of the connector shell to the module package. One side of the package remained open to allow the optical coupler's connector ferrule socket to protrude.

Similar packages were molded for the receiver and transmitter. Differences included modifying the receiver package to house one chip instead of two, changing the inner lead design of the leadframe, and having longer inner leads with wirebonds spanning from the inner leads to the heat sink. The transmitter, which held two chips, had wirebonds from the heat sink to the driver's chip, and longer inner leads for the multiple wirebond connections. The special inner leads provided 30% less inductance than a standard-width lead for powering the final driver stage.

6.2.2 OPTICAL COUPLER

The plastic, injection-molded optical coupler consisted of a transparent array lens and a sealed opaque housing. The purpose of the optical coupler was to couple light from the vertical-cavity surface-emitting laser (VCSEL) array (20 µm diameter each on 500-µm centers) to the fiber array (200 µm core diameter step index) and then to the photodiode array. This entire subassembly was then placed in the cavity of the premolded module package.

The Jitney array lens consisted of a backbone structure on which 22 optical elements were premolded and evenly spaced. Only the central 20 optical elements were optically active, as shown in Figure 6.2 [2]. A detail of a single element, with ray tracing, is shown in insert B. A segment of the lens array is depicted in insert A.

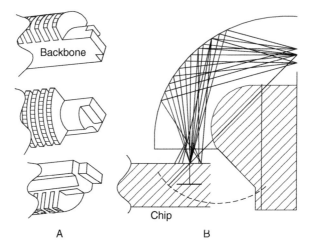

FIGURE 6.2 Molded Jitney array lens.

This design had to be manufacturable, and the design had to isolate the external environment from the optoelectronic chips. The molding contract was developed by Pitney Bowes in Danbury, Connecticut, and Apex Machine Tool Co. in Farmington, Connecticut [4]. Each optical element had an input and output face perpendicular to one another and a metallized plane parallel to the sidewall. The individual elements had elliptically shaped surfaces and were metallized by a vapor deposition process. Copper deposition was preferred over aluminum because it had a higher reflectivity, but it showed reduced reflectivity after 1100 hours of humidity and stress tests. Before the deposition, adhesion was increased by a preliminary bake followed by a plasma treatment. A special apparatus was constructed to hold up to 300 lenses during the deposition. Deposition was performed on the curved and back sidewall of each optical element by orientating the optical element down toward the evaporation source. Overall film deposits were 1.5 μm thick on the elliptical side, with smaller thickness on the sidewall.

The design of the optical coupler allowed preassembly so that it could be mounted onto the leadframe pins in the modules. A liquid crystal polymer housing included contour surfaces to aid in the epoxy seal between the nonoptical portions and the optically active chips and array lens. The epoxy was applied with syringe injection and cured at low temperatures. The result was a housing that held the array lens in alignment and positioned it properly with respect to the package.

6.2.3 Chips

Advances in OE chip technology have been a boost to the Jitney program. Costs can be lowered if yield and wafer testability can be increased. The chips included a laser driver, a laser VCSEL, and a receiver optoelectronic integrated circuit (OEIC).

The laser driver array chip was manufactured on IBM's standard CMOS IC manufacturing line [5]. The nominal size was 5×13 mm. The 2-byte-wide electrical

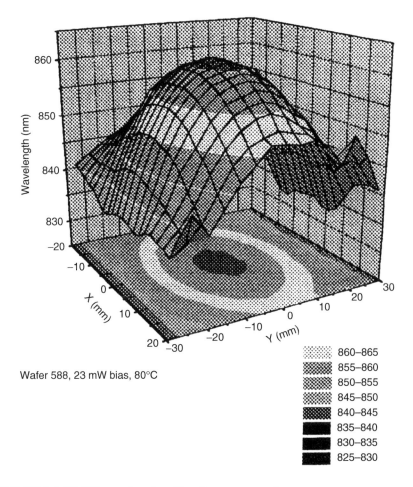

Wafer 588, 23 mW bias, 80°C

▒	860–865
▓	855–860
▓	850–855
▓	845–850
■	840–845
■	835–840
■	830–835
■	825–830

FIGURE 6.3 VCSEL wavelength uniformity across a wafer.

interface operated each line with an IEEE-Std. 15963-1994 low-voltage differential signaling (LVDS) logic receiver. Depending on user selection, the driver could be operated in either synchronous mode (2-byte-wide transmission) or asynchronous mode (many parallel lines). Standard IC tools and processes were used to place and attach the driver chip to the heat sink.

The monolithic 20-element VCSEL chip was nominally 1×13 mm, with devices on a 500-μm pitch, and exhibited excellent uniformity, as shown in Figure 6.3 [3]. The VCSEL chip emitted at a wavelength equal to 850 ± 10 nm, with $I_{th} = 3.75 \pm 0.5$ ma. This wavelength was compatible with the laser standard of 780 to 850 nm for optical networks. The chip design included two ±20-μm edges cut with the IC chip dicing tool to permit the chip to be passively aligned with the transmitter module. Gold wires located along one edge of each chip supplied the current for the VCSEL laser array. After exiting the VCSEL base, the return current passed through a silver epoxy layer and returned through wirebonded gold wires from the heat sink.

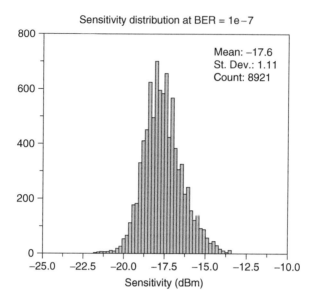

FIGURE 6.4 Sensitivity uniformity across OETC receivers.

The GaAs receiver chip was designed to give high yield and uniform sensitivity, as illustrated in Figure 6.4 [3]. The chip uses metal semiconductor metal (MSM) photodiodes integrated with metal semiconductor field effect transistor (MESFET) pre- and postamplifiers, digital logic, and off-chip drivers. The chip made at Vitesse Semiconductor used 1.25 μm H-GAS-III process. The nominal size was 4 × 13 mm, and the chip employed special on-chip dynamic control circuits to adjust logic decision levels under the presence of long run lengths of NRZ data. Figure 6.5 shows the schematic of the receiver functions [3]. Alignment could be done passively because the photodetector active area dimensions, 300 × 300 μm, were large. The chip outputs were lost value of service (LVOS) logic levels, with a power consumption of 3.7 W and an output skew of less than 100 psec. Similar to the VCSEL chip, the standard IC chip was diced with two reference edges cut to within ±10 μm for chip placement for passive alignment.

Screening of all three chips included direct current (DC) and alternating current (AC) measurements. The nominal time to test channels was 4 seconds for VCSEL chips and 20 sec for each channel of the transmitter and receiver chips. Rapid screening advances at the wafer level could further lower costs.

6.2.4 JITNEY CABLE

As with other components, the design of the Jitney cable assembly coupled optimum performance with low-cost plastic parts. Approximately ±20 μm was budgeted for mechanical misalignment between the fiber and optical coupler. The large core 20-fiber ribbon cable relaxed alignment tolerances for the plastic parts and had the connectors molded on each end. The designers took advantage of an existing 3M

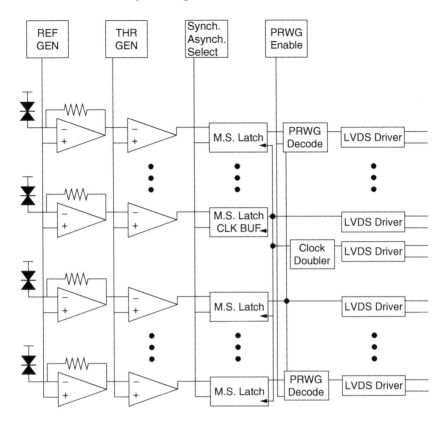

FIGURE 6.5 Receiver chip functions.

cable assembly and made a modified TECS™ fiber. There are two varieties, depending on application use. The first cable assembly, used for applications less than 45 m, had a pure silica core with a 200-μm diameter wrapped with low-index fluoropolymer cladding (225 μm) and coated with an acrylate (300 μm). The numerical aperture (N.A.) measured 0.4, with a propagation loss of 5 dB/Km. A more expensive fabrication using chemical vapor deposition (CVD) improved the performance of the cable. This premium fiber ribbon cable used a graded index central core of 175 μm surrounded by silica (200 μm), followed by a similar cladding (225 μm), and finally by a strong acrylate coating (300 μm). Tests showed that the cable had a N.A. of 0.2. When this premium cable was excited by the VCSEL, the measured bandwidth (3 dB electrical) of a 30-m length was 660 MHz. This could be compared to the previous cable, with a N.A. of 0.2, which could be excited by a VCSEL at 470 MHz over a 200-m distance. Results thus were more than adequate to be able to transmit data at a 500 Mbit/sec/channel data rate to a targeted distance of 100 m.

The fiber ribbon assembly illustrated in Figure 6.6 [6] helped eliminate the labor-intensive process of adding the connector parts after the cable assembly. The simplified process involved feeding reels of 20 TECS™ fiber into a comb

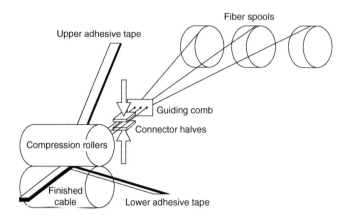

FIGURE 6.6 Fiber ribbon assembly tool.

structure that spaced the fibers (500 μm from center to center). Connector ferrules were laminated into the fiber array by a robotic system. The result was a continuous multifiber ribbon cable with face-to-face connector ferrules at preselected distances. The ferrule assemblies were cut and optically finished before being snapped into a molded plastic connector body. Tests on various cable lengths from 5 to 100 m showed no significant degradation in fiber properties.

6.2.5 PASSIVE ALIGNMENT CONSIDERATIONS FOR PACKAGING

Standard pick-and-place tools having chip placement accuracies of approximately ±30 μm were not suitable for the Jitney interconnect. Therefore, a passive chip placement tool was designed for the Jitney application, as shown in Figure 6.7 [2]. The diagram shows a table that moved uniaxially along a horizontal track. A pick-and-place head controlled by an electromagnetic actuator moved vertically. A bridged structure over the table supported the vertical actuator. The table was designed to allow three mechanically defined positions: the first for load/unload, the second for chip pickup, and the third for chip placement. Different corner plates for the receiver and transmitter modules fixed the chip with pins. Epoxy was applied to the heat sink, and a hinged chuck secured the package to the plate. Three directional micrometers defined the package position. Displacements from target positions for operational chips and packages showed large displacements ranging from ±2 to ±30 μm, depending on chip type. Some increased displacements were attributed to package-to-package variation of pins and variation in the housing fit to the package. However, results of the tool performance indicated that chip placements were adequate for module construction.

The chips were connected to the leadframe with standard gold-ball wiring, using an existing PQFP tool. The premolded package required surface decontamination to achieve strong wirebonds. Oxygen ashing and nitrogen curing done after the chip attachment provided the best adhesion. Finally, the optical coupler was

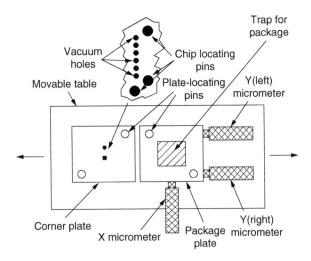

FIGURE 6.7 Passive alignment chip placement tool.

mounted into the package by matching the pins into the housing holes. Vertical positioning of the optical coupler resulted when the foot of the array lens rested on the heat sink. A bead of epoxy provided an airtight cavity for the light paths. Additional epoxy provided by a globbing procedure protected the wirebonds and leadframe from the environment.

6.2.6 ALIGNMENT STUDIES

A package was designed with cost in mind that allowed maximum allowance to misalignment. The alignment errors in the laser-to-array lens caused signal and bandwidth losses, which were attributed to beam shape degradation [7]. The bandwidth-distance of the fiber to more than 12 MHz-km was achieved with a low numerical aperture (< 0.2). Large misalignments of the laser to the fin contributed less than 0.5 dB and were small in comparison to losses associated with the transmitter and receiver. To measure mechanical misalignments, a test setup was built. A testing stage holding an array lens, a laser, and a fiber allowed positioning of the chip and fiber to the array lens by the use of two microscopes. This apparatus provided measurements of the sensitivity to misalignment. Under normal alignment, the array lens resulted in a 2.5-dB insertion loss from the laser to the fiber and a 1.8-dB insertion loss from the fiber to the detector.

The normalized coupling efficiency (decibels) was measured as a function of laser misalignment in the X and Y directions. A 2-dB loss was recorded if the laser was displaced up to ±120 μm in the X direction and ±42 μm in the Y direction while holding the fiber normal. If the laser was held normal, a 2-dB loss was measured for fiber-to-array lens misalignment up to ±65 μm in the X or Y direction. Receiver-to-lens displacements resulted in only a 1-dB penalty when the receiver was ±75 μm in X and ±123 μm in Y. A similar loss was recorded for fiber-to-lens misalignments.

An investigation of all alignment-error contributions showed that the total loss values of the transmitter and receiver were 4.8 and 4.3 dB, respectively. These limits were within the budgeted goals. Systematic errors stemmed from the bowing of the array lens in the housing, run out, or offset errors in the connector ferrule and array lens, and chip placement difficulties with setting the package-plate micrometers. Random errors included variations in the fiber geometry, lens alignment in the housing, and chip placement tool repeatability.

6.3 TESTING

Unique challenges were presented for test applications having an optical component that are both mixed signal and parallel while collecting data at an extremely high rate. Again, costs affect the testing methodologies for the Jitney optical bus to be cost competitive with copper bus technology. There are two suggested routes that one may implement depending on the type of data sought. The optoelectronic bus could be tested via the mixed-signal approach or by an all-electrical functional test. Both approaches have advantages and disadvantages, depending on the stage of manufacturing development the bus product is in. The mixed-signal approach looks at the mixed optical and electrical input and output signals of the optical bus components. This is especially useful at the wafer level because the information gathered can be used for screening, sorting, and perfecting designs. Extensive data were gathered by testing every component of the optical bus including the VCSELs, the laser driver, OEIC receiver chips, and both the receiver and transmitter packaged modules [8]. One drawback of mixed-signal testing is that it requires a great deal of resources invested in specialized testing equipment (multiple light sources, calibrated optoelectronic amplifiers, etc.) or time expended in serial testing of single channels. However, depending on what stage of development the optical components are in, it may be possible to eliminate some tests if the design is perfected, thus lowering overall component costs.

In addition, electrical function tests or digital testing would be more effective if the optical bus were in production because of the automated technology. Commercial digital testers function at data rates as high as 1 Gbps NRZ (no return to zero) and can provide up to several hundred pins.

Mixed-signal testing provides the quality of information needed during the design stage. Every semiconductor component generated from the Jitney project has had extensive tests run. Test results were available for the chips, both laser driver and OEIC receiver; VCSELs; and receiver and transmitter modules. The data were used not only to evaluate the components but also to sort the components for packaging.

6.3.1 COMPONENT TESTING

6.3.1.1 Laser Driver

Tests of the laser driver chip included AC electrical characterization at the wafer level. Designed to receive LVDS inputs and drive an output source [5], the chip had several variables that needed to be tested. The design included a linear feedback shift register (LFSR), which generated a pseudo-random bit (PRBS) pattern that limited optical transmission lengths and was used as a self-test generator. The built-in self-test (BIST) was a real time saver. The LFSR generated a 2^15 -1 pseudo-random bit corresponding to the polynomial $x^{15} + x^{14} + 1 = 0$. The bit sequence was sent to the output latches of the driver and then to the output current driver. If the LFSR and the drivers were operating, a 2^15-1 PRBS signal was output from the Jitney laser driver chip. A bit error rate tester (BERT) was used to check all channels [8].

One test performed included output signal characterization in both asynchronous and synchronous modes, using a Tektronix CSA11801 digitizing oscilloscope. A database of device characteristics was generated, including duty cycle and minimum and maximum output currents. The sensitivity of the driver chip channels was tested, using bit error rate in synchronous mode. The data were derived from one wafer of driver chips and represented 2318 driver channel measurements per plot. A total of 14,640 channels were tested, yielding valuable data for evaluating the effects of process variation on performance. The modulation current amplitude data in Figure 6.8 [8] allowed examination of device uniformity across the wafer.

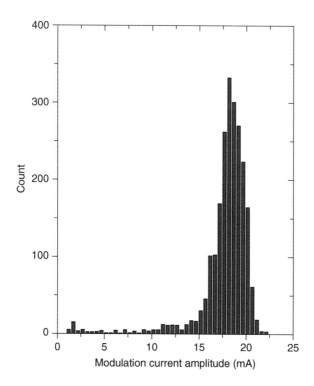

FIGURE 6.8 Driver chip modulation current amplitude.

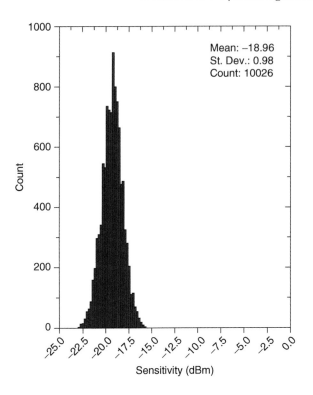

FIGURE 6.9 Receiver channel sensitivity distribution.

6.3.1.2 OEIC Receiver

OEIC receiver chips were designed with ease of testability as a criterion. Characteristics measured included optical input sensitivity and electrical output. A bit error rate of $1e^{-7}$ was favored over an error rate threshold of $1e^{-9}$ because of the reduced time required to measure the sensitivity of a channel. The $1e^{-9}$ threshold required 25 min per channel, in comparison to 15 sec to test the average $1e^{-7}$. Data were collected on approximately 20,000 receiver channels, and characterization was used for sorting, design evaluation, and process monitoring [9]. Figure 6.9 [8] represents the mean sensitivity of 10,026 receiver channels tested, and statistical analysis showed a mean sensitivity of −18.96 dBm ± 0.98 dB.

6.3.1.3 VCSEL

At the wafer level, testing involved obtaining the light current voltage (LIV) curves for each device and also the optical spectra of a subset of devices. The LIV curves yielded many useful parameters including threshold current, on-state resistance, and output power at a fixed current bias. The data were valuable for screening the VSCELs and affecting future designs. Data such as the optical power output for the wafer shown in Figure 6.10 [8] were used to match VCSEL

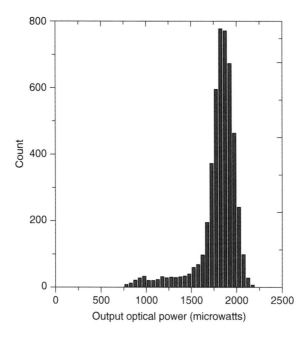

FIGURE 6.10 Output power distribution for VCSEL wafer.

arrays to laser driver chips. Optical spectral data gave mean wavelength and spectral width. Only devices with a wavelength below 860 nm could be used in links with GaAs-based receivers, and design considerations were influenced by a skew dependence on the laser wavelength [9]. A three-dimensional contour plot of the VCSEL emission wavelength across a wafer similar to Figure 6.3 was generated from 495 sample points. In this plot, several of the lasers tested failed the maximum 860-nm wavelength, indicating a need for further refinement of the laser growth process.

6.3.2 MODULE CHARACTERIZATION TESTING

Automated testing was carried out with custom computer software. A schematic of a test station for a receiver module is shown in Figure 6.11 [8]. Module level testing was similar to wafer level tests, but channels were measured serially, rather than in parallel. The testing involved the examination of waveforms and amplitude sensitivity. Comparison of the module sensitivity data to the corresponding wafer-level data helped in determining the packaging effect on device performance.

6.4 JITNEY TESTBEDS

The feasibility of the passively aligned parallel optical interconnections was demonstrated and the OETC modules evaluated on several different testbeds,

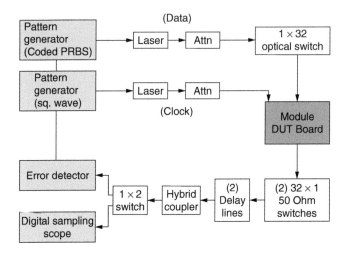

FIGURE 6.11 Diagram of a receiver module automated test station.

including an ESCON testbed, an IBM AS/400 system, and an IBM RISC Power-Parallel switch testbed. A brief overview will be given of each.

6.4.1 OETC PROTOTYPE TESTBED

Tests of the performance of parallel optical links were run on OETC prototype modules [11]. The 32-channel VCSEL array transmitter operating at 850 nm was interconnected with a fiber ribbon cable to a 32-channel MSM photodiode receiver. The testbed choices included one readily available for fiber-optic evaluation, the ESCON, and an IBM System/390 CMOS-based Parallel Transaction Server Model 9672. The testbed card accepted four independent data streams from the bus and then serialized each into a 200 Mbit/sec stream of encoded data. The data stream was then inputted into the ESCON optical transceivers. The testbed mapped four ESCON channels to four channels on the OETC. Results of a parallel optical link used in a mainframe with clustered CMOS processors indicated that the OETC module could run error free in asynchronous mode up to 240 m at 200 Mbit/sec. [12] The successful demonstration of the optical link demonstrated several advantages. First, space savings on the card were achieved because the 32-channel OETC occupied the same space as two ESCON transceivers. Second, there was a reduction in cable configuration allowing stacking of cards, and the denser card packaging and smaller module size reduced electromagnetic interference and noise.

Another reported IBM testbed involved mounting modules on a testcard with a port-adapter IC module supporting a version of the IEEE standard scalable coherent interface (SCI), as shown in Figure 6.12 [6]. Results indicated low bit-error rate (BER) (33 hours between CRC frame error) and a 10 times reduction in cable bulk. In addition, the flexible cable allowed cost-effective installation.

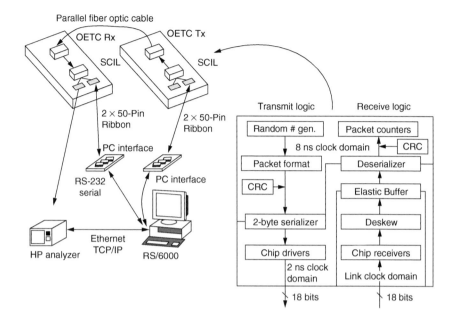

FIGURE 6.12 IBM SCI testbed layout.

6.4.2 IBM AS/400 Testbed

This testbed was configured by the connection of two IBM AS/400 high-end commercial computers [6,12]. The testing took part in two stages. First was the design, implementation, and debugging of the shared memory cluster system. The IBM I/O hub was extended to support a 500 Mbit/sec/channel data rate. Second, the Jitney parallel fabric interconnection was added to the testbed. The interconnection fabric used a 10-bit-wide IEEE 1596.3 compatible interface, allowing the interconnect to be added transparently.

The physical packaging and interconnections of the two-node Jitney testbed are shown in Figure 6.13 [13]. A production-level 1063 Mbit/sec serial OptiConnect link provided a baseline for cluster performance measurements. The I/O tower, although not required for the shared memory cluster interconnections, provided a cluster switch for the OptiConnect cluster. A shared memory cluster system resulted when the I/O hub modules were connected to the Jitney link. The Jitney technology demonstrated that it could function in an AS/400 production computer running a production warehousing application over 30 m of step index fiber.

6.4.3 Power Parallel Switch Testbed

The IBM RISC PowerParallel system is based on RS/6000 workstations interconnected through a switch fabric [13]. Each node has its own processor, memory, and I/O running its own AIX operating system. A message-passing programming

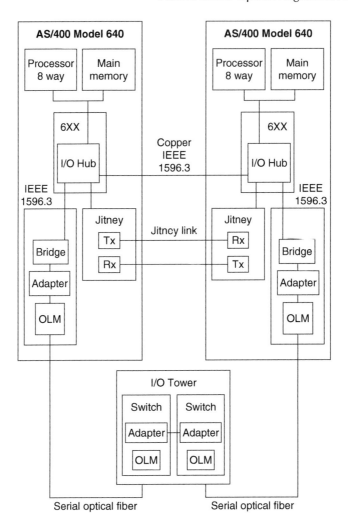

FIGURE 6.13 Jitney AS/400 testbed configuration.

module allows parallelism. Nodes are attached to a cross-point switch via high-bandwidth interconnect. Large systems have node-to-switch and switch-to-switch links. The maximum size tested was 512 nodes. The testcard used was designed for a harsh environment in which the back wall was sealed to contain any electromagnetic interference (EMI) and lacked air flow. Such was the case with the copper cable connections in the PowerParallel hardware. The transmitter and receiver were placed on opposite sides of the testcard to minimize height. To examine the various user modes of the Jitney module, a series of dip switches were included. Results of wrap tests on a single switch port on several test cards indicated no errors on test times from several hours to overnight. Testcards were run from a variable 3.0–3.6-V switch power supply. Hot plugability of the testcard was accomplished by moving the testcard from one switch port to the next while the card and switch were powered.

6.5 CONCLUSION

The Jitney multimode optical interconnect successfully demonstrated the feasibility of a passively aligned, low-cost, high-performance optical link. The plastic leadframe package designed by manufacturing and system design engineers emphasized ease of assembly. The success of these prototypes demonstrated control of transfer plastic molding and high alignment-pin dimensional and positional accuracy, plus the injection molding technology of the optical coupler fins [2]. Cost-saving measures included using existing volume production applications and relaxed alignment-tolerance optics. In the mid-1990s, estimated costs indicated that the Jitney approach would be 10 to 100 times less expensive than telecommunications-based parallel connections. The transmitter/receiver costs would be approximately $125, with an estimate of $7/m for the cable [6]. Functional testing of links using high-volume production could further lower costs.

REFERENCES

1. J.R. Broomall and H. Van Deusen, "Extending the Useful Range of Copper Interconnects for High Data Rate Signal Transmission," Proc. 47th Electronic Components and Technology Conference, San Jose, CA, May 1997, pp. 196–203.
2. M.S. Cohen et al., "Packaging Aspects of the Jitney Parallel Optical Interconnect," Proc. 48th Electronic Components and Technology Conference, Seattle, WA, May 1998, pp. 1206–1215.
3. J. Crow et al., "The Jitney Parallel Optical Interconnect," Proc. 46th Electronic Components and Technology Conference, Orlando FL, May 1996, pp. 292–300.
4. D. Smock, "Super Molding Makes Tiny Lens," *Plastics World,* 5 p. 9, May 1997.
5. P. Xiao et al., "A 500 Mbit/sec, 20-Channel CMOS Laser Diode Array Driver for a Parallel Optical Bus," 1997 IEEE International Solids State Circuits Conference, Digest of Technical Papers San Francisco, CA, pp. 250–251, February 1997.
6. J. Crow, "Parallel Fiber Optical Bus Technology — A Cost Performance Breakthrough," 22nd European Conference on Optical Communications, ECOC'96 Oslo, Vol. 2, Sept. 1996, pp. 47–54, 1996.
7. B. DeBaun et al., "Direct VCSEL Launch into Large Core Multimode Fiber: Enhancement of the Bandwidth*Distance Product," Proc. SPIE, Vol. 3003, pp. 142–152, 1997.
8. K. Stawiasz and D. Kuchta, "Automated Testing Methodologies for Low Cost, Parallel Optical Bus Components" Proc. 47th Electronic Components and Technology Conference, May 1997, pp. 217–224.
9. J. Choi et al., "High Performance High Yield, Uniform 32 Channel Optical Receiver Array," Optical Fiber Communication, Vol. 2, pp. 309–310, 1996.
10. P. Pepeljugoski et al., "Modeling and Simulation of the OETC Optical Bus," Proc. IEEE Laser and Electro-Optics Society, San Francisco, CA, Oct. 1995, Vol. 1, pp. 185–186.

11. Y.M. Wong et al., "Technology Development of High Density 32 Channel 16 Gibit/s Optical Data Link for Optical Interconnect Applications for the Optoelectronics Technology Consortium (OETC)," *J. Lightwave Technol.* 13, pp. 995–1016, 1995.

12. C. DeCusatis and T. Quinn, "Optoelectronics Technology Consortium 32-Channel Parallel Fiber Optic Transmitter/Receiver Array Testbed," *Optics and Photonics News,* Vol. 7, Issue 12 pp. 33–34, 1996.

13. D.M. Kuchta et al., "Low Cost 10 Gigbit/s Optical Interconnects for Parallel Processing," Proceedings of the 5th International Conference on Massively Parallel Processing, Las Vegas, NV, June 1998, pp. 210–215.

7 Mechanical Methods for Free Space Passive Alignment

Nagesh Basavanhally

CONTENTS

7.1 INTRODUCTION

To date, the use of serial optical solutions has been well established in long-distance communication technology. Because of market needs and the cost of optical components, its introduction into the realm of short-distance interconnections, such as optical interconnection between cabinets, has been rather sluggish. However, with the recent increase in demand for advanced services such as entertainment video, high-definition television, video teleconferencing, parallel processing systems, Internet access, and so forth, attention is being focused on short-distance interconnection technology. The above-mentioned services need systems that call for high throughput, with bandwidths in the giga- to terahertz range, and a high density of channels. It has been argued [NLN92] that such systems will require parallel optical data links, based on array technology, to meet performance and cost expectations. Parallel fiber optic data links that use transmitter and receiver arrays are being introduced into the marketplace (e.g., PARA-OPTIX™ transceiver modules by Xan3D Technologies) to satisfy the bandwidth-hungry systems. Such optical links are suitable for board-to-board, rack-to-rack, or cabinet-to-cabinet interconnections.

With the emergence of surface-active devices such as self-electro-effective device (SEED) [LCD94, GOO95] modulator arrays, vertical cavity surface-emitting lasers (VCSEL) [LCS95, CHA00], and microelectromechanical systems [APC03], systems with a very large number [KBW03, NFK04, KNK03] of interconnection channels are gradually becoming a reality. Handling such large optical interconnection channels using the better known technology of guided wave optics (fiber and waveguides) is rather impractical from a packaging and economical manufacturing point of view. In contrast, free space optics (unguided) provides the most practical solution for handling such systems. This is because free space optics offer both spatial parallelism and three-dimensional structure. In addition to the obvious advantages of optics over electrical interconnection (see, e.g., MBL97), free space optics also allow the implementation of interconnect topologies with large throughputs and lower energy requirements. Furthermore, free space optical systems are being developed to include functionalities such as channel equalization, blocking filters, and wavelength switching in wavelength division multiplexed systems [NTG04, MAR03]. Figure 7.1 depicts the gradual evolution of optical interconnection technology.

With free space optical interconnection technology gradually making its way into system demonstrators, such as digital free space switches [HCT95, YYH96, MAR03], free space optical back planes [NEJ95, SZH96, BAR97], computers [NEF94, NEF96], and large optical cross connect switches [KBW03, KNK03], attention needs to be focused on assembling these systems to withstand the harsh environments in which they will be used. When a multistage system is required, it is always advantageous to use a modular design approach. Modular design allows one to assemble and test various stages separately before assembling them to form an entire system.

Typical free-space optical systems require broad integration of passive bulk optical components, such as lenses, mirrors, beam splitters, and so forth, and custom components such as fiber arrays, microlens arrays, and so on. In addition, these systems need integration of components such as lasers (optical power supply), microelectromechanical system arrays, liquid crystal pixel arrays, and surface-active devices—smart pixels, VCSELs, computer-generated holographic elements (CGH) and detector arrays. The development of these devices has been mainly responsible for the emergence of the free space optical systems. Typically, the bulk optical components have a major dimension of less than 25 nm.

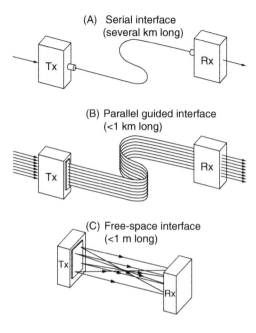

FIGURE 7.1 Evolution of fiber interconnection technology.

The following sections elaborate on the assembly issues and different passive assembly techniques that can be used in integrating free space optical systems.

7.2 ASSEMBLY CHALLENGES

7.2.1 ALIGNMENT TOLERANCE

The most challenging issues that an optomechanics designer encounters relate to aligning various components of the system to the required tolerances and maintaining them during system operation. Unlike an optoelectronic package, in which the distance between components is small (e.g., the gap between a laser and a single-mode fiber or a waveguide is on the order of a few microns, and the required lateral alignment between them is <1 μm), in a free space system the range of distance between components can vary from tens of microns to 10 to 15 cm. Because of these large distances, it is difficult to maintain precision alignment between the components. For example, consider the free space interconnect demonstrator board-to-board relay [BAR97] shown in Figure 7.2. The optical design requires that between the smart pixel array and the lenslet array (LA2) the lateral alignment needs to be < ±1 μm over a distance of 886 μm to keep the optical losses resulting from misalignment at < 1%. Furthermore, an alignment tolerance of ±5 μm between the lenslet arrays LA1 and LA2 over a distance of ≈7 mm, and 220 μm between the stage 1 and stage 2 over a distance of 14 cm, is required.

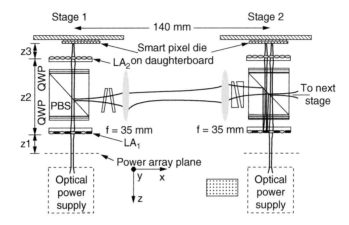

FIGURE 7.2 Board-to-board relay in a free space optical backplane demonstrator. From [BAR97], copyright 1997 OSA.

Free space optical components can be aligned using "active techniques," "passive methods," or a combination of the two. In active techniques, a signal launched through the system by an active component, which is an integral part of the system, is used for aligning the components. The final assembly of components is achieved by seeking the desired output while monitoring the input signal. In passive methods, components are aligned either by using passive "mechanical techniques" or by using an "aligner" that uses fiducials or reference marks on the surfaces of the components. The aligner method is necessarily used for aligning various stages within a system because of the large distances between components (2 to 15 cm). In most cases, the alignment-aiding components [ROB79], such as the ones shown in Figure 7.3, are

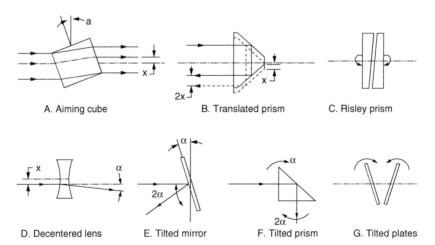

FIGURE 7.3 Some of the beam steering aids used in free space optical systems. Adapted from [Rob79], copyright 1979 SPIE.

incorporated into the system. In any fine-adjusting mechanism, it is desirable to have a vernier action: small output movements for large input movements. This can be effectively accomplished using decentered weak lenses, cubes, and Risley prisms. Furthermore, their mounts can be designed to be almost insensitive to thermal and vibrational factors. Mirrors and prisms are the least desirable elements, as the movements are usually doubled. In any case, passive alignment techniques are less cumbersome but will require components made with utmost precision or have "mechanical aids" built into them with the highest precision.

The concept of mechanical tolerance can be best understood with the following example. Figure 7.4 is a subassembly of a fiber array with two microlens arrays assembled using mechanical techniques. The fiber spacer (substrate) and the microlens arrays have mechanical aids built into them for passive alignment purposes. One can study the tolerance build-up in the assembly either by using the square root of sum of the squares of the tolerances of each part or by using the Monte Carlo analysis (see Chapter 11), which uses measured part tolerance data. In this particular example, one needs to take into account the following errors and tolerances to estimate the final assembly error:

1. Fiber data
   a. Fiber core eccentricity
   b. Fiber diameter tolerance
2. Fiber-locating substrate (A)
   a. Fiber hole and ball-mounting hole location tolerance with reference to mask alignment mark
   b. Fiber hole diameter tolerance

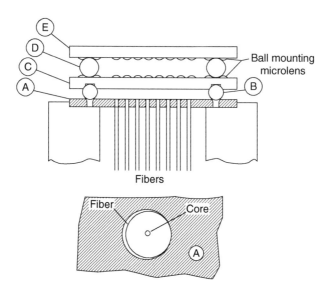

FIGURE 7.4 Fiber array subassembly.

    c. Ball-mounting hole diameter tolerance
3. Alignment ball (B, E)
    a. Ball diameter tolerance
4. Microlens array (C)
    a. Ball-locating socket diameter tolerance (backside)
    b. Double-side mask alignment tolerance
    c. Microlens array location tolerance with reference to mask alignment mark
    d. Ball-mounting microlens location error
5. Microlens array (D)
    a. Microlens location with reference to ball-mounting microlens location error

Table 7.1 lists the expected assembly accuracy of such an assembly. Although all the errors that can occur in a subassembly have been listed above, some extremely small errors, such as lateral misalignment caused by the tilt of a microlens array

---

TABLE 7.1
Misalignment Calculation for a Fiber Assembly Module

Component	Error (μm)	1 Sigma Misalignment	Comments
Fiber diameter	1.00	0.333	Standard commercial fiber
Core offset	0.60	0.200	Based on 0.6-μm concentricity
Hole-to-hole location on fiber spacer	0.50	0.167	Lithography error
Fiber hole diameter	2.00	0.333	Etching error
Ball diameter tolerance	2.00	—	Lateral misalignment resulting from tilt caused by ball diameter (negligible)
Ball-locating socket diameter	1.00	—	Lateral misalignment resulting from tilt caused by socket diameter (negligible)
Double-sided mask alignment	0.70	0.233	
Microlens array location with respect to mask alignment mark	0.20	0.067	
Ball-mounting location error	0.20	0.067	
Microlens error	0.20	0.067	
Square root of sum of squares		0.598	
Total misalignment (3 sigma)	1.79		

caused by variation in ball lens diameter, can be neglected for all practical purposes. Walking through such a tolerance exercise in conjunction with optical tolerance analysis will help determine the best alignment technique to pursue to achieve the needed performance of a free space interconnect system.

In addition to misalignment caused by mechanical effects, as noted above, it might be necessary to consider misalignment caused by thermal effects when thermally dissimilar materials are used in the assembly. This effect can be considerable if the coefficients of expansion of the materials are not well matched [PMO94].

7.2.2 COMPONENT MOUNTS

7.2.2.1 Kinematic Design

One of the key requirements in any optical system is that the components are mounted in a repeatable manner without causing any distortion. Hence, it is desirable not to overconstrain the system, as overconstraint will likely cause the components to distort or misalign because of any small external perturbation, which may be a result of either change in temperature or vibration.

Kinematic design is the discipline of mounting components such that they are located rigidly without any stress or strain. Every rigid body has six degrees of freedom—three translations and three rotations—respectively, along and around three mutually perpendicular axes. The theory of kinematic design states that a rigid body has 6-N degrees of freedom, where N is the number of constraints (or contact points). It can be seen from this statement that a rigid body with more than six contact points is overconstrained. A classic example of kinematic mount is shown in Figure 7.5, in which the three spheres that belong to the top part are located in three V-grooves of the bottom part. With each ball in a V-groove providing two constraints, for a total of six constraints, the parts, when put together, are uniquely positioned.

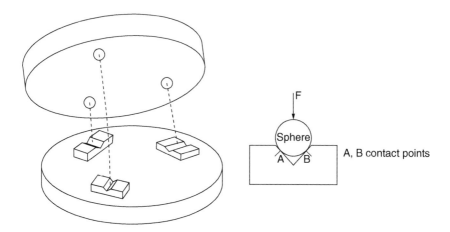

FIGURE 7.5 Typical example of a kinematic design showing six degrees of freedom constraint.

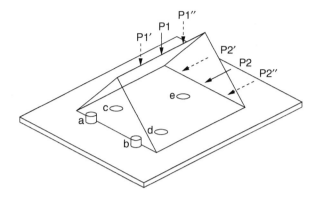

FIGURE 7.6 Precision mount for prism From [Dur68]. Reprinted with permission from *Machine Design*, May 1968. A Penton Publication.

From a kinematic design point of view, only point contacts are required to support the component. However, it is well known that point contacts result in high Hertzian contact stresses, and hence small area contact pads are preferred. Further, as only three points are required to define a plane, any scheme that uses more than three support points will either not uniquely constrain the component or will end up in distorting the component. Figure 7.6 and Figure 7.7 show examples of schematics of kinematic mounts for right-angle prisms [DUR68].

Positioning bulk optical components using cylindrical mounts in V-grooves is a common technique in optomechanical design. Mounting cylinders in V-grooves (four degrees of constraint) allows for easy adjustment of axial and rotational positioning during assembly process. These adjustments are sometimes required for focusing, changing magnification, or positioning the image. If the cylinders are made out of

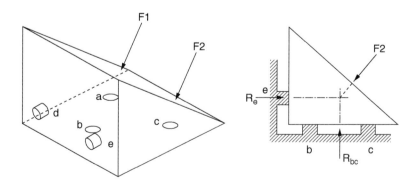

FIGURE 7.7 Mounting principle showing applied force distribution on support pads. From [Dur68]. Reprinted with permission from *Machine Design*, May 1968. A Penton Publication.

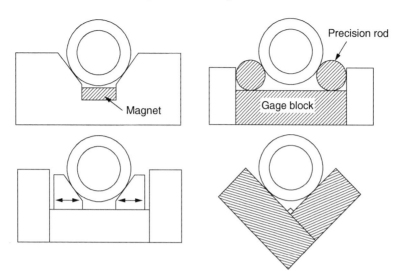

FIGURE 7.8 A few methods of mounting cylinders in Vs.

magnetic material (e.g., 416 ss), magnets attached at the bottom of the V-grooves can be used to temporarily hold the cylinder in place during component positioning. Such a positioning technique has been successfully used in Lucent Technologies' free-space optical switching demonstrators [MTB92, MCT93, MCL94]. The most important process that affects this type of mount is the machining tolerance of both the cylinder and the V-mount. Monolithic Vs are constructed from a single, continuous piece of material and are stiffer and more stable than fabricated ones [GOO97]. However, the use of assembled Vs, for example, precision rods or gage slugs, simplifies the fabrication because of the noncriticality of the V-angle.

Figure 7.8 shows a few methods of mounting cylinders in Vs. For rotationally symmetric optics, the lateral positioning (orthogonal to the optical axis) is very critical. Because both the cylinder and V dimensions determine the lateral position, some sort of lateral adjustment is invariably needed before permanently securing the cylinder. One common method is the use of spacers on the V-surfaces to shift the cylinder laterally. In microoptics, precision V-grooves can be machined in silicon, using the lithographic process. Because silicon is a crystalline material, precise and smooth etched surfaces can be realized. This will be discussed further in Section 7.3.

7.2.2.2 Deflection Control

Because deflection can compromise system performance, deflection control is very important in optomechanical design. Deflection occurs mainly as a result of external forces, including self-weight, clamping forces, vibration, material instability, and temperature changes. It is often necessary to ensure, by analytical means, that the structural design does not affect system performance. Support points and clamping forces should be carefully chosen such that they do not distort the structure [RSB96]. For example, in a beam with a uniform cross section, equal and parallel deflection at the end points can be achieved by placing the supports at the Airy points. For a beam of length L with simple supports, the Airy points are located at a distance 0.2113 L from the ends of the beam [VUK92].

Choosing the right material to minimize deflection is critical for maintaining the alignment between components in the system. In addition, the factors that determine the material choices for mechanical mounts include mechanical constraints, environmental considerations, material properties, and manufacturing cost. Typical properties used for comparing materials are listed in Table 7.2. Table 7.3 lists figures of merit used in the design of optical systems (see, e.g., [PAQ97, YOD86]). The specific stiffness E/ρ, where E is the modulus of elasticity and ρ is the density, is used when mass or deflection resulting from self-weight is critical to the system. This figure of merit is commonly used when comparing materials for deflection when the thickness is constant. Common structural materials—aluminum, steel, glass, molybdenum, titanium—have similar specific stiffness values. Materials such as silicon and graphite epoxy (GY70 × 30) have higher stiffness values. However, beryllium and SiC stand out in this category. The structural material that is most commonly used in system demonstrators is Al 6061-T6 [BAR97, RSB96, MTB91] because of its relatively low cost, known machinable characteristics, and a specific stiffness that closely matches that of glass. Furthermore, Al 6061 can be heat treated to reduce microcreep, which can be detrimental to the long-term stability of the structure.

For systems used in switching, it is desirable to mount the free space optical interconnect system in an electronic cabinet. Although these cabinets are designed to withstand earthquakes from the viewpoint of strength, they can sway and deflect by a considerable amount because of external disturbances. Furthermore, they have cooling fans that can impart vibration to the system. In such situations, it is necessary to determine the effect of vibration on the critical optical alignment. In most cases, vibration isolation mounts are required. A figure of merit used for comparing the resonant frequency of a design is $(E/\rho)^{1/2}$. The larger this number, the higher the natural frequency is. As listed in Table 7.3, SiC and beryllium also outperform other materials in this category.

Considering the thermal aspects, high thermal conductivity (k), thermal diffusivity ($D = \rho C_p$, C_p = specific heat), and low coefficient of thermal expansion (α) are desired. The figure of merit for steady-state and transient thermal distortions are, respectively, k/α and D/α. Higher values of these figures of merit are a better choice for material selection. Figure 7.9 compares the figures of merit for various structural materials that can be used in designing component mounts and fixtures. It can be seen that in thermally challenging designs, Be and SiC provide the least thermal distortion with the most dimensional stability. However, graphite epoxy (see Table 7.3) outshines all materials when thermal distortion is the main consideration.

Although beryllium has a superior figure of merit for use as a structural material, it has serious drawbacks, including high cost and highly toxic dust that is generated during machining operations.

TABLE 7.2
Structural Material Properties

Design Requirement	Density (ρ) gm/cc Low	Young's Modulus (E) GPa High	Poisson's Ratio Small	0.2% Yield Stress MPa High	Coefficient Th. Expansion (α) ppm/°C Low	Thermal Conductivity (K) W/m°C High	Specific Heat (Cp) J/gm°C High	Thermal Diffusivity k/ρCp m²/sec × 10⁻⁶ High
Aluminum 6061-T6	2.7	68	0.330	276	22.50	167.00	0.90	68.72
Aluminum 5083-O	2.66	71	0.330	145	22.60	120.00	0.90	50.13
Steel 1010 mild-annealed	7.86	200	0.280	180	12.20	60.00	0.45	16.96
SS 17-4 PH-H900	7.8	200	0.272	1170	10.40	22.20	0.46	6.19
SS 304-annealed	8	193	0.270	241	17.3	16.20	0.50	4.05
SS 416-annealed	7.8	215	0.283	275	9.90	24.90	0.46	6.94
Cu OFHC-annealed	8.94	117	0.343	195	16.50	391.00	0.38	115.09
Titanium 6Al-4V	4.4	114	0.342	427	8.70	21.90	0.67	7.43
Beryllium:I-70	1.85	287	0.043	276	11.40	216.00	1.92	60.81
Molybdenum	10.22	324	0.293	600	5.00	140.00	0.28	48.92
Invar 36	8.05	141	0.259	276	1.00	10.40	0.52	2.48
Super Invar	8.13	148	0.260	—	0.30	10.50	0.51	2.53
Zerodur[a]	2.53	91	0.240	—	0.05	1.64	0.76	0.85
CERAFORM SIC	2.95	364	0.140	—	2.44	172.00	0.67	87.02
Silicon	2.33	131	0.260	—	2.60	135.20	0.71	81.73
Fuzed silica	2.19	72	0.170	—	0.50	1.40	0.75	0.85
Graphite/epoxy GY70x30	1.78	93	—	—	0.02	35.00	—	—
Electroless Ni	8	140	0.410	—	14.00	5.00	0.46	1.36
Electroplated Al	2.7	69	0.330	—	22.70	234.00	0.90	96.30

Note: Adapted primarily from references [PAQ97, EWW97, HOP97] and as noted.

[a] Data from Schott Glass Tech, Inc.

TABLE 7.3
Figure-of-Merit for Material Comparison

Figure of Merit (Arbitrary Units)	E/ρ	$\sqrt{E/\rho}$	K/α	D/α
Design Criteria	Deflection	Natural Frequency	Steady-state Thermal Distortion	Transient Thermal Distortion
Preferred Value	High	High	High	High
Aluminum 6061-T6	25	5.0	7.4	3.1
Aluminum 5083-O	27	5.2	5.3	2.2
Steel 1010 mild-annealed	25	5.0	4.9	1.4
SS 17-4 PH-H900	26	5.1	2.1	0.6
SS 304-annealed	24	4.9	0.94	0.23
SS 416-annealed	28	5.3	2.5	0.7
SS420	29	5.4	3.1	0.9
Cu OFHC-annealed	13	3.6	23.7	7.0
Beryllium:I-70	155	12.5	18.9	5.3
Molybdenum	32	5.6	28.0	9.8
Invar 36	18	4.2	10.4	2.5
Super Invar	18	4.3	35.0	8.4
Zerodur	36	6.0	32.8	17.1
CERAFORM SIC	123	11.1	70.5	35.7
Silicon	56	7.5	52.0	31.4
Fuzed silica	33	5.7	2.8	1.7
Graphite/epoxy GY70x30	52	7.2	1750.0	—
Electroless Ni	18	4.2	0.4	0.1
Electroplated Al	26	5.1	10.3	4.2

Note: Copyright 1997 SPIE.

7.2.2.3 Fine Motion

As mentioned in Section 7.2.1, fine adjustments are inevitable for achieving needed alignment accuracy. This is because the optical tolerances needed are much finer than those achievable by bulk machining procedures. It is good practice to design the system with the perspective of having to use minimal adjustment during the final tuning. There are a number of mechanisms available for fine adjustment of optical systems [TUT67, REI86, VUK92], such as differential screws (e.g., a micrometer), a preloaded spring (e.g., bowed leaf spring), a differential lever, worm and gear, and so forth. These mechanisms involve sliding movement between parts within and are thus prone to stick-slip action. Furthermore, adjustments using such mechanisms can experience drift as a result of thermal and vibrational loads.

In a well-designed free space optical system, the translational and rotational adjustments needed are, respectively, less than 1 mm and 3° to 4°. Hence, the use of

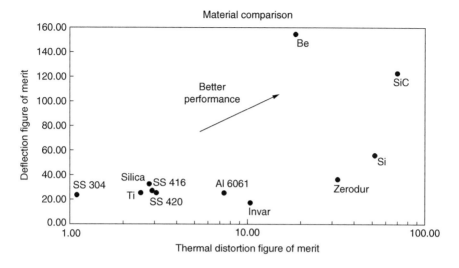

FIGURE 7.9 Comparison of figures of merit for various structural materials.

flexures is a suitable choice for tuning the system. The advantages of flexures include freedom from friction, no stick-slip effect, not requiring any lubrication, stability, repeatability, working very well in adverse environments (e.g., in dust), and miniaturization. The commonly used flexure designs are single and double strips for translational and rotational motion, parallel spring guides [REI86], and crossed springs for pivot bearings (e.g., Lucas Free Flex®, Lucas Aerospace Power Transmission Corp., Utica, New York). Some of these designs are illustrated in Figure 7.10. Typical materials used in flexure designs are beryllium copper, carbon, stainless steel, aluminum, invert, and titanium. The figure of merit used for comparing the flexure material is S_y/E, where S_y is the yield stress and E the modulus of elasticity. Table 7.4 lists the figure of merit for some of the commonly used spring materials.

The fabrication of flexures usually requires special machine tools such as a wire electrical discharge machine. Therefore, the main impediments to wider use of flexure are the cost and difficulty of fabrication. Figure 7.11 illustrates the use of dual flexures for fine adjustment of an optical grating in two orthogonal directions [BRT04]. Such grating mounts are typically used in optical wavelength division multiplexed systems.

7.2.2.4 Long-Term Stability

Environmental effects such as temperature, vibration, and humidity provide the most difficult challenges to an optomechanical designer. Long-term stability of an assembly will depend on such factors as the material used for fixing the components (e.g., creep and relaxation, specific stiffness, thermal expansion coefficient, etc.), athermalization, and attachment methods. Choosing the right material for mounting optical components, which are usually made from optical glass, is the key to successful system assembly.

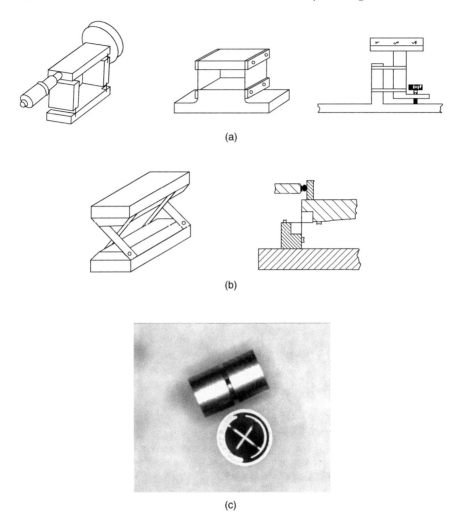

(a)

(b)

(c)

FIGURE 7.10 (a,b) Flexure design examples. From [REI86] copyright 1986 SPIE. (c) Free Flex®, crossed spring pivot bearing. Reprinted with permission from Lucas Aerospace Power Transmission Corp.

Temperature changes often introduce stresses and shift to the focal length of optical elements. Reducing the effect of temperature changes in an optical assembly is termed athermalization. An optomechanical designer always seeks such a system design, though it is not easily realized in practice. One way of reducing the thermal effect is to use materials with matched thermal coefficients of expansion. This means that one would be forced to choose materials such as Zerodur®, ULE®, invar, beryllium, titanium, composite materials, and so forth, which are expensive and difficult to machine. It is best to limit the use of such materials to local areas

TABLE 7.4
Comparison of Flexure Materials

Material	S_y (MPa)	E (GPa)	S_y/E ($\times 10^{-3}$)
Aluminum 6061-T6	276	68	4.06
SS type 304	241	193	1.25
SS type 420[a]	1524	200	7.60
SS 17-4 PH H900	1170	200	5.85
Titanium 6AI-4V	427	114	3.75
Invar 36	276	141	1.96
Beryllium copper-BW25 1/2 H	586	131	4.47

[a] Material used in Free Flex®.

where it is absolutely necessary. A second approach would be to allow for free movement caused by thermal expansion, while maintaining the required centration.

An example of an athermalization mount for a beam splitter is shown in Figure 7.12 [LIP68]. As shown, the beam splitter is kinematically mounted with five contact points (three in the yz plane and two in the xz plane), which allows for movement along the z-axis without introducing any error. The beam splitter is spring loaded directly opposite the constraint points to keep it in position with the proper compressive load. The springs allow for thermal expansion of the beam splitter (shown in dotted line) without affecting the light path to each image (X and Y detectors). Figure 7.13 shows another example that uses materials with two different coefficients of expansion for achieving athermalization. Allowing the longer tube to

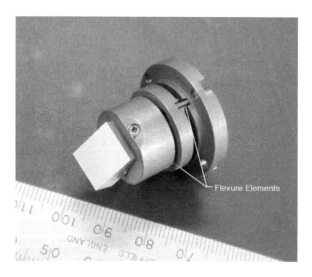

FIGURE 7.11 A dual-flexure optical grating mount.

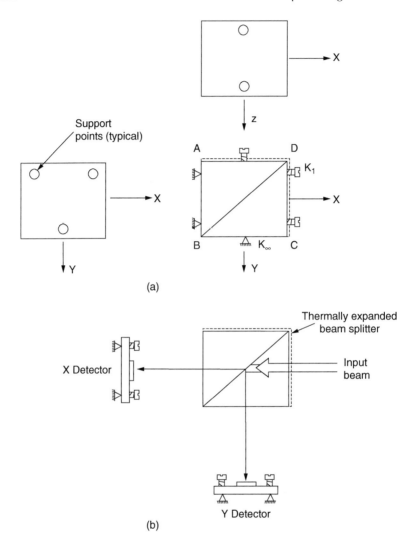

FIGURE 7.12 Beam splitter mount showing minimized effect of thermal expansion on light ray path. From [LIP68] copyright 1968 OSA.

expand less by using a material with lower coefficient of expansion enables maintenance of the relative distance between the two optical elements [VUK92]. The condition that maintains a constant relative distance is given by $\alpha_1 d_1 = \alpha_2 d_2$, where α and d are, respectively, the coefficient of expansion and length of the tubes. Typical materials used for such applications include aluminum ($\alpha = 23.6$ ppm/°C) and stainless steel ($\alpha = 9.9$ ppm/°C) [FRI80]. It should be noted that this type of compensation is not suitable when thermal gradients are present in the system because α, for most materials, is a function of temperature.

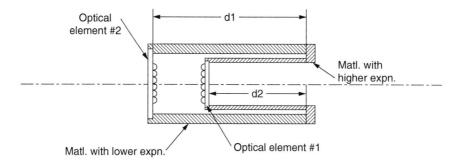

FIGURE 7.13 Thermal expansion compensation using dissimilar materials.

To avoid thermal gradients, it is necessary to conduct the heat away from internal heat generators and to shield the system from external heat sources. Typically, free space optical systems are mounted on baseplates with good thermal conduction properties. Aluminum is a fairly good choice from the point of view of spreading the heat away from the internal heat generators. However, because of its high thermal expansion, it can sometimes cause misalignment in the system components. Furthermore, if such a system has to be mounted in a cabinet full of electrical circuit packs, the heat generated by the electronics will have to be managed to maintain the baseplate at a relatively constant temperature.

Materials such as polymer composites (a mixture of polymers and fillers) or graphite composites (graphite and epoxy) that can be engineered to have good thermal stability, an excellent strength to weight ratio, and good dynamic damping are worthy of consideration. These materials, which are typically used to fabricate precision structural bases (e.g., Anocast™, made by Rockwell Automation, Inc. Shirley, NY), can be cast with metallic inserts and fillers, which could then be used for mounting components and removing heat from internal heat generators.

7.3 MECHANICAL "BUILDING BLOCKS" FOR MICRO-PASSIVE ALIGNMENT

For passive alignment at the microassembly level, it is absolutely necessary to have parts fabricated with the highest precision possible. This makes it necessary to use materials that can be micromachined using lithographic techniques. Optical elements used in photonic component packaging fit this requirement and can be ingeniously put to use as mechanical elements for alignment purposes. The following section describes a few of these "building blocks."

7.3.1 SILICON AND GLASS SUBSTRATES

Single-crystal silicon is an excellent mechanical material [PET82] and is widely used in the fabrication of miniature components. It is an excellent mechanical

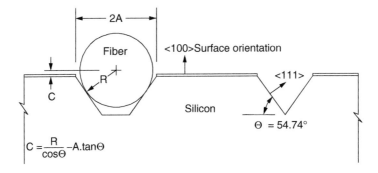

FIGURE 7.14 Fiber or ball lens position in V-groove etched in <100> silicon. From [MBL97] copyright 1997 John Wiley & Sons, Inc.

building block available for precision assembly because of its micron-level machining capability, good thermal distortion characteristics (see Figure 7.9), and higher specific stiffness compared to fused silica. The common etch features include V-grooves and pyramidal cavities in <100>-oriented silicon using the anistropic wet-etching process (see Figure 7.14). Vertical-walled shallow grooves on the order of 10 to 15 μm can also be etched, using the reactive ion etching process for providing mechanical stops or alignment features [HSM96]. The V-grooves, as described in Section 7.2.2.1, are usually used for locating cylinders (precision glass fiber or precision pins) and the cavities for locating spheres (precision glass ball lenses). Using a glass fiber as the cylinder and a well-established etching process, the center of the fiber can be located within a micron from the surface of the silicon. Many fiber array and array connector products, such as MACII (MACII™ connectors are products of Berg Electronics, St. Louis, MO) use etched silicon V-grooves for locating both fibers and alignment pins.

Epoxy is usually used for the attachment of fibers and ball lenses to the silicon V-grooves and pyramidal cavities. Other attachment processes such as soldering or ALO bonding [COU93] can also be used, as the silicon surface can easily be metallized. ALO bonding technology, (developed at Lucent Technologies, Murray Hill, NJ) uses an aluminum thin-film-coated substrate to form a solid-state bond with oxide components, such as glass fibers and ball lenses.

Figure 7.15 shows a cross-sectional view of a demonstration prototype using light-emitting diodes and CMOS photoreceivers. The alignment between the transmitter and receiver array has been realized using optical glass fiber and a stack of silicon spacers with V-grooves etched on both sides [CSC94].

Glass substrates are typically used in fabricating such optical elements as microlens arrays, holographic elements, and so forth. Vertical-walled precision grooves and cavities (up to depths of 10 to 15 μm) can be etched in these substrates, using lithographic techniques and reactive ion etching. Sometimes, mechanical aids such as using a set of three microlenses to form a tripod to locate microspheres or notches to locate microcylinders (e.g., fiber stubs), which are located lithographically

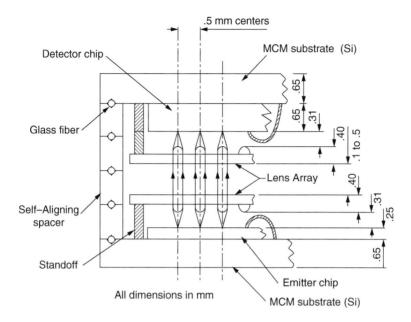

FIGURE 7.15 A demonstration prototype showing the use of silicon spacers. From [CSC94] copyright 1994 JOM.

using a single mask, can be machined along with the main optical elements. Figure 7.16 shows such a mechanical aid machined during the fabrication of optical elements [BAS94a, BAS94b].

7.3.2 MACHINED CERAMICS

Parts fabricated using ceramic have been used in photonic packaging for a long time. Ferrules made out of ceramics or glass with a bore maintained within 0.5 μm

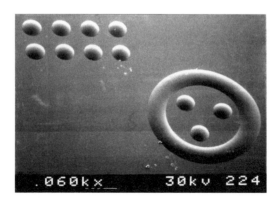

FIGURE 7.16 A tripod etched in quartz substrate to support a sphere used for alignment.

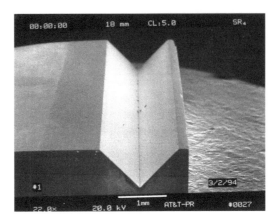

FIGURE 7.17 Machined V-groove in SiC material. Machined by NGK Insulators Ltd.

tolerance are typically used in optical fiber connectors. The ferrules are also available with chamfered ends for attaching ball lens, which then can be passively aligned with the fiber. Although this is an optical assembly, it can be used as a precision mechanical building block.

Figure 7.17 shows a ceramic block with a deep precision V-groove machined in it. Such V-grooves can be used for locating alignment pins. Also, shallow V-grooves, which are typically used for locating fibers in one-dimensional fiber arrays, are mechanically ground (using technology developed by NGK Insulators Ltd., Japan) to a precision of < 1 µm. The 72-wide linear fiber array, described in Section 7.4.3, has been fabricated using this technology. This machining versatility allows for the creation of a wide range of V-grooves without incurring extra setup costs.

7.3.3 OPTICAL ELEMENTS—BALL LENS AND GLASS FIBER

From a kinematic assembly point of view, ball lenses and optical glass fiber stubs respectively are perfect-precision microspheres and cylinders. Precision (grades) ball lenses are available up to a few millimeters in diameter in size and are typically made out of different optical glasses—ruby and sapphire. Optical glass fibers, which are mainly used for guiding light through them, are made with diameter held to ±1 µm. Although the typical diameter of long-haul communication fibers is 125 µm, they are also available with diameters of 140 and 200 µm.

7.3.4 PRECISION GAGE PINS AND BLOCKS

Precision gage blocks have been used in the machine tool industry for a number of decades. These gage blocks are available in dimensional increments of 0.5 µm over a large range of dimensions. Typically, they are made from hardened steel,

TABLE 7.5
Precision Building Blocks

Material	Machined/Formed Shape	Machining Process	Tolerance	Typical Use
Silicon substrate	V-grooves and pyramidal cavities	Anisotropic wet etching	±0.5 μm	Fiber array, spacer, aligners, etc.
Silicon and glass substrates	Vertical walled cavities, spherical surface	Reactive ion etching	±0.2 μm	Spacer, mounting spheres, etc.
Ceramic substrate	V-grooves	Grinding	±0.5 μm	Fiber array, spacers, aligners, etc.
Glass, ruby, sapphire	Precision spheres		±1.0 μm	Spacers and aligners
Glass	Glass fibers	Drawing	±1.0 μm	Spacers and aligners
Hardened steel, chromium carbide, etc.	Gage pins and blocks	Grinding	±0.5 μm	Constructed Vs and spacers

chromium carbide (Croblox®, L. S. Starrett Company, Ltd., Athol, Massachusetts), ceramic, and tungsten carbide. Chromium carbide with Rockwell C 71-73 wears extremely well and is stable over a long period of time. Table 7.5 concisely lists the use and tolerances of these building blocks.

7.4 MODULE ASSEMBLY

In any system that requires fine alignment, it is advantageous to use modular design because it allows one to test the modules separately before assembling them to form an entire system. Further, it allows for interchangeability and quick replacement when a module malfunctions either because of the failure of an active device or because of mechanical misalignment within a stage. An example of a highly modularized free space optical system [MCL94] is shown in Figure 7.18. This five-stage free space optical switching network demonstrator system consists of an input fiber bundle module, which includes a single-mode fiber array, microlens array, and hybrid lens; optical power supply module that has a laser diode, collimating lens, external cavity grating, and Brewster telescope; objective lens module; field effect transistor self-electro-effective device (SEED) smart pixel array module; and output fiber array module, which includes a multimode fiber array, and hybrid lens.

FIGURE 7.18 Photonic switch demonstrator: laser units are to the left, field-effect transistor self-electrooptic-effect devices are on the right, and the input fiber bundle is in front. Courtesy of Lucent Technologies. From [MCL94] copyright 1994 OSA.

All these modules are assembled, aligned, and tested external to the system and before assembly on the system base plate, permitting easy placement and removal.

7.4.1 OPTICAL POWER SUPPLY MODULE (LASER PEN)

The optical power supply shown in Figure 7.19 consists of a semiconductor laser diode, a collimating lens, a binary phase grating to split the beam into a large number of equal intensity clock beams, and a Brewster telescope to linearly polarize and circularize the beams [MSC92]. All these components have been assembled

FIGURE 7.19 Optical power supply module assembly. Courtesy of Lucent Technologies. From [FCL94] copyright 1994 OSA.

and mounted on a subplate to allow for simple laser replacement. Alignment of the laser pen has been accomplished using a radially shearing interferometer, which enables the alignment of the collimating lens' optical axis to the laser diode facet. This alignment is needed to minimize the lens' off-axis aberration.

7.4.2 SURFACE ACTIVE DEVICE ASSEMBLY

As mentioned in Section 7.2.1, the alignment between the surface-active devices such as smart pixel or VCSEL and the first optical element such as a microlens array is very critical. The alignment requirement is dictated by the optical power, which has to be transported to or from the device, and the cross talk between the adjacent devices. Monolithic integration of surface emitting lasers (SELs), drivers, and micro-optic lenslets is one way of lithographically aligning and forming the lenslet on the device. Figure 7.20 shows a hybrid approach, in which a lenslet array is aligned and attached to a surface-active device [BBB96]. The microlens array serves both as an optical element and an optical window. The design uses flip-chip solder self-alignment assembly to provide both optical alignment and electrical integration in one assembly step.

The solder self-alignment technology is finding wide use in optoelectronic packaging for alignment of devices with optical fiber [BAS93, LEB94]. Using a proper metallization scheme, such as Ti:Pt:Au or Ti:Ni:Au, a well-controlled solder deposition process, and a well-established reflow process, alignment accuracy of < 1 μm can be achieved. The solder self-alignment process has been addressed elsewhere in this book.

Because the device performance is affected by uniformity of temperature across the chip, it is imperative that a good heat sink be well attached to the backside of the chip. In the example shown in Figure 7.20, a metal tub has been soldered, using a low-temperature solder. The low-temperature solder is chosen based on the solder melting temperature hierarchy, so that the solder reflow process does not disturb the chip alignment.

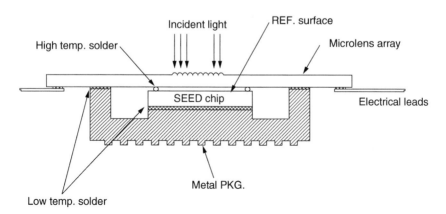

FIGURE 7.20 Two-dimensional surface active device package. From [BBB96] copyright 1996 IEEE.

7.4.3 Fiber Array Assembly

Fiber arrays are essential within a free space optical interconnection system for bringing in the optical signals from the outside and transmitting them out after signal manipulation. Typically, these arrays are two dimensional in nature and are spaced with a pitch tolerance that can vary from a couple of microns for single-mode fibers to several microns for multimode fibers. In addition to the fiber spacing, the path of the optical beams emerging from the fiber array with respect to the optical axis of the system is also critical. The location of the light beam emerging from the array depends on the fiber core concentricity and on the location of the fiber. Reference [MBL97] lists the different approaches that have been used in fabricating prototypes for fiber arrays. The following paragraphs illustrate three of the approaches used for passive fabrication of fiber arrays.

Figure 7.21 is the photograph of an 8×8 polarization-maintaining and absorption-reducing fiber array used in NTT's exciton absorption reflection-switch module [YYH96]. A stack of microglass ferrules—a building block discussed earlier—is used to form an array. The fibers are inserted into the microglass ferrules, and arranged in a square pattern, using zirconia plates and brass frames. The ferrules are held together using ultraviolet curable adhesive, and their ends are polished to complete the array fabrication. The fiber spacing is dictated by the ferrule diameter, which in this example is 250 μm. The average resulting pitch tolerance resulting from the combined variation in ferrule outside and inside diameters and from fiber diameter and core eccentricity is claimed to be ±3.1 μm. The average fiber axis deviation from the optical axis is 4°.

Figure 7.22 illustrates another method of fabricating a large fiber array, by placing linear arrays, with alignment pins precisely located with respect to fibers, next to each other [MBL97, BAS95a]. The alignment pins are used for plugging the individual arrays into a cage that uses precisely located V-grooves. The advantage

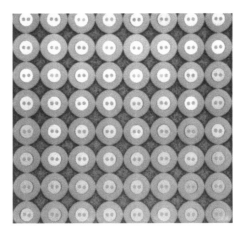

FIGURE 7.21 An 8×8 PANDA fiber array assembled using stacked microferrules. Courtesy of NTT, reprinted with permission from [YYH96], copyright 1996 IEEE.

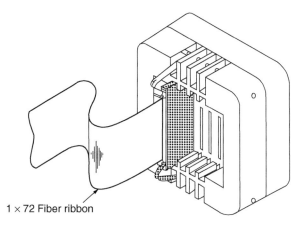

1 × 72 Fiber ribbon

FIGURE 7.22 A "plug-in"-type fiber array.

to using this approach is the ability to quickly replace a linear array in case of fiber breakage. However, the distance between arrays is dictated by the thickness of the individual arrays. The linear fiber array shown uses a 72-V-grooved ceramic chip. Figure 7.23 illustrates the accuracy of a 2 × 72 SM (single mode) fiber array that can be achieved in such a plug-in type fixture [BAS95b].

Figure 7.24 shows yet another method of fabricating a very large fiber array (19 × 19 SM fibers) with a silicon microlens array aligned and attached to it. The fiber array is fabricated by inserting fibers through a stack of substrates with precision-etched holes [BSC04]. The stack helps minimize the beam pointing error, maintaining flatness across the array and providing a rigid base for attaching the fibers. A fiber spacing deviation of less than 2 μm and a beam pointing error of less than 0.3° over the entire array have been reported. Furthermore, these fiber arrays have passed the Telecordia (Telecordia Technologies, Piscataway, NJ) environmental test requirements.

7.4.4 MICROLENS ARRAY ASSEMBLY

Microlens arrays are an important element that has been used in several system demonstrators for optical interconnections. Typically, they are used with other optical elements such as fiber arrays, surface-active device arrays, and conventional lenses for hybrid-lens-based optical interconnect systems. As explained earlier, for passive alignment schemes, it is necessary to have alignment aids built into the microlens array substrate. Figure 7.25 illustrates a few of the different techniques [BAS92, BAS94a, BAS94b] that could be used to align microlens arrays. Figure 7.25b shows SEM photographs of a tripod stand formed out of three microlenses and an alignment sphere (0.8 mm in diameter) mounted on the tripod for alignment purposes. The epoxy used for bonding the sphere (not shown) is restricted from spreading onto the optical path by the dam formed by the ring around the tripod stand. Earlier, Figure 7.4 illustrated (see Section 7.2.1) a microsphere-locating socket etched in

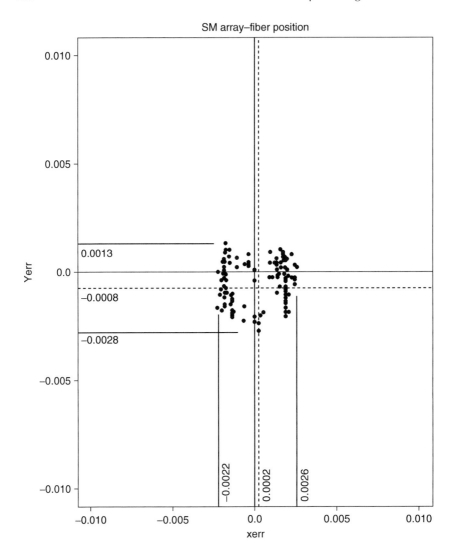

FIGURE 7.23 Dimensional accuracy of a 2 × 72 "plug-in"-type fiber array.

the microlens array for alignment purposes. Solder self-alignment — use of surface tension forces for alignment (see Section 7.4.2) — is yet another technique that could be used for aligning microlens arrays.

7.5 PASSIVE-ALIGNER APPARATUS

As described in Section 7.2.1, a passive aligner is needed for aligning optical components in the system either because of the large distance between the components or because of the components not lending themselves to alignment

FIGURE 7.24 19×19 single-mode fiber array with microlens attached.

schemes such as solder self-alignment, mechanical stops, and so on. The following paragraphs, through examples, illustrate the use of passive aligners in optical component alignment.

Figure 7.26 shows an optical component aligner for the passive assembly of components, used in NTT's exciton absorption reflection-switch (EARs) module [YYH96]. The aligner consists of a beamsplitter and a plane mirror assembly (half-mirror assembly), microscopes with charge-coupled device (CCD) cameras, vacuum chucks for holding the optical components, and manual six-axis micropositioning devices. After mounting the components on the vacuum chucks, a laser interferometer and microscope #2 are used to make the two components planar with respect to each other. Lateral alignment is achieved by viewing the fiducials (alignment marks) on both of the components at the same time through the half-mirror assembly and microscope #1. The two components are then attached to each other by moving the microscope aside and bringing the components together. The reported positioning accuracy of this aligner is ±2 μm. This type of aligner could also be used as a view port to align the optical components, which are further apart and attached to a base plate.

Figure 7.27 shows an optical arrangement used in the Karl Suss flip-chip bonder. Both the chip holder and the arm of the flip-chip bonder are made of zerodur to minimize the misalignment resulting from temperature excursions. Both lateral and planar alignments are made by viewing the alignment surfaces of the components. This type of bonder, which is used for aligning laser chips onto V-grooved silicon substrates that hold optical fibers, has an alignment and bonding accuracy of less than 1 μm.

Figure 7.28 shows a setup used in a free-space optical backplane demonstrator for aligning bulk turning mirrors in the baseplate [BAR97]. The optical system uses four mirrors in a closed loop ring system. As can be seen, the base

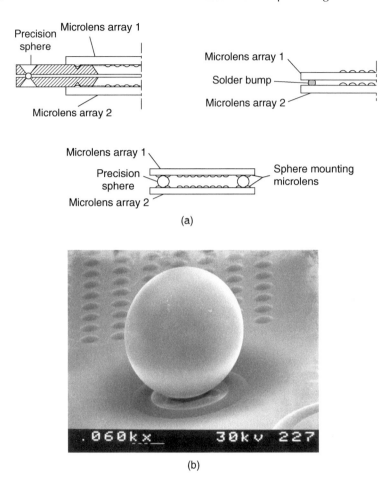

FIGURE 7.25 (a) Examples of aligning microlens arrays. From [BAS95] copyright 1992 ASME. (b) A ball lens mounted on a tripod formed by microlens for alignment purposes.

plate has slots and apertures in strategic places for beam observation during the alignment process, and the alignment method uses a reference beam launched through a pellicle holder that holds a 70/30 transmittive-reflective pellicle. First, the reference beam is adjusted (within a few minutes of arc and 50 μm off axis) to be coaxial with the optical axis on the baseplate, using customized apertures in the baseplate. Then mirror 1 is installed in its holder and adjusted, so that the beam reflected from it toward mirror 2 is parallel to and on axis. This procedure is repeated for mirrors 2 and 3. For the final mirror 4, the reflected beam R is observed, and the misalignment is reduced by adjusting the mirror until the angle β is minimized. Figure 7.29a shows the observed reference beam T and the reflected beam R (projected onto a screen located 3 m from the baseplate) during the alignment procedure. This picture is taken by

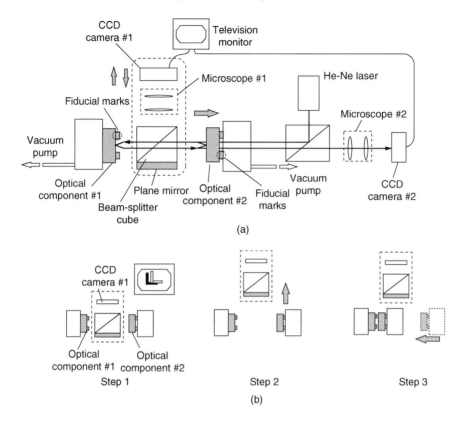

FIGURE 7.26 NTT's optical component aligner showing the assembly steps. From [YYH96] copyright 1996 IEEE.

deliberately misaligning mirror 4 to illustrate the alignment procedure. The actual alignment in the demonstator is claimed to be much tighter (with $\beta < 0.05°$) and is comparable to the one shown in Figure 7.29b. This misalignment is small enough that the final system tuning can be performed using Risley beam steering prisms.

7.6 SUMMARY

Assemblies of free space interconnect components pose challenges that are different when compared to packaging of optoelectronic devices that are used in guided wave interconnects. This is because of the range of distances involved and the degree of alignment accuracies needed between the components. A modular design approach is advantageous because it allows for pretesting, interchangeability, and quick replacement. Optomechanical design requires a good understanding of the critical optical tolerances that are required to maintain the optical integrity, and the type of environment in which the system is used.

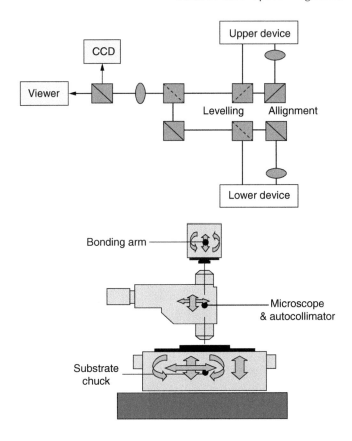

FIGURE 7.27 Alignment mechanism used in Karl Suss flip-chip bonder. Schematics courtesy of Karl Suss, Inc.

This understanding will enable the designer to make the proper material choice and design component mounts that will withstand various external loads such as temperature excursions and vibration. Deflection analysis, rather than strength, is critical in optomechanical design. Hence, analysis based on stiffness is of paramount importance. In this chapter, an attempt has been made, with the help of a few examples, to explain the techniques and building blocks that are available for use as mechanical aids in free-space alignment.

ACKNOWLEDGMENTS

I would like to acknowledge Anthony Serafino and Laurence Watkins for, respectively, taking the scanning electron microscopy photographs and providing critical comments during the preparation of this manuscript. I am indebted to Frederick McCormick for reviewing the final draft.

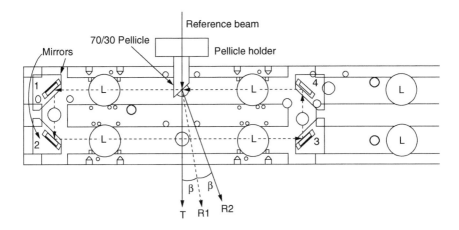

FIGURE 7.28 Setup for mounting and aligning turning mirrors on a baseplate. From [BAR97] copyright 1997 OSA.

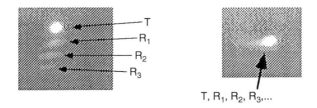

FIGURE 7.29 Results of aligning turning mirrors. (a) Mirror 4, appreciably misaligned, and (b) mirror 4 in its optimum alignment. From [BAR97] copyright 1997 OSA.

REFERENCES

[APC03] V.A. Aksyuk et al., "Beam-steering micromirrors for large optical cross-connects," *Journal of Lightwave Technology,* 21:3, pp. 634–642 (March 2003).

[BAR97] G.C. Boisset, M.H. Ayliffe, B. Robertson, R. Iyer, Y.S. Liu, D.V. Plant, D.J. Goodwill, D. Kabal, and D. Pavlasek, "Optomechanics for a four-stage hybrid-self-electro-optic-device-based free-space optical backplane," *Applied Optics,* **36**:29, pp. 7341–7358 (October 1997).

[BAS92] N.R. Basavanhally, "Opto-mechanical alignment and assembly of 2D-array components," in *Manufacturing Aspects in Electronic Packaging,* EEP Vol. 2, PED-Vol.60 (New York: ASME, 1992).

[BAS93] N.R. Basavanhally, "Applications of soldering technologies for optoelectronic component assembly," in *Advances in Electronic Packaging,* EEP Vol. 4–2 (New York: ASME, 1993).

[BAS94a] N.R. Basavanhally, "Alignment and Assembly Method," U.S. Patent No. 5,281,301 (1994).

[BAS94b] N.R. Basavanhally, "Optical Fiber Alignment Techniques," U.S. Patent No. 5,346,583 (1994).

[BAS95a] N.R. Basavanhally et al., "Evolution of fiber arrays for free space interconnect applications," *OSA Spring Topical Meetings,* Salt Lake City, UT, March 1995.

[BAS95b] N.R. Basavanhally, unpublished data (1995).

[BBB96] N.R. Basavanhally, M.F. Brady, and D.B. Buchholz, "Optoelectronic packaging of two-dimensional surface-active devices," *IEEE Trans Components, Packaging, and Manufacturing Tech, Part B: Advanced Packaging,* 19:1, pp. 107–115 (1996).

[BRT04] N.R. Basavanhally, D.A. Ramsey, and H. Tang, "Optical Grating Mount," U.S. Patent No. 6,757,113 B1 (2004).

[BSC04] N. Basavanhally, J. Sherman, and D. Cluff, "Fiber arrays for large optical cross-connect switches; a design and reliability study," *17th Annual Meeting of the IEEE Lasers and Electro-Optics Society,* **2,** Puerto Rico, pp. 659–660 Nov. 2004.

[CHA00] C.J. Chang-Hasnain, "Tunable VCSEL," *IEEE J Selected Topics in Quantum Electronics,* **6:**6 pp. 978–987 (December 2000).

[COU93] A. Coucoulas et al., "ALO bonding: a method of joining oxide optical components to aluminum coated substrates," *Proc 43rd Electronic Components and Technology Conference,* Orlando, FL, May 1993, pp. 470–481.

[CSC94] R.F. Carson, P.K. Seigal, D.C. Craft, and M.L. Lovejoy, "Future manufacturing techniques for stacked MCM interconnections," *Journal of Metals,* pp. 51–55 (1994).

[DUR68] D.S.L. Durie, "Stability of optical mounts," *Machine Design,* pp. 184–190 (May 1968).

[EWW97] M.A. Ealey et al., "CERAFORM SiC: roadmap to 2 meters and 2Kg/m² areal density," *Proceedings of SPIE,* **CR67,** pp. 53–70 (July 1997).

[FRI80] I. Friedman, "Thermo-optical analysis of two long focal-length aerial reconnaissance lenses," *Proceedings of SPIE,* **216,** pp. 146–155 (1980).

[GOO95] K.A. Goosen et al., "GaAs MQW modulators integrated with silicon CMOS," *IEEE Photonics Technology Letters* 7, pp. 360–362 (1995).

[GOO97] D.S. Goodman, "Cylinders in Vs — An Optomechanical Methodology," *Proceedings of SPIE* 3132, pp. 196–217 (1997).

[HCT95] H.S. Hinton, T.J. Cloonan, F.A.P. Tooley, F.B. McCormick, and A.L. Lentine, "Free-space digital optical interconnections," *Proceedings of IEEE,* **82,** pp. 1632–1649 (1995).

[HOP97] M. R. Howells and R.A. Paquin, "Optical substrate materials for synchrotron radiation beam lines," *Proceedings of SPIE,* **CR67,** pp. 339–372 (July 1997).

[HSM96] H. Han, J.E. Schramm, J. Mathews, and R.A. Boudreau, "Micromachined silicon structures for single mode passive alignment," *Proceedings of SPIE,* **2691,** San Jose, CA, 1996, pp. 118–123.

[KBW03] M. Kozhevnikov et al., "Compact 64 × 64 Micromechanical Optical Cross Connect" *IEEE Photonics Technology Letters,* **15:**7, pp. 993–995 (July 2003).

[KNK03] J. Kim et al., "1100 × 1100 port MEMS-based optical crossconnect with 4-dB maximum loss" *IEEE Photonics Technology Letters,* **15:**11, pp. 1537–1539 (November 2003).

[LEB94] Y.C. Lee and N. Basavanhally, "Solder engineering for optoelectronic packaging," *Journal of Metals,* 46:6, pp. 46–50 (June 1994).

[LCD94] A.L. Lentine et al., "Field-effect transistor self-electrooptic-effect device (FET-SEED) electrically addressed differential modulator array," *Applied Optics,* **33:**14, pp. 2849–2655 (May 1994).

[LCS95] K.L. Lear, K.D. Choquette, R.P. Schneider Jr., S.P. Kilcoyne, and K.M. Geib, "Selectively oxidized vertical cavity surface-emitting lasers with 50% power conversion efficiency," *Electronics Letters,* **31,** pp. 208–209 (February 1995).

[LIP68] M.L. Lipshutz, "Optomechanical considerations for optical beam splitters," *Applied Optics,* **7**:11, pp. 2326–2328 (November 1968).

[MAR03] D.M. Marom, "Wavelength selective 1 × K switching systems," *2003 IEEE/LEOS International Conference on Optical MEMS,* Naikoloa, HI, August 2003, pp. 43–44.

[MBL97] A.R. Mickelson, N.R. Basavanhally, and Y.C. Lee, eds., *Optoelectronic Packaging* (New York: Wiley, 1997).

[MTB91] F.B. McCormick et al., "Optomechanics of a free-space photonic switching fabric: the system," *Proceedings of SPIE,* **1533,** San Diego, CA, July 1991, pp. 97–114.

[MTB92] F.B. McCormick et al., "Design and tolerancing comparisons for S-SEED-based free-space switching fabrics, *Optical Engineering,* **31**:12, pp. 2697– 2711 (December 1992).

[MCT93] F.B. McCormick et al., "Six-stage digital free-space optical switching network using symmetric self-electro-optic-effect devices," *Applied Optics,* **32**:26, pp. 5153–5171 (September 1993).

[MCL94] F.B. McCormick et al., "Five stage free-space optical switching network with field-effect transistor self-electro-optic-effect-device smart-pixel arrays," *Applied Optics,* **33**:8, pp. 1601–1618 (March 1994).

[MSC92] F.B. McCormick et al., "Fabrication and testing issues in free-space digital optical switching and computing," *Proceedings of SPIE,* **1720** pp. 553–512 1992.

[NEF94] J.A. Neff, "Optical interconnects based on two-dimensional VCSEL arrays," *Proceedings of the First International Workshop on Massively Parallel Processing Using Optical Interconnections* (Piscataway, NJ: IEEE Computer Society Press, April 1994).

[NEF96] J.A. Neff, "Thermal considerations of free space optical interconnects using VCSEL based smart pixel arrays," *Proceedings of SPIE,* 2691 pp. 150–161 (February 1996).

[NEJ95] J.A. Neff and K. Johnson, "Optical computers: improving medicine, manufacturing, and the military," *Photonics Spectra,* **29**:11 (November 1995).

[NFK04] D.T. Neilson et al., "256 × 256 Port optical cross-connect subsystem," *IEEE Journal of Lightwave Technology* **22**:6, pp. 1499–1509 (June 2004).

[NTG04] D.T. Neilson et al., "Channel equalization and blocking filter utilizing microelectromechanical mirrors," *IEEE Journal of Selected Topics in Quantum Electronics,* **10**:3, pp. 563–569 (May/June 2004).

[NLN92] R.A. Nordin et al., "A systems perspective on digital interconnection technology," *IEEE Journal of Lightwave Technology,* **10**:6, pp. 811–827 (June 1992).

[PAQ97] R.A. Paquin, "Advanced materials: an overview," *Proceedings of SPIE,* **CR67** pp. 3–18 (July 1997).

[PET82] K.E. Peterson, "Silicon as Mechanical Material," *Proceedings of IEEE,* **70**:5, pp. 420–457 (1982).

[PMO94] S.K. Patra, J. Ma, V.H. Ozguz, and S.H. Lee, "Alignment issues in packaging for free-space optical interconnects," *Optical Engineering,* **33**:5, pp. 1561–1570 (May 1994).

[REI86] R.S. Reiss, "Fine adjustments for optical alignment," *Proceedings of SPIE,* **608,** p. 68 (1986).

[RSB96] D.J. Reiley, J.M. Sasian, and M.L. Beckman, "Optomechanical design of a robust free-space optical switching system," *Proceedings of SPIE,* 2691 pp. 84–90 (February 1996).

[ROB79] D.A. Roberts, "Integrating Nd:YAG lasers into optical systems," *Proceedings of SPIE,* **193**, pp. 121–128 (1979).

[SZH96] T.H. Szymanski and H. Scott Hinton, "Reconfigurable intelligent optical backplane for parallel computing and communications," *Applied Optics,* **35**:8, pp. 1253–1268 (March 1996).

[TUT67] S.B. Tuttle, "How to achieve precise adjustment," *Machine Design,* pp. 227–229 (February 16, 1967).

[VUK92] D. Vukobratovich, "Principles of optomechanical design," in *Applied Optics and Optical Engineering,* R.R. Shannon and J.C. Wyant, eds. Vol. XI (San Diego: Academic Press, 1992, pp. 239–283).

[YOD86] P.R. Yoder, *Opto-Mechanical Systems Design* (New York: Marcel Dekker, 1986).

[YYH96] M. Yamaguchi, T. Yamamoto, K. Hirabayashi, S. Matsuo, and K. Koyabu, "High-density digital free-space photonic-switching fabrics using exciton absorption reflection-switch (EARS) arrays and microbeam optical interconnections," *IEEE Journal of Selected Topics in Quantum Electronics,* **2**:1 pp. 47–54 (April 1996).

Section 2

Visual Passive Alignment

8 Solder-Bump and Visual Passive Alignment Technologies for Optical Modules

Yasufumi Yamada, Tsuyoshi Hayashi,
Yuji Akahori, Hideki Tsunetsugu,
and Kuniharu Katoh

CONTENTS

8.1 INTRODUCTION

The application of optical communication technology is now being extended from
trunk lines to subscriber systems and interconnections between computers, and
so it is becoming increasingly important to reduce the cost of optical modules.
The component assembly process is the dominant factor governing the cost of
conventional optical modules. A conventional optical module based on microoptics consists of a number of components including laser diodes (LDs), photodiodes (PDs), lenses, mirrors, and fibers. These optical components are assembled
by an active alignment procedure that requires monitoring via the activation of
an LD and a PD. Although this procedure ensures micron or submicron order
alignment accuracy, it is complicated and time-consuming.

In recent years, many studies have been conducted attempting to simplify the
LD-to-fiber alignment through the use of a passive alignment technique, in which
alignment is achieved without the need to monitor the output power from the fiber.
Most of the reported approaches use a Si optical bench and a mechanical contact
alignment technique. The position of the LD is determined by micromachined guiding
structures such as notches or pedestals formed both on the Si optical bench and on
the LD. V-grooves for aligning the optical fibers are also formed on the Si optical
bench. The LD-to-fiber alignment can be achieved simply by bringing the guiding
structures into contact and by placing fibers in the V-grooves. Therefore, the mechanical contact alignment technique is now used for assembling LDs or PDs on a Si
optical bench.

In addition, there are two other types of passive alignment methods: the
solder-bump alignment method and the index alignment method. These methods

are expected to have a wider range of applications than the mechanical contact alignment method. Recently, the solder-bump alignment technique was used to simplify the optical alignment procedure in state-of-the-art optical modules. The index alignment technique has been successfully used to provide optical hybrid integrated circuits based on a silica-based planar lightwave circuit (PLC). We describe these passive alignment techniques and explain why these techniques are necessary. We then describe the development of an optical module that employs the solder-bump technique, and finally, we report on applications of index alignment to optical hybrid integrated circuits.

8.2 COMPARISON OF THREE PASSIVE-ALIGNMENT METHODS

This section compares the three passive alignment technologies mentioned above in terms of optical component fabrication, to clarify the importance of the solder-bump and the index alignment methods.

8.2.1 MECHANICAL CONTACT ALIGNMENT

With the mechanical contact alignment, the major factor determining the horizontal alignment accuracy is the accuracy with which the guiding structures are formed. The vertical alignment accuracy depends mainly on how precisely the thickness of the semiconductor layer is controlled during deposition, and on the accuracy of the core height of the optical fiber when it is placed on a substrate. Therefore, this method requires a sophisticated micromachining technique with an accuracy of better than ± 1 μm both vertically and horizontally. As long as Si is used as the substrate, accurate guiding structures can be easily formed via anisotropic etching or reactive ion etching (RIE).

However, it is difficult to use this technique with substrates other than the Si optical benches. Nowadays, optical component technology uses not only the Si optical bench but also other substrates including silica waveguides on Si (SiO$_2$/Si) and ceramics. SiO$_2$/Si is considered to be the most promising substrate for optical hybrid integration. Because this waveguide has an overcladding layer of typically ~30-μm-thick glass, deeper etching is required to form guiding structures on SiO$_2$/Si. The need for deep etching makes it difficult to form a guiding structure with better than ± 1 μm accuracy.

Furthermore, the fusion of electrical and optical technologies has been progressing rapidly, and the desire to push performance levels even higher will soon generate the need for hybrid optoelectronic (OE) devices that integrate optical chips alongside electrical chips. From this point of view, ceramic substrates will also be useful, as well as SiO$_2$/Si substrates, as ceramics have been widely applied for packaging conventional electronic chips and photonic chips because of their superior electrical characteristics. However, if ceramic substrates are used, accurate guiding structures cannot be easily formed using wet chemical etching or RIE. Therefore, we must develop new passive alignment techniques that do not require micromachining of the substrate to reduce the cost of various kinds of optical module.

8.2.2 Solder-Bump Alignment Method

The chip-bonding technique using solder bumps, which is regarded as a type of flip-chip bonding, offers a number of advantages over other chip-bonding techniques, such as wire bonding or tape-automated bonding. The technique's most attractive feature for optical devices is its ability to align chips with circuit boards spontaneously. This is known as the self-aligning effect (Figure 8.1) [1–4]. Even if there is a large misalignment after chip placement (Figure 8.1a), the surface tension of the molten solder moves the chip toward the desired position during the solder reflow stage (Figure 8.1b), and the chip is thus bonded accurately without the need for positioning adjustments (Figure 8.1c). Solder bumps have practical potential for two-dimensional alignment with a tolerance in the micrometer or submicrometer range (ideally, the technique has the potential to eliminate misalignment completely). This technique is thus expected to emerge as an alternative to conventional positioning techniques that use, for example, precision mechanical systems and piezoelectric actuators. In addition, solder can be used for mounting microoptoelectronic devices on many kinds of substrate, including those made of Si, ceramic, and metal. This means that the solder bump technique can be applied not only to future hybrid OE modules but also to conventional microoptic modules.

Other attractive features are as follows: terminals can be arranged over the entire area of a chip, so more chip input/outputs (I/Os) are possible, and these

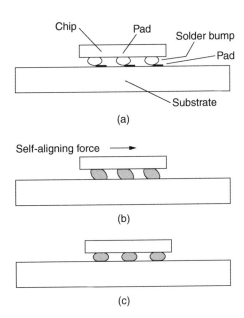

FIGURE 8.1 Schematic showing principle of self-aligning effect with solder bumps.

terminals can be simultaneously bonded in one solder reflow operation even if they are distributed over many chips, and the interconnection length is in the micrometer range, so we obtain superior electrical characteristics at high frequencies. Other bonding techniques often involve interconnection lengths measured in millimeters, and terminals must be arranged peripherally and bonded sequentially.

8.2.3 INDEX ALIGNMENT METHOD

The index alignment technique uses alignment marks formed both on the substrate and on the LD. These marks are usually lithographically formed metal patterns. The horizontal position of an LD is determined by aligning these marks with the help of a high-precision manipulator. The vertical alignment is accomplished simply by placing the LD on the substrate.

This technique was first demonstrated by M. S. Cohen et al. in 1991, as shown in Figure 8.2 [5]. Fiducial marks were formed lithographically both on an LD array and a V-grooved Si substrate. The fibers were aligned on the substrate, which worked as a fiber carrier. When aligning the LD with the fiber, an alignment plate with fiducial marks both for the fiber carrier and for the LD is used to align the fiber carrier and the LD. The fiber carrier and the LD are held in position, placed on a substrate, and then attached to it with SnPb solder. Because the height of LD and fiber has already been adjusted, this procedure ensures LD-to-fiber alignment without the need for activating the LD and monitoring the fiber output power. A single-mode fiber-to-LD optical coupling loss of 11 dB was achieved by using this procedure. This value was only 0.7 dB higher than that achieved by active alignment.

Figure 8.3 shows another type of index alignment, which was developed by K. Kurata et al. [6] (see Chapter 9). This technique uses an infrared light to observe

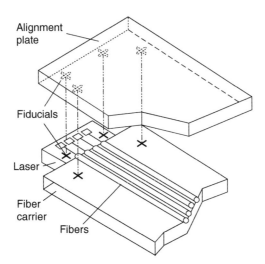

FIGURE 8.2 Concept of index alignment.

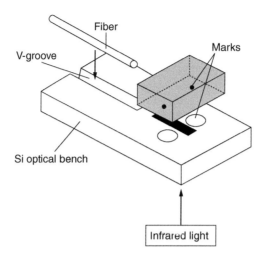

FIGURE 8.3 Index alignment using infrared light.

the alignment marks. By employing the infrared light from the underside of the Si bench, marks both on the Si bench and the LD surface can be aligned. Then, the LD is placed on the Si substrate and bonded in position with AuSn solder. This procedure realized an average alignment accuracy of better than ±1 μm.

The notable advantage of index alignment over mechanical contact alignment is that the former does not require a precise etching process. The only process necessary is the fabrication of alignment marks on the substrate and OE device surfaces. This means that index alignment techniques can be applied to any kind of substrate and OE device and are therefore expected to have a wide range of applications.

The above comparison makes it clear that both the solder-bump and index alignment methods are necessary to reduce the cost of optical modules that use substrates other than the Si optical bench.

8.3 OPTICAL MODULES USING MICROSOLDER BUMP TECHNIQUE

8.3.1 PERFORMANCE OF MICROSOLDER BUMPS

8.3.1.1 Formation Technique and Mechanical Performance

8.3.1.1.1 Formation

Figure 8.4 shows the basic process of solder bump formation and reflow bonding [7]. The microsolder bumps are fabricated on the chip as follows. First, solderable Ti/Pt/Au base metal is formed lithographically, with a total thickness of the metallization of about 0.3 mm. The base metal is used for the microsolder bumps

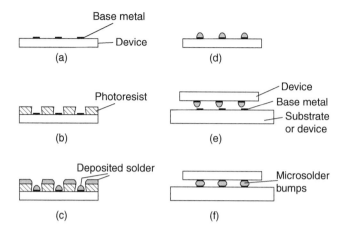

FIGURE 8.4 Basic process of microsolder-bump formation and reflow bonding.

(Figure 8.4a). The pattern for the solder bumps is formed by using a thick photoresist. It is essential to make an overhanging photoresist pattern (Figure 8.4b). Next, the solder is deposited on the base metal by evaporating the InPb solder (Figure 8.4c). Solder bumps are formed by lifting off the photoresist (Figure 8.4d). The chip is set face down onto the substrate, and the solder bumps on the chip are aligned to the substrate electrodes (Figure 8.4e). Finally, the solder bump interconnections are completed by the reflow process (Figure 8.4f).

8.3.1.1.2 Self-Alignment Accuracy and Shear Strength

The specimens consisted of a 1-mm-square Si chip bonded to a 3-mm-square Si or glass substrate with an array of microsolder bumps [8]. The glass substrate was used to evaluate alignment accuracy; the Si substrate was used to evaluate shear strength. Various numbers of bumps (8 to 60), all with a diameter of 26 μm and of various pitches (70 to 350 μm) were located symmetrically around the perimeter of the chip. A marker to measure the alignment accuracy was located at the center of each chip. Three types of solder were used: pure indium, 50% InPb, or 70% InPb solder.

The self-aligning effect of the microsolder bumps is as follows. A chip with solder bumps was aligned face-to-face with the substrate and preattached to the substrate, using slight pressure. The solder bumps were interconnected by reflowing the solder. The chip was flip-chip bonded onto the substrate, and the correct position was achieved by the self-alignment effect of molten solder bumps. A resin flux was used to attain nonoxidizing conditions; the bonding temperature was about 200°C, and the bonding time was 3 min.

The configuration we used to measure the alignment accuracy is shown in Figure 8.5. The bonding process was observed through the glass substrate, and the centers of both the chip and the substrate were marked to provide quantitative data on the alignment accuracy. First, the position of the substrate was determined

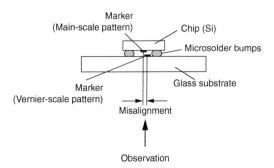

FIGURE 8.5 Configuration for measuring alignment accuracy.

by focusing on the substrate marker (vernier-scale pattern) and the fixed origin. The position of the chip was then determined by focusing on the chip marker (main-scale pattern); the chip alignment accuracy was determined by measuring the misalignment. A pattern with the same 3-μm width for the main scale and vernier scale was formed concentrically on the Si chip and glass substrate. The pitches of the two patterns were 6.0 and 5.9 μm. We could thus measure the misalignment and its direction with 0.1-μm accuracy by observing the coincidence position (measuring point) of the two patterns; that is, by counting the number of coincidence points from the original point. The marker patterns were formed simultaneously with the base metal used for the microsolder bumps, so their centers were located at the centers of the microsolder bumps formed around the perimeter of the Si chip. The marker patterns were about 500 μm in diameter. A photograph of the chip pattern through the glass substrate after bump bonding is shown in Figure 8.6. This shows that the Si chip was flip-chip bonded onto the glass substrate very accurately because of the self-alignment effect.

The measured misalignment is shown in Figure 8.7. The alignment accuracy increased with an increasing number of microsolder bumps. With the more than 20 microsolder bumps, the average misalignment was less than 0.2 μm, and the maximum misalignment was less than 0.5 μm. These results show that microalignment using 26-μm-diameter microsolder bumps is suitable for mounting laser diodes, which require a mounting accuracy of better than 1 μm.

The shear strength of microsolder bumps against thermal stress is shown in Figure 8.8. Each 1-h heat cycle consisted of 30 min at 55°C and 30 min at 125°C. These results show that InPb microsolder bumps provided strong and stable flip-chip bonding even after 1000 heat cycles.

8.3.1.2 Frequency Response

Photographs of the bump-connected coplanar waveguide (CPW) line and CPW through-line test samples are shown in Figure 8.9 [9]. A CPW with 50-Ω characteristic impedance was formed in a dummy chip and in a chip carrier. Each CPW line was connected by microsolder bumps in series with repeated interconnections.

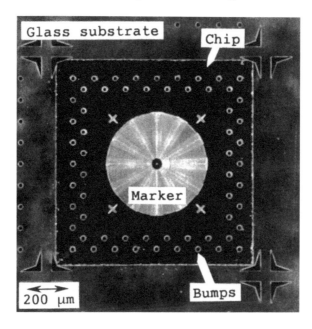

FIGURE 8.6 Photograph of chip pattern through glass substrate after bump bonding.

The width of the signal line was 60 μm, and the gap between the signal line and ground plane was 40 μm. Microsolder bumps with two different heights, 20 and 30 μm, were used. The dummy chip and chip carrier were about 0.5 × 1 mm and 2 × 4 mm, respectively. The pitch between two bumps on a CPW line was 250 μm.

Both the dummy chip and chip carrier had InP substrates. Plasma-deposited SiN was used as the insulation layer and a solder dam, and electron-beam vapor-deposited

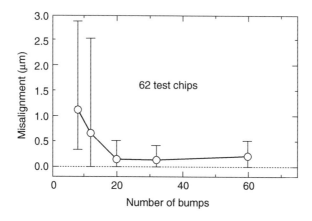

FIGURE 8.7 Alignment accuracy after microsolder-bump bonding.

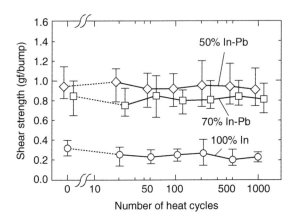

FIGURE 8.8 Dependence of shear strength on number of heat cycles for each solder material.

Ti/Au was used for the CPW lines. A contact hole in the SiN insulation layer between each CPW line and the base metal was formed by reactive ion etching, and the Ti/Pt/Au base metal was formed by electron-beam vapor deposition. To investigate the effect of the bump connections, we also made the same measurements using a CPW through-line without bump connections (Figure 8.9b). A signal was applied to the CPW line at one end of the chip carrier by an RF probe, and the output signal passing through the CPW line and two microsolder bumps on the dummy chip was detected at the other end of the chip carrier by a spectrum analyzer.

Figure 8.10 shows the measured S-parameters of bump-connected CPW lines with 20- and 30-μm-diameter microsolder bumps, as well as of that of the CPW through-line. Comparing the two bump-connected CPW lines, the insertion losses (S21) were less than 3 dB and had no resonance mode; in contrast, the return losses (S11) were more over 10 dB in the range from direct current (DC) to 60 GHz, which was the frequency limitation of the spectrum analyzer. The dependence of the frequency response on the diameter of the microsolder bump was slight. S11 for the 30 μm-diameter bumps was about 3 dB lower than that for the 20 μm-diameter bumps. Comparing the bump-connected CPW lines with the CPW through-line, S21 was almost the same for both; however, S11 for the CPW through-line was 20 dB, which was about 10 dB lower than that for the bump-connected CPW lines.

The frequency response was computed using three-dimensional electromagnetic field analysis [9]. The microsolder bumps were approximated as a cubic model, with a 30-μm side for the 30-μm-diameter microsolder bumps. In the bump-connected CPW line, S21 was less than 3 dB and had no resonance mode; S11 was more than 10 dB over the range from DC to about 90 GHz. These results are nearly equal to those for the measured frequency responses, indicating that the bandwidth performance of this microsolder bump interconnection is DC to 90 GHz. The frequency response of a CPW through-line with a dummy chip located 30 μm

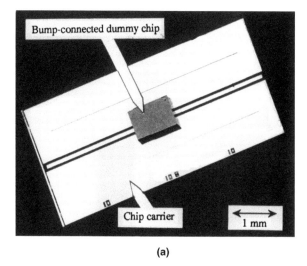

(a)

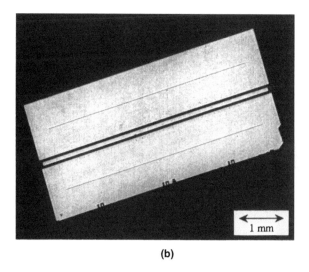

(b)

FIGURE 8.9 Fabricated test sample: (a) bump-connected CPW line and (b) CPW through-line.

above the through-line without a bump connection was also investigated. The frequency responses were similar to those for the 30-μm-diameter bump-connected CPW line. This implies that the primary limitation on the frequency response is not the bump connection but the decrease in the characteristic impedance resulting from the gap between the chip carrier and the dummy chip. Therefore, considering the impedance matching between the chip and the chip

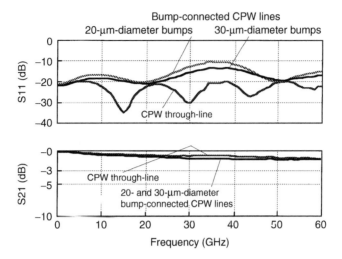

FIGURE 8.10 Measured S-parameters of bump-connected CPW line and CPW through-line.

carrier, the frequency response of the microsolder bump connection has a return loss of more than 20 dB, from DC to over 100 GHz.

8.3.1.3 Optical Modules Using Ferrule-Integrated Platform

This section describes a new receptacle microoptic module that requires no optical-axis adjustment [10,11]. In this scheme, a photonic device to a fiber is basically coupled by simple butt joining, which is accomplished automatically by solder-bump bonding the device onto a ceramic platform with an optical fiber glued by a ceramic ferrule. (In this section, this is called a ferrule-integrated platform/chip-carrier.)

In conventional modules, there are two major problems in simplifying the fabrication flow. One is the complicated assembly of the module. The optical fiber is reinforced by a holder, and the photonic device is packaged using a transistor-outline or butterfly package. A lens system is also needed. Complex and troublesome optical-axis adjustment is performed during the final assembly of these parts. The other problem occurs in installing the pigtail module on the circuit board. The latest high-throughput mounting machines cannot be used for mounting an optical module because the fiber pigtail is easily broken. The optical modules must be installed separately after the other electrical parts have been automatically mounted. In addition, the fiber pigtail occupies a large dead space on the circuit board.

In addition, the fiber pigtail occupies a large dead space on the circuit board. To overcome these problems, the new module structure illustrated in Figure 8.11 has been developed.

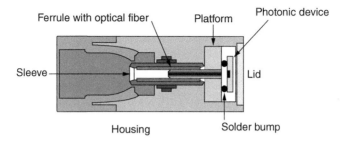

FIGURE 8.11 Schematic cross-section view of new receptacle optical module using ferrule-integrated chip carrier (platform).

Figure 8.12 shows the fabrication flow. The complexity in assembly of the optical modules mentioned as the first problem can be solved by changing the fabrication flow to make it sequential as shown in Figure 8.12a to Figure 8.12d:

a. An optical fiber is glued into a ferrule.
b. It is integrated into a platform.
c. A photonic device is connected to the platform. By using solder bumps, the optical axis of the photonic device and that of the platform's ferrule are automatically arranged.
d. The final assembly simply requires installation of the platform into the module housing. The platform's ferrule is automatically aligned at a predetermined position in the sleeve inside the module housing.

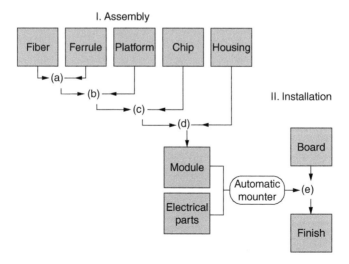

FIGURE 8.12 Fabrication flow of new receptacle module and circuit board.

FIGURE 8.13 New chip carrier with optical ferrule.

The second problem in the installation of the optical modules on the circuit board can be solved by the receptacle layout of the new module. Receptacle optical modules accommodate peripheral electrical parts very well because they have no fiber pigtail. Therefore, optical modules and peripheral electrical parts can be mounted using one mounting machine as shown in Figure 8.12e.

Figure 8.13 shows a sample of the ferrule-integrated chip carrier as the platform. The 1.25-mm-diameter zirconia ferrule follows the miniature universal coupling (MU) optical connector standard International Electrotechnical Commission ([IEC] standard, IEEE P1355). A 10/125-μm single-mode fiber is in the ferrule. The ferrule endface was polished using the advanced physical contact (AdPC) polishing technique [12]. This guarantees good optical characteristics at the ferrule-to-ferrule contact point. The other endface inside the alumina ceramic chip carrier was simply plane polished. The metallized pattern for solder-bump bonding the photonic device is on the 4.5 × 4.5-mm top of the chip carrier. Low insertion loss of less than 0.2 dB (almost the same value as that of the original MU connector [13]) and fabrication accuracy below 5 μm (cumulative distribution of 50%) were achieved.

Figure 8.14 shows a prototype simplex MU-type receptacle PD module. Solder bumps (Sn/Pb = 3/7) 100 μm in diameter were used for bonding the positive intrinsic negative (PIN) PD chip onto the chip carrier. The electric connection of the PIN PD was accomplished without wire connections, and optical coupling was achieved at the same time. Final assembly of the module was completed by simply installing the chip carrier with a PD into a housing and sealing the lid. The chip carrier's ferrule was arranged automatically at a predetermined position in the sleeve inside the module housing, as shown in Figure 8.11. The appearance of the receiver module conforms to the MU receptacle, so an MU-type optical plug could easily be connected by pushing it on, and physically connecting it to the ferrule of the chip carrier. No adjustment processes were required. The DC characteristics of the

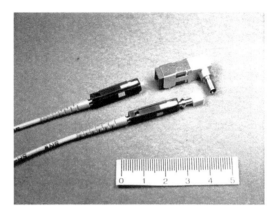

FIGURE 8.14 Prototype receptacle receiver: chip carrier with PD chip and MU sleeve is installed in a receptacle housing where chip carrier's ferrule and MU optical connector's ferrule are physically connected.

module are shown in Figure 8.15. The photocurrent generated by the light from the MU optical plug confirmed the optical coupling between the PIN PD and the fiber. The frequency characteristics are shown in Figure 8.16.

8.3.2 HIGH-SPEED OPTICAL MODULES CONTAINING SOLDER-BUMP-BONDED HYBRID OPTOELECTRONIC INTEGRATED CIRCUITS

8.3.2.1 Requirements for Packaging High-Speed Optical Devices

A high-speed communication transmission system has been demonstrated experimentally [14]. For this purpose, photoreceivers that use a broadband PD and a high-speed preamplifier have been developed. However, these photoreceivers use

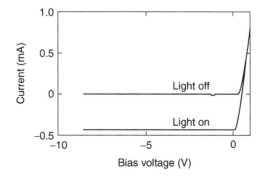

FIGURE 8.15 DC characteristics of prototype receptacle receiver.

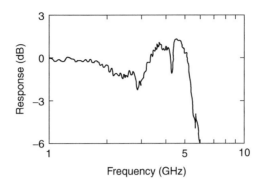

FIGURE 8.16 Frequency characteristics of prototype receptacle receiver.

conventional bonding wire for the electrical interconnections between the PD electrode, the preamplifier electrode, and the ceramic package terminal. As a consequence, the parasitic capacitance and inductance of these interconnections degrade the electrical performance of the photoreceiver. To reduce these parasitic elements, a packaging technology is required for high-speed photoreceivers.

Recently, flip-chip interconnection technology, which uses gold bumps, was developed to eliminate the inductive elements of the interconnection between the PD electrode and the preamplifier electrode [15]. However, this technology generally requires very careful alignment between chips because of the small size of the gold bumps; in addition, thermocompression and thermosonic bonding must be considered because they may damage the PD.

We have applied coherent transmission to a high-speed photoreceiver that uses microsolder bumps [16,17]. This photoreceiver can cover the frequency range from DC to about 20 GHz, and its use of microsolder bumps eliminates parasitic elements in the interconnection between the PD and the preamplifier.

Because microsolder bumps provide the shortest connection to a device (about 20 to 50 mm) between the twin PIN-PDs and the preamplifier, they can eliminate parasitic elements. They thus have the potential to drastically improve electrical performance. Furthermore, to achieve high electrical performance, it is important to reduce the capacitance of the twin PIN-PD junction areas by using a microsolder bump connected directly to the surface of the twin PIN-PDs junction areas. With regard to electrical interconnection between preamplifier and ceramic package, it is very important to achieve an impedance-matched interconnection [18].

This section introduces two novel photoreceiver packaging techniques using microsolder bump and impedance-matched film (IPF) carrier.

8.3.2.2 Application to Ultra-High-Speed Photoreceiver

8.3.2.2.1 Structure and Features

The new photoreceiver's structure is shown in Figure 8.17. It consists of a PD package and an optical fiber unit. The PD package contains a back-illuminated twin InGaAs p-i-n PD (PIN-PD), a GaAs metal semiconductor field effect transistor

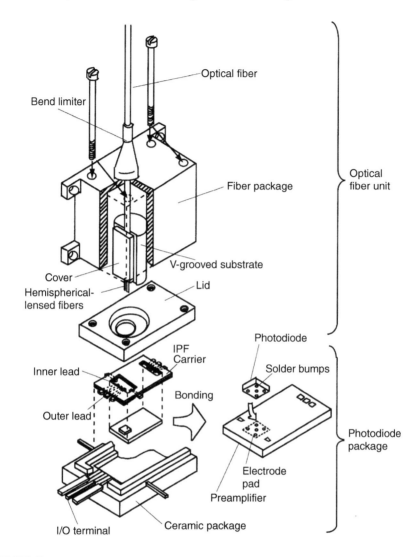

FIGURE 8.17 Detailed structure of photoreceiver.

(MESFET) preamplifer, an IPF carrier, and a ceramic package. The optical fiber unit contains a V-grooved substrate with hemispherical-ended fibers, a fiber package, and a bend limiter. The photoreceiver has the following features:

1. Microsolder bump interconnection: The bumps eliminate parasitic elements in the interconnection between PD electrodes and preamplifier electrodes. Solder having a low melting point is used to make the bump connections; as a result, no mechanical stress or thermal damage is inflicted on the PD during the reflow bonding process.

2. IPF carrier: This carrier forms a coplanar waveguide on a polyimide film and provides an impedance-matched interconnection between the preamplifier electrode and the ceramic package terminals.
3. Hemispherical-lensed fibers: These provide high-efficiency optical coupling.

8.3.2.2.2 Design

The photoreceiver is composed of two parts (a PD package and an optical fiber unit) for easy assembly. To achieve high-speed signal transmission, it is necessary to minimize the parasitic elements and impedance at the electrical interconnections. Because microsolder bumps can provide the shortest means for a connecting device's pad (i.e., their diameters are less than 50 μm), they can significantly reduce parasitic elements. Microsolder bumps are connected directly to the surface of the small PD junction area. To achieve 10-Gbit/sec transmission, a minimum PD junction diameter of 20 μm [19–21] and 26-μm-diameter microsolder bumps were selected.

For the interconnection between preamplifier electrodes and ceramic package terminals, a new type of interconnection technology using an IPF carrier was used. The IPF carrier consists of an impedance-matched coplanar waveguide on a polyimide film. The coplanar waveguide is implemented in the signal plane with a center strip and parallel adjacent ground areas. To obtain 50-ohm characteristic impedance matching, and to achieve a fine IPF carrier, the typical waveguide structure should be designed with a width of 90 μm and have 50-μm spacing [22, 23].

To achieve perfect optical coupling between fiber and PD, we used a hemispherical-ended fiber. Therefore, a 28-μm-radius hemispherical-lensed fiber was used.

8.3.2.2.3 Fabrication and Assembly Process

The flow of fabrication and assembly of the photoreceiver is given below. The first stage is microsolder bump formation and bonding. Subsequent stages are IPF carrier formation and bonding, followed by optical fiber unit mounting.

The solder bump formation and reflow bonding process used for the photoreceiver were explained in Section 8.3.1.1. Figure 8.18 shows an array of 26-μm-diameter solder bumps. The IPF carrier was fabricated as follows. First, a copper adhesion layer is formed on a polyimide film. Second, photoresist applied to the surface of the copper adhesion layer is exposed and developed to form a conductive layer mask. Then a copper conductive layer is electroplated onto the copper adhesion layer and the photoresist is removed. Third, another photoresist layer is applied to the surface of the conductive layer to form a bump mask at the distal ends of the inner and outer leads. Gold bumps are plated, and the photoresist is removed. Fourth, the polyimide film is etched photolithographically to form bonding windows for the PD. Finally, the unnecessary copper adhesion layer is etched away, and the coplanar IPF carrier is finished. A PD package with an IPF carrier is assembled as explained in the following steps. The inner leads with bumps are bonded thermosonically to the preamplifier electrodes, and the IPF carrier with a preamplifier is inserted into a ceramic package. The outer leads with bumps are connected to the ceramic package terminals by thermocompression bonding. Figure 8.19 shows the fabricated PD package.

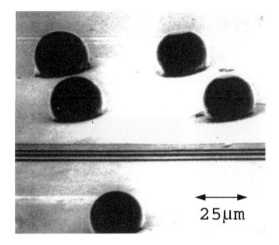

FIGURE 8.18 Twenty-six-micrometer-diameter solder bumps.

The optical fiber unit is mounted as follows. First, the lid is attached to the ceramic package. Second, hemispherical-lensed fibers are aligned precisely and fastened onto the V-grooved substrate. Third, the optical fiber unit is mounted precisely onto the package in such a way that light input from the fiber end is directed to the junction area through the rear of the PD. Figure 8.20 shows a photoreceiver ($7 \times 10 \times 15$ mm) assembled by the above techniques.

8.3.2.2.4 Photoreceiver Characteristics
The photoreceiver combines a back-illuminated twin InGaGa PIN-PD and a broadband GaAs MESFET preamplifier to form a balanced optical receiver. To measure

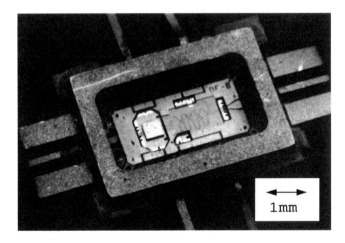

FIGURE 8.19 Photograph of fabricated PD package.

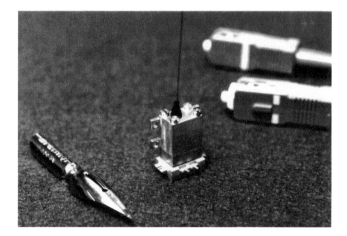

FIGURE 8.20 Photograph of photoreceiver.

the photoreceiver frequency response, optical heterodyne signals of 1.55-μm wavelength were input into the rear of the PD, and the output of the preamplifier was detected with a spectrum analyzer. The twin PIN-PD was reverse-biased to cancel the local intensity noise power. Figure 8.21 shows the measured frequency response results. The measured gain was 3 dB at more than 20 GHz. These results show that the packaging technology will be very useful for future high-speed optical transmission systems.

8.3.2.3 Application to Wideband Optical 90-Hybrid Balanced Receiver

8.3.2.3.1 Structure and Features
The proposed packaging configuration is shown in Figure 8.22. The receiver has two optical input ports (signal input and local oscillator power) and two electrical

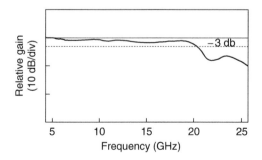

FIGURE 8.21 Frequency response of photoreceiver.

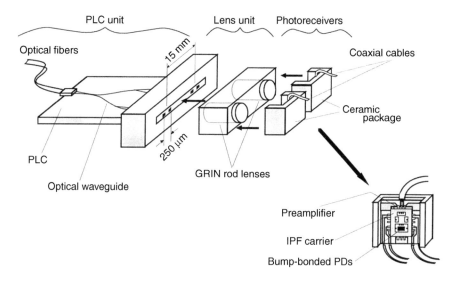

FIGURE 8.22 Basic construction of receiver module.

output ports. It contains a silica-based PLC with a 1:1 optical tunable coupler and two polarization beam splitters (PBSs), two graded-index (GRIN) rod lenses, and two photoreceivers, each having a broadband GaAs MESFET preamplifier and back-illuminated twin InGaAs PIN-PDs. The silica-based PLC was designed to ensure that an optical 90-hybrid is formed and that the lengths of all the optical paths are the same. Four optical lights injected from the PLC are connected to the two twin PIN-PDs through the two rod lenses.

The receiver has the following features:

1. Microsolder bump interconnection (see Section 8.3.2.2)
2. IPF carrier (see Section 8.3.2.2)
3. Silica-based PLC: designed to ensure that the optical 90-hybrid uses the polarization states of the transmitter and local oscillator and that all of the optical paths have the same length
4. GRIN rod lenses: used for wide-tolerance optical interconnections between the PLC and the twin PIN-PDs for easy fabrication and to achieve the same optical length from the PLC to the twin PIN-PDs

8.3.2.3.2 Design

A silica-based PLC [24, 25] is used for the optical 90-hybrid to get an accurate optical path length. Its core is 8×8 μm, and the cladding is 50 μm thick. Its two PBSs are achieved by controlling the birefringence of an amorphous silicon (a-Si) film that applies stress to the waveguide. The optical signal and the local oscillator power are input to the coupler through an optical fiber. The output optical polarization states are controlled by the thermo-optic phase shifters. The port pitch is set at 250 μm to match that of the twin PIN-PD junction area.

To achieve perfect coupling efficiency between the PLC and the photoreceivers and easy alignment and assembly of the lens unit and photoreceivers, two GRIN rod lenses [26], which have a graded refractive index distribution, are used for the optical interconnections. GRIN rod lenses that are 1.8 mm in diameter and 8.1 mm long are used for this module.

To obtain high-speed signal transmission, we must minimize parasitic elements and impedance mismatching at the electrical interconnections. Because microsolder bumps provide the shortest connection to a device between the twin PIN-PDs and the preamplifier, they can eliminate parasitic elements by using a microsolder-bump connected directly to the surface of the twin PIN-PDs junction areas. For uniform bump interconnection, the diameter of the microsolder bumps is 26 μm.

An IPF carrier was used for the interconnections between the preamplifiers and ceramic packages. To get 50-Ω-characteristic impedance matching, and to make the IPF carrier compact, we designed the waveguide to be 90 μm wide, with 50-μm spacing. To minimize degradation in the optical coupling resulting from external mechanical vibration of the module, the two RF output ports of the strip-lines were connected to the two output connectors by flexible coaxial cables.

8.3.2.3.3   Fabrication and Assembly

The flow of fabrication and assembly is as follows. First, the lens unit with two GRIN rod lenses mounted in the lens holder is attached to the metal frame of the PLC end. Then the photoreceivers are mounted one by one onto the lens unit, while the electrical output of the twin PIN-PDs is monitored. Finally, the PLC unit with the lens unit and two photoreceivers is mounted onto the module case, and the connector panel is attached.

Two 15-mm pitch V-grooves are formed in the lens holder, and a GRIN rod lens is inserted into each groove. Next, a lens cover is placed over each lens; both ends of the lens holders are formed so as to maintain a distance of 0.5 mm between the lens ends and the lens holder end. Finally, the lens covers and the lenses are fixed using ultraviolet resin.

The photoreceiver assembly process is shown as follows. Each photoreceiver consists of twin PIN-PDs, a preamplifier, an IPF carrier, and a ceramic package. The twin PIN-PDs, on which 26-μm-diameter microsolder bumps have been formed, are flip-chip bonded onto the preamplifier, the inner leads on the IPF carrier are bonded to the electrode pads of the preamplifier, and the preamplifier with the IPF carrier is inserted into the ceramic package; the outer leads on the IPF carrier are connected to the electrode pads of the ceramic package. Figure 8.23 shows a photoreceiver fabricated using an IPF carrier.

After the lens unit and photoreceivers have been mounted, the PLC unit is set in the module case, and the connector panel and module cover are attached to the case. Figure 8.24 shows the fabricated receiver module. It is 16 × 9 × 3 cm, excluding the optical fiber.

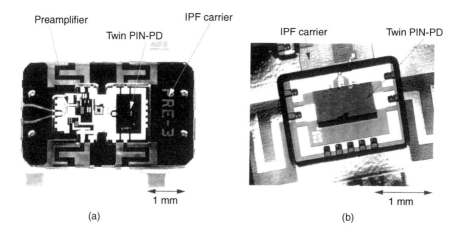

(a) (b)

FIGURE 8.23 Photoreceiver fabricated using IPF carrier.

8.3.2.3.4 Receiver Characteristics

External cavity laser diodes were used as optical sources. A multielectrode distributed feedback (DFB) laser diode with an external cavity was used as the local oscillator (wavelength: 1.55 μm). The photoreceiver combined back-illuminated twin InGaAs PIN-PDs and a broadband GaAs MESFET preamplifier. The outputs of the preamplifier were detected using a spectrum analyzer.

Figure 8.25 shows the measured frequency responses of the PIN-PDs, the preamplifier, and the receiver after subtracting the losses of the test fixture. The frequency response of the receiver exceeded those of the preamplifier and PIN-PDs because of the peaking effect at 14 GHz resulting from the improved DC bias circuits of the

FIGURE 8.24 Photograph of fabricated receiver module.

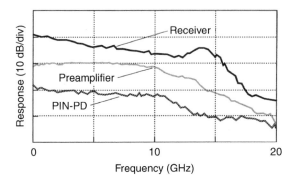

FIGURE 8.25 Frequency response of receiver.

IPF carrier. As a result, this receiver achieved 10-Gbit/sec binary phase shift keying (BPSK) homodyne detection, showing that a receiver fabricated using our new packaging technique gives excellent performance.

8.4 PLC HYBRID-INTEGRATED MODULES USING INDEX ALIGNMENT TECHNIQUE

This section describes successful application of an index alignment method that enables the realization of a hybrid-integrated optical circuit based on a PLC platform.

8.4.1 PLC PLATFORM

8.4.1.1 Basic Structure and Fabrication Process of PLC Platform

Figure 8.26 shows the basic configuration of a PLC platform with a silica-on-terraced-silicon (STS) structure [27]. The platform consists of a PLC region and

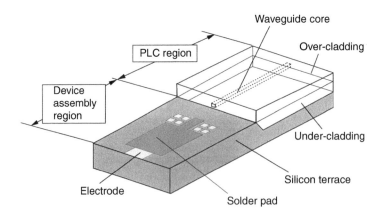

FIGURE 8.26 Basic structure of PLC platform.

a device assembly region. In the PLC region, an embedded-type silica waveguide is formed on the ground plane of the terraced silicon substrate. Because an embedded-type silica-waveguide has the same structure as a conventional PLC, various kinds of circuits can be fabricated on the platform.

In the device assembly region, the silicon terrace is used as both the alignment plane and the heat sink for the OE chips. A Au electrode and a AuSn solder pad are formed on a thin passivation layer deposited on the Si terrace. The AuSn solder pad is formed as a thin film 2 to 3 μm thick. The height from the solder surface to the waveguide core center is designed to be identical to the height of the active layer of the OE chip. Au alignment marks are also formed on the terrace to enable the passive alignment of the OE chips. Therefore, OE chips can be integrated simply by aligning marks on the silicon terrace and on the OE chips themselves and by placing the OE chips on the silicon terrace.

Figure 8.27a to Figure 8.27e shows the PLC platform fabrication process [27]. First, as shown in Figure 8.27a, an undercladding layer is deposited by flame hydrolysis deposition (FHD) on a chemically etched Si substrate. Second, as shown in Figure 8.27b, a thin layer for OE device height adjustment is deposited

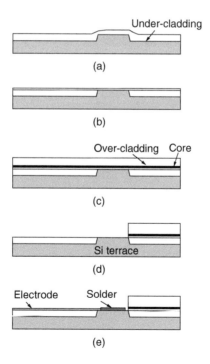

FIGURE 8.27 Planar lightwave circuit platform fabrication process: (a) deposition of undercladding layer and flattening, (b) deposition of OE-device height adjustment layer, (c) deposition of core and overcladding layers, (d) RIE process to form Si terrace, (e) lift-off process to form electrode, marks, and solder film.

on a flattened surface by polishing. Third, as shown in Figure 8.27c, core and overcladding layers are formed. Then, as shown in Figure 8.27d, a Si terrace is formed by RIE and a thin SiO_2 passivation layer (about 0.5 μm thick) is formed on the Si terrace. After that, electrodes and alignment marks are formed simultaneously by lifting off the evaporated Au layer. Finally, a thin AuSn solder film (about 2 to 3 μm thick) is evaporated and patterned using the lift-off method shown in Figure 8.27e [28]. The AuSn film consists of three layers of Au/Sn/Au with a composition of 80 wt% Au and 20 wt% Sn.

8.4.1.2   Index Alignment for PLC Platform

The index alignment method used on PLC platforms is as follows, and is shown in Figure 8.28 [28]: Alignment marks are formed on both the chip and the PLC surface. During the alignment, the OE chip is held in a face-down position by a quartz arm. Horizontal alignment is carried out by aligning the marks on the PLC platform with those on the OE chip. The marks are observed with an infrared (IR) microscope as dark images by passing an infrared ray through the Si terrace and the OE chip. After the alignment, the OE chip is placed on the solder pad. Vertical alignment is accomplished at the same time, as the height of the center of the waveguide core from the solder pad surface is designed to be the same as that of the active layer of the OE chip.

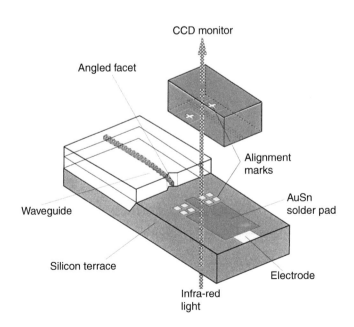

FIGURE 8.28 Index alignment method used on PLC platforms.

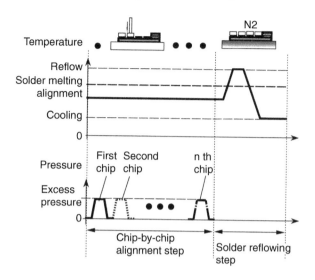

FIGURE 8.29 Multichip hybrid integrated procedure.

8.4.1.3 Multichip Integration on PLC Platform

A two-step assembly method has been developed to enable both multichip bonding and high-density integration on a PLC platform [29]. The method consists of a chip-by-chip alignment step and a simultaneous solder reflowing step.

Figure 8.29 shows the two-step assembly method. In the chip-by-chip alignment step, the platform is first placed on a heater and heated to the alignment temperature T_1, which is well below the solder melting temperature. The first OE chip is then held face down by a quartz arm, and marks on both the chip and the silicon terrace are aligned by observing them with an infrared microscope, as mentioned in Section 8.4.1.2. Then the chip is pressed down on the silicon terrace with excess pressure P. This procedure allows the first chip to be placed in the optimum position on the PLC platform. The second chip is similarly placed by repeating the same procedure.

In the solder reflowing step, the platform is heated to the solder reflow temperature T_2 in an N_2 atmosphere without excess pressure, and all the chips are bonded simultaneously onto the PLC platform. This procedure has two advantages. First, it prevents the solder pads from being oxidized regardless of the number of chips and their integration density because the temperature of the platform can be kept below the solder melting temperature during the chip-by-chip alignment step. Second, the bonding time for multichip integration is shorter with this procedure than with the local heating method because the solder reflow for all the chips can be performed with only one heating step.

We should mention here one major difference between the two-step assembly method and the solder bump method. In the solder bump method, an OE chip is aligned with a waveguide by making use of the self-alignment effect of molten solder. However, this self-alignment effect requires a complex structure to make it possible to form thick solder bumps on the platform. In contrast, the two-step assembly method uses solder layers thin enough to suppress the self-alignment effect. This consequently simplifies the solder pad structure on the PLC platform. Therefore, the two-step assembly method is more suitable for reducing the module assembly cost.

8.4.2 APPLICATIONS OF MULTICHIP INTEGRATION ON PLC PLATFORM

8.4.2.1 Optical Wavelength Division Multiplexing Transceiver Module

An optical wavelength division multiplexing (WDM) transceiver module has been fabricated to demonstrate the potential of the two-step assembly method [30,31]. Figure 8.30 shows the module configuration. The module consists of a 1.3/1.55 μm WDM circuit, a 1.3-μm bidirectional transceiver circuit, and three optical devices: a LD as a transmitter, a monitor photodetector (M-PD) for automatic power control of the LD, and a receiver photodetector (R-PD). As the WDM circuit, a thin-film filter was inserted into a groove formed in the platform [32]. The transceiver circuit consisted of a Y-branch circuit formed by using a silica-waveguide [33] with a refractive index difference of 0.45%. The waveguide facet was slanted at 8° to reduce reflection [34].

The LD was integrated at one end of the Y-branch silica waveguide circuit, and the M-PD was bonded just behind the LD. The R-PD was bonded at the other end of the Y-branch. The optical chip bonding area was only 2.0 × 1.3 mm.

A spot size converter integrated-LD (SS-LD) [35] was used as a transmitter LD. The spot size of the output light was converted to match the silica-waveguide. A significant advantage of the SS-LD is that it has a low intrinsic coupling loss with a waveguide of about 2 dB and a 1-dB down tolerance of ±2 μm. A waveguide

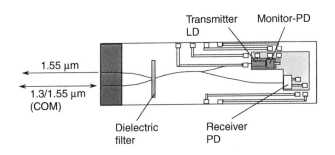

FIGURE 8.30 Configuration of 1.3/1.55-μm WDM transceiver circuit.

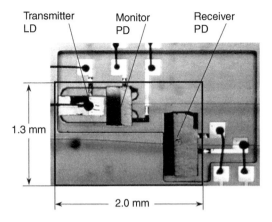

FIGURE 8.31 Microscope observation of assembled optical devices.

PD (WG-PD) [36] was used for both the R-PD and the M-PD. Because the WG-PD can be integrated on a PLC platform by the same method used for LD integration, the device assembly method is simplified.

These chips were integrated on the PLC platform, using the two-step assembly method. First, the SS-LD, M-PD, and R-PD were passively aligned and placed in the optimum position. In this step, the PLC platform was held at an alignment temperature that was well below the melting point of AuSn solder, 280°C. Then the platform was heated to 300°C in an N_2 atmosphere. Three chips were bonded to the platform simultaneously. This procedure made it possible to obtain both multichip and high-density integration of the LD, M-PD, and R-PD on the small area of the PLC platform, as shown in Figure 8.31.

The module exhibited a responsivity as high as 0.41 A/W and an output power of 0 dBm at an injection current of 60 mA. The insertion loss for the 1.55 μm port was 1.0 dB. These characteristics are good enough for practical fiber-to-the-home applications.

As shown in Figure 8.32, the OE chips of the module were encapsulated in transparent silicone resin and then covered with a black epoxy resin conventionally used in the Si-IC industry, to realize both low cost and high reliability [37]. This epoxy resin protects the chips from humidity and mechanical stress from the outside. The mechanical stress originating from the epoxy resin is released by the interior silicone. Various reliability tests were performed on the modules. Storage tests at 85°C and 85% humidity resulted in no observable degradation in the optical output power of the module, even after more than 4000 h of storage. The threshold current of the tested SS-LD scarcely changed, and the dark current of the receiver WG-PD remained at less than 1 nA at 25°C during the test. The stable operation of the modules was confirmed using high-humidity/high-temperature tests and also using temperature cycling tests (−40° to +85°C).

FIGURE 8.32 Photograph of encapsulated WDM transceiver circuit.

8.4.2.2 External Cavity LD Module

A multiwavelength light source with an external cavity was fabricated as another demonstration of the two-step assembly method [38]. Figure 8.33 shows the configuration of the hybrid integrated four-channel multiwavelength light source. It was composed of silica-waveguides with UV written gratings and four SS-LDs. The gratings were written in the waveguides by excimer laser irradiation through phase masks. The Bragg wavelength interval of these gratings was 2 nm. Each SS-LD was mounted on silicon terraces via the two-step assembly procedure. The rear and front facets of the SS-LDs were high reflection (HR) and antireflection (AR) coated, respectively. These coatings provided a low-loss laser cavity between the LD rear facet and the grating. The fabricated module had a channel-to-channel distance of 3 mm, and the substrate was 12×15 mm.

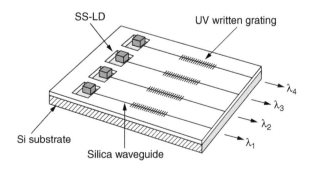

FIGURE 8.33 Configuration of four-channel multiwavelength light source.

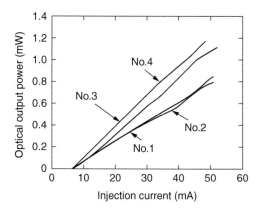

FIGURE 8.34 Relationship between LD injection current and output power.

Figure 8.34 shows the relationship between the LD injection current and the output power from a single-mode fiber connected to the module. A threshold current of 8 mA and an output power of more than 0.8 mW were obtained at an injection current of about 50 mA, resulting from the low coupling loss of 4 dB between the SS-LD and the waveguide. The coupling loss corresponds to an LD-PLC alignment accuracy of better than ±2 μm.

Figure 8.35 shows the output spectra of the LD module. These spectra confirm the realization of a four-wavelength LD module with simultaneous single-mode oscillation and an oscillation wavelength interval of about 2 nm. Figure 8.36 shows the measured oscillation wavelength vs. the substrate temperature. The wavelength is

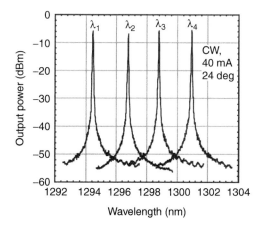

FIGURE 8.35 Output spectra of external cavity LD module.

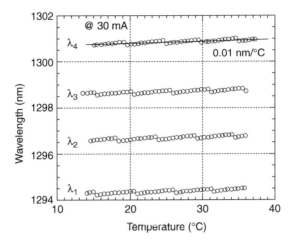

FIGURE 8.36 Oscillation wavelength vs. substrate temperature.

locked to the reflection peak of the grating in all four lasers. The average temperature dependence is as low as 0.01 nm/°C, which coincides with the temperature dependence of the silica waveguide. The small steps observed every 5°C are a result of mode hopping. In this case, the oscillation mode jumps to the neighboring mode because of the refractive index change in the LD. Mode hopping degrades the receiver sensitivity at the photoreceiver. However, the imposed power penalty resulting from these steps was measured and found to be as low as 1.5 dB for a 1 Gbit/nonreturn zero (NRZ) signal.

Figure 8.37 shows the bit-error-rate (BER) characteristics at 2.488 Gbit/sec, pseudo random bit stream (PRBS) 27-1. The substrate temperature was 22°C, at

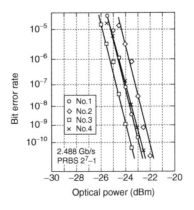

FIGURE 8.37 Bit error rate for four wavelengths at 2.488 Gbit/sec.

which temperature none of the LDs suffered from mode hopping. The amplitude of the injection current was 20 mA. As shown in this figure, a BER of 10^{-9} was obtained for all the LDs [39]. The operation speed of the LD module was limited by a resonance resulting from the parasitic inductance at the bonding wires and the electrical circuit outside the module. The bandwidth will be improved by using wiring technology for high-speed operation and an appropriate chip carrier.

8.4.3   HIGH-SPEED PLC PLATFORM

8.4.3.1   Approach for High-Speed PLC Platform

The platform configuration mentioned in the previous section has the advantages of a simple structure suited to cost reduction and an easy alignment process that uses the silicon terrace as the height reference for the silica waveguide core. However, the silica insulation film between the solder and the silicon terrace has to be thin enough to provide precise vertical direction alignment. Therefore, the thickness of this silica insulation layer is set at only 0.5 μm. The use of the thin insulation film results in a parasitic capacitance between the solder and the silicon substrate. The large loss tangent of the silicon substrate also increases the electrical propagation loss in the coplanar transmission lines. These drawbacks, caused by the use of the silicon terrace, are crucial when the OE modules handle high-frequency electrical signals. To reduce the detrimental effects, electrodes formed on the undercladding layer are used in high-speed PLC platforms. In this configuration, the silicon terrace is also used as a height reference. Therefore, solder bumps are employed to interconnect the flip-chip-bonded OE devices and the PLC platform. This section describes assembly technologies using solder bumps for the high-speed applications.

8.4.3.2   Hybrid Integration Technologies
for High-Speed Applications

A cross section of the assembly region of a high-speed PLC-platform is shown in Figure 8.38. In this configuration, vertical alignment is achieved by bringing the OE devices into mechanical contact with the silicon terrace. The horizontal position is defined by the index alignment technique used with hybrid integration for low-cost applications. PLC platforms for high-speed applications are formed as follows: first, a silica undercladding layer is formed on the terraced silicon substrate by flame hydrolysis deposition. Then the surface is flattened by mechanical polishing so that the surface of the undercladding layer and the silicon substrate can be matched. After flattening the surface, the core and overcladding layers are deposited on the substrate by FHD. Then the overcladding layer on the assembly region is removed by RIE. In this step, the etching depth is controlled so that it is about 10 μm below the top surface of the silicon terrace. This depth defines the solder bump height after the flip-chip bonding of an OE device. Finally, electrodes and solder film are formed on the undercladding.

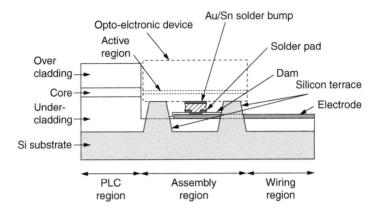

FIGURE 8.38 Cross-section of high-speed PLC platform.

The structure of the assembly region is formed by the following steps: gold electrode evaporation and lift off, silica glass sputtering followed by contact-window opening using RIE, evaporation of the Ti/Pt/Au solder pad and lift off, and Au/Sn eutectic multilayer solder film evaporation and lift off. The silica glass formed in the second step acts as a dam that collects the solder on the solder pad during the reflow process. Figure 8.39a shows the cross-sectional view of the solder film.

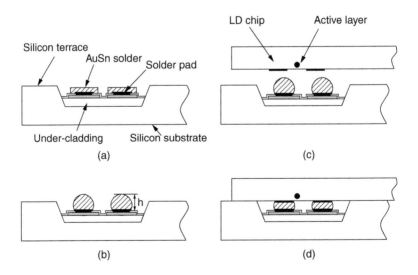

FIGURE 8.39 Hybrid-integration process on PLC platform: (a) cross-sectional view of assembly region before integration; (b) solder reflow at temperature over 280°C, using resin-based flux; (c) index alignment of OE device; (d) flip-chip bonding at temperatures above 300°C.

The OE device is assembled on the PLC platform by the procedure shown in Figure 8.39b to Figure 8.39d. For Figure 8.39b, the solder film was reflowed using resin-based flux. For Figure 8.39c, after removing the flux, the OE device was aligned with the silica waveguide in the horizontal direction, using the index alignment technique, and in Figure 8.39d, the solder bumps were reflowed again and the OE device was flip-chip bonded. In this step, position of the OE device in the vertical direction is defined by the mechanical contact between the device and the silicon terrace.

Because the initial height of the solder surface is set lower than that of the silicon terrace, as shown in Figure 8.39a, the height of the solder bumps after the first reflow has to be controlled during these assembly steps so that electrical connection can be made by the flip-chip-bonding step. Therefore, one of the requirements for the solder bumps on the PLC platform is good controllability of the solder height during the first reflow.

The solder bump height is controlled by the thickness and diameter of the multilayer solder film and the diameter of the solder pad. The solder film diameter (d_s) and thickness (h) define the mass of the solder bump. The solder pad diameter (d_m) defines the bump diameter after the reflow. As the solder bump height after the reflow is defined by the mass of the solder and the diameter of the bump, its height can be controlled by controllong the diameter ratio of the multilayer solder film and the solder pad. Figure 8.40 shows the measured solder bump height after the reflow at different diameter ratios (d_s/d_m). In these measurements, the thickness of the solder film was set at 4 μm. This graph shows the linear relationship between the solder height and the diameter ratio. For example, when the gap between the undercladding layer surface and the OE device is 10 μm, the graph shows that the 4-μm-thick solder film with a diameter ratio (d_s/d_m) of 1.5 provides sufficient solder bump height to interconnect the PLC platform and the OE device.

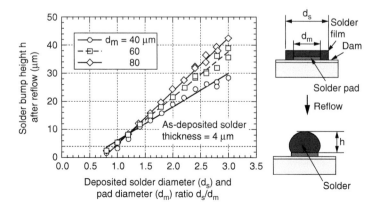

FIGURE 8.40 Solder bump height after reflow.

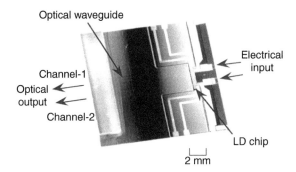

FIGURE 8.41 Photograph of fabricated transmitter module.

8.4.3.3 10-Gbit/sec Hybrid Integrated Transmitter Using Solder Bump Technology

Several hybrid modules for the high-speed applications have been realized using a PLC platform [40–42]. The 10-Gbit/sec transmitter array is one such module. Figure 8.41 shows a photograph of a submodule following the LD array chip assembly. The PLC platform was 14×12 mm. The submodule was hybrid integrated with a 1.55-μm InGaAsP/InP LD array. The LD array consisted of a buried heterostructure of 1.55-μm laser diodes embedded in a high-resistance InP epitaxial layer on a semiinsulating substrate. This structure provides low parasitic capacitance, which gives a wide bandwidth over 10 GHz and also provides good electrical isolation between the channels. Both p- and n-type electrodes were formed on the same surface of the LD chip to facilitate flip-chip bonding. The distance between the channels was 1.6 mm. The assembly region had eight AuSn solder bumps, four of which were used to disperse the heat generated in the active regions of the LD array. The multilayer solder film was designed to be 4 μm thick, with a diameter of 90 μm. The solder pad diameter was 50 μm. On the basis of the data in Figure 8.40, this design will provide a solder bump height of about 15 μm. The platform also used coplanar lines formed on the undercladding layer as the signal lines to achieve a wide operation bandwidth [43]. In the assembly process, the solder bumps were reflowed first within the resin-based flux at a temperature above 280°C. Then the LD chip was flip-chip bonded in an N_2 atmosphere and at a temperature above 300°C.

Figure 8.42 shows the measured continuous wave (CW) light output power against injection current characteristics of the submodule. The optical power was 0.92 and 0.82 mW for channels 1 and 2, respectively, at an injection current of 50 mA. The estimated total losses were 8 and 9 dB, respectively. These values were calculated by comparing the differential efficiencies of the fabricated module and the discrete laser diode. The intrinsic coupling loss caused by the mode-field mismatch between the silica waveguide and the LD was estimated to be 7 dB. Other losses resulting from the coupling loss between the silica waveguide and

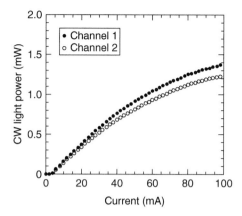

FIGURE 8.42 CW light against injection current characteristics of transmitter module at room temperature.

the fiber, and the propagation loss at the silica waveguide, totaled around 0.6 dB. Therefore, the excess losses caused by the misalignment of the LD chip were 0.4 and 1.4 dB for channels 1 and 2, respectively. The low excess loss reveals that flip-chip-bonding technology using the index alignment technique and a silicon terrace is promising for application to high-speed PLC platforms.

Figure 8.43 shows the frequency response of the module. The measurements were performed using an on-wafer probing technique and a lightwave component analyzer. The response shows that each channel has a wide 3-dB bandwidth of 10 GHz. The bandwidth was identical to that of the discrete LD chip. Therefore, the assembly technologies using the solder bumps on the undercladding layer and the wiring technology using coplanar lines did not have a detrimental effect on the operation bandwidth of the discrete OE device.

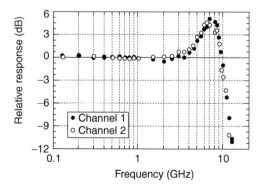

FIGURE 8.43 Frequency response of hybrid integrated transmitter module.

8.5   CONCLUDING REMARKS

This chapter describes applications for solder-bump alignment and an index alignment method for passive alignment. Unlike mechanical contact alignment methods, they do not require the micromachining of the platform or of the semiconductor optical device. Therefore, these techniques have a wide range of applications.

The solder-bump technique uses the surface tension of molten solder to align an optical device. This technique has been employed to provide a microoptics-type PD module that eliminates the need for optical-axis adjustment. This indicates that the solder bump technique can reduce the assembly cost of a conventional optical module. The solder-bump technique also made it possible to construct high-speed PD modules which operated at over 10 Gbit/sec. This is because the solder bumps reduce the parasitic capacitance between the PD electrode and the substrate.

The index alignment technique uses alignment marks formed lithographically both on a platform and on a semiconductor device. This technique made it possible to realize optical multichip hybrid integration on a silica-PLC platform. By employing this technique, a low-cost WDM transceiver module has been developed that consists of a 1.3 to 1.55-μm WDM circuit, a 1.3-μm bidirectional transceiver circuit, and three optical devices: an LD and two PDs. The same technique was also used to realize a four-channel external-cavity LD module that is applicable to dense-WDM networks. The solder-bump technique was also used with a PLC platform, together with index alignment, to realize high-speed optical hybrid integrated circuits. A transmitter module was fabricated using a high-speed PLC platform. The module showed a very wide bandwidth of 10 GHz and a low excess loss of below 1.6 dB. Therefore, this approach makes it possible to realize low-cost and highly functional optical modules for future optical networks.

REFERENCES

1. T. Hayashi and T. Ohsaki, "Accurate optical coupling using solder-bump bonding," Institute of Electronics, Information, and Communications Engineers of Japan, National Conf. Rec., S9-6, vol. 2, p. 338, 1987 (in Japanese).
2. T. Hayashi, "An innovative bonding technique for optical chips using solder bumps that eliminate chip positioning adjustments," IEEE Trans. Components, Hybrids, and Manufacturing Technology, vol. 15, No. 2, April 1992.
3. H. Tsunetsugu, T. Hayashi, and K. Katsura, "Micro-alignment technique using 26-mm diameter microsolder bump and its shear strength," Proceedings of '95 Japan International Electronic Manufacturing Technology Symp., pp. 52–55, December 1995.
4. H. Tsunetsugu, T. Hayashi, K. Katsura, M. Hosoya, N. Sato, and N. Kukutsu, "Accurate, stable, high-speed interconnections using 20-to 30-μm-diameter microsolder bumps," IEEE Trans. Comp. Packaging, and Manuf. Technol.-Part A, vol. 20, no. 1, pp. 76–82, March 1997.
5. M. S. Cohen, M. F. Cina, E. Bassous, M. M. Oprysko, and J. L. Speidell, "Passive laser-fiber alignment by index method," IEEE Photon. Technol. Lett., vol. 3, pp. 985–987, 1991.

6. K. Kurata, K. Yamauchi, A. Kawatani, H. Tanaka, H. Honmou, and S. Ishikawa, "A surface mount type single-mode laser module using passive alignment," Proc. Electronic Components and Technology Conference 45th, p. 759, 1995.

7. K. Katsura, T. Hayashi, F. Ohira, S. Hata, and K. Iwashita, "A novel flip-chip interconnection technique using solder bumps for high-speed photoreceivers," *J. Lightwave Technol.*, vol. 8, no. 9, pp. 1323–1327, 1990.

8. H. Tsunetsugu, T. Hayashi, K. Katsura, M. Hosoya, N. Sato, and N. Kukutsu, "Accurate, stable, high-speed interconnections using 20-to 30-μm-diameter microsolder bumps," IEEE Trans. Comp. Packaging, Manuf. Technol.-Part A, vol. 20, no. 1, pp. 76–82. March 1997.

9. N. Kukutsu and R. Konno, "Super absorption boundary condition for guided waves in the 3-D TLM simulation," *IEEE Microwave Guided Wave Lett.*, vol. 5, no. 9, pp. 299–301, September 1995.

10. T. Hayashi and H. Tsunetsugu, "Optical module with MU connector interface using self-alignment technique by solder-bump chip bonding," Proc. 46th Electronic Components and Technology Conference, pp. 13–19, May 1996.

11. T. Hayashi and H. Tsunetsugu, "New receptacle optical modules using ferrule-integrated chip carrier with solder-bump-bonded photonic device: singlemode-fiber-based high-speed PIN PD receivers," submitted to IEEE Trans. CPMT.

12. Y. Ando, S. Iwano, R. Nagase, K. Kanayama, and E. Sugita, "Advanced optical connectors for single mode fibers," *NTT Rev. Japan*, vol. 3, no. 3, pp. 110–121, 1991.

13. R. Nagase, E. Sugita, S. Iwano, K. Kanayama, and Y. Ando, "Miniature optical connector with small Zirconia ferrule," *IEEE Photon. Technol. Lett.*, vol. 3, no. 11, pp. 1045–1047, November 1991.

14. N. Takachio, K. Iwashita, S. Hata, K. Onodera, K. Katsura, and H. Kikuchi, "A 10 Gbit/sec optical heterodyne detection experiment using a 23 GHz bandwidth balanced receiver," IEEE Trans. Microwave Theory Techniques, vol. 38, no. 12, December 1990.

15. R. S. Sussman et al., "Ultra-low-capacitance flip-chip-bonded GaInAs PIN photodetector for long-wavelength high-data-rate fiber-optic systems," *Electron. Lett.*, vol. 21, no. 14, pp. 593–595, 1985.

16. H. Tsunetsugu, K. Katsura, T. Hayashi, F. Ishitsuka, and S. Hata, "A new packaging technology using microsolder bumps for high-speed photoreceivers," IEEE Trans. Components, Hybrids, Manuf. Technol., vol. 15, no. 4, pp. 578–582, 1992.

17. H. Tsunetsugu, M. Hosoya, S. Norimatsu, N. Takachio, Y. Inoue, and S. Hata, "A packaging technique for an optical 90°-hybrid balanced receiver using a planar lightwave circuit," IEEE Trans. Comp. Packaging, and Manuf. Technol.-Part B, vol. 19, no. 3, pp. 569–574. August 1996.

18. H. Tomimuro, F. Ishitsuka, N. Sato, and M. Muraguchi, "A new packaging technology for GaAs MMIC modules," IEICE Trans., vol. E74, pp. 1209–1213, May 1991.

19. J. E. Bowers, C. A. Burras, and R. J. McCoy, "InGaAs PIN photodetectors with modulation response to millimeter wavelengths," *Electron. Lett.*, vol. 21, no. 18, pp. 812–814, 1985.

20. J. Schlafer, C. B. Su, W. Powazinik, and R. B. Lauer, "20 GHz bandwidth InGaAs photodetector for long-wavelength microwave optical links," *Electron. Lett.*, vol. 21, no. 11, pp. 469–471, 1985.

21. J. E. Bowers and C. A. Burras, "Ultrawide-band long-wavelength p-i-n photode-tectors," *J. Lightwave Technol.*, vol. LT-5, no. 10, pp. 1339–1350, 1987.

22. C. P. Wen, "Coplanar waveguide: a surface strip transmission line suitable for nonreciprocal gyromagnetic device applications," IEEE Trans. Microwave Theory Techniques, vol. MTT-17, no. 12, pp. 1087–1090, December 1969.

23. E. Jahnke and F. Emde, *Tables of Functions with Formulae and Curves*, 4th ed, New York: Dover, 1945.

24. M. Kawachi, "Silica waveguides on silicon and their application to integrated-optic components," *Opt. Quant. Electron.*, vol. 22, pp. 391–416, 1990.

25. M. Okuno, N. Takato, M. Kawachi, and A. Sugita, "Polarization beam splitter switch with controlled silica waveguide birefringence on Si substrate," *J. Light-wave Technol.*, vol. 12, no. 4, pp. 625–633, 1994.

26. K. Kawano, M. Saruwatari, and O. Mitomi, "A new confocal combination lens method for a laser diode module using a single-mode fiber," *J. Lightwave Technol.*, vol. LT-3, p. 739, 1985.

27. Y. Yamada, A. Takagi, I. Ogawa, M. Kawachi, and M. Kobayashi, "Silica-based optical waveguide on terraced silicon substrate as hybrid integration platform," *Electron. Lett.*, vol. 29, pp. 444–445, 1993.

28. T. Hashimoto, Y. Nakasuga, Y. Yamada, H. Terui, M. Yanagisawa, K. Moriwaki, T. Tohmori, Y. Suzuki, and M. Horiguchi, "Hybrid integration of a laser diode chip on a planar lightwave circuit platform by passive alignment method," MOC '95, Tech. Digest, vol. D5, pp. 66–69, 1995.

29. Y. Nakasuga, T. Hashimoto, Y. Yamada, H. Terui, M. Yanagisawa, K. Moriwaki, Y. Akahori, Y. Tohmori, K. Kato, S. Sekine, and M. Horiguchi, "Multi-chip hybrid integration on PLC platform using passive alignment technique," Proc. 46th Elec-tronic Components and Technology Conference, pp. 20–25, 1996.

30. Y. Yamada, S. Suzuki, K. Moriwaki, Y. Hibino, Y. Tohmori, Y. Akatsu, Y. Nakasuga, T. Hashimoto, H. Terui, M. Yanagisawa, Y. Inoue, Y. Akahori, and R. Nagase, "Application of planar lightwave circuit platform to hybrid integrated optical WDM transceiver/receiver module," *Electron. Lett.*, vol. 31, pp. 1366–1367, 1995.

31. N. Uchida, Y. Yamada, Y. Hibino, Y. Suzuki, and N. Ishihara, "Low-cost hybrid WDM module consisting of a spot-size converted laser diode and a waveguide photodiode on a PLC platform for access network systems," IEICE Trans. Electron., vol. E80-C, pp. 88–97, 1997.

32. Y. Inoue, T. Oguchi, Y. Hibino, S. Suzuki, M. Yanagisawa, K. Moriwaki, and Y. Yamada, "Filter-embedded wavelength-division multiplexer for hybrid-inte-grated transceiver based on silica-based PLC," *Electron. Lett.*, vol. 32, pp. 847–848, 1996.

33. S. Suzuki, T. Kitoh, Y. Inoue, Y. Yamada, Y. Hibino, K. Moriwaki, and M. Yanag-isawa, "Integrated optics Y-branching waveguide with an asymmetric branching ratio," *Electron. Lett.*, vol. 32, pp. 735–736, 1996.

34. I. Ogawa, Y. Yamada, M. Yasu, and K. Moriwaki, "Reduction of waveguide facet reflection in optical hybrid integrated circuit using sawtoothed angle facet," *IEEE Photon. Technol. Lett.*, vol. 7, pp. 44–47, 1995.

35. Y. Tohmori, Y. Suzuki, H. Oohashi, Y. Sakai, Y. Kondo, H. Okamoto, Y. Kadota, M. Mitomi, Y. Itaya, and T. Sugie, "High temperature operation with low-loss coupling to fiber for narrow-beam 1.3 μm lasers," *Electron. Lett.*, vol. 31, pp. 1838–1840, 1995.

36. K. Kato, A. Kozen, M. Yuda, Y. Muramoto, K. Noguchi, Y. Akatsu, O. Nakajima, and J. Yoshida, "Low-cost low driving-voltage waveguide p-i-n photodiode for optical hybrid integration," Proc. OECC'96, pp. 410–411, 1996.

37. M. Fukuda, F. I. chikawa, H. Toba, J. Yoshida, Y. Yamada, Y. Inoue, K. Kato, H. Sato, and T. Sugie, "Highly reliable plastic packaging for laser diode and photo-diode modules used for access network," *Electron. Lett.*, vol. 33, pp. 2158–2159, 1997.

38. T. Tanaka, H. Takahashi, T. Hashimoto, Y. Yamada, and Y. Itaya, "Fabrication of hybrid integrated four-wavelength laser composed of UV written waveguide grating and laser diodes," Proc. OECC'97, pp. 500–501, 1997.

39. H. Takahashi, T. Tanaka, Y. Akahori, T. Hashimoto, Y. Yamada, and Y. Itaya, "A 2.5 Gbit/sec, four-channel multiwavelength light source composed of UV written waveguide gratings and laser diodes integrated on Si," Proc. ECOC '97, pp. 3/355–3/358, 1997.

40. S. Mino, T. Ohyama, H. Hashimoto, Y. Akahori, K. Yoshino, Y. Yamada, K. Kato, M. Yasu, and K. Moriwaki, "High frequency electrical circuits on a planar light-wave circuit platform," *IEEE J. Lightwave Technol.*, vol. 14, no. 5, pp. 806–811, 1996.

41. T. Ohyama, S. Mino, Y. Akahori, M. Yanagisawa, T. Hashimoto, Y. Yamada, Y. Muramoto, and H. Tsunetsugu, "10 Gbit/s hybrid integrated photoreceiver array module using a planar lightwave circuit platform," *Electron. Lett.*, vol. 32, no. 9, pp. 845–846, 1996.

42. Y. Akahori, T. Ohyama, M. Yanagisawa, Y. Yamada, H. Tsunetsugu, Y. Akatsu, M. Togashi, S. Mino, and Y. Shibata, "A hybrid high-speed silica-based planar lightwave circuit platform integrating a laser diode and a driver IC," Proc. ECOC '97, vol. 3, pp. 339–362, 1997.

43. S. Mino, T. Ohyama, Y. Akahori, Y. Yamada, M. Yanagisawa, T. Hashimoto, and Y. Itaya, "10 Gbit/s hybrid-integrated laser diode array module using a planar lightwave circuit (PLC) platform," *Electron. Letters*, vol. 32, no. 24, pp. 2232–2233, 1996.

9 Passive Alignment for Surface Mount Packaging and a Low-Cost Plastic Packaged Optical Module as an Application of the Passive Optical Alignment Method

Kazuhiko Kurata and Kimio Tatsuno

CONTENTS

9.1 PASSIVE ALIGNMENT FOR SURFACE MOUNT PACKAGING

Recently, the application of optical communications has expanded from trunk lines to subscriber loops [1] and local area networks (LANs). Various types of low-cost optical modules such as LD/PD (laser diode/photodetector diode) modules or hybrid integrated PLC (planar light wave circuit) modules are required. To obtain low-cost modules, reduction of assembly and packaging cost is very important, because it accounts for 70% of the total module cost. For drastic cost reduction of optical modules, development of new packaging techniques based on a passive alignment technique is strongly required. To meet such requirements, a new packaging technique has been developed. The key techniques of packaging are based on passive alignment and effective optical coupling. A LD is passively mounted on a Si substrate, and a fiber is self-aligned to the LD by placing it on a Si V-groove. Alignment marks are patterned on the LD bottom surface and on the Si surface, using a photolithography technique. The LD is mechanically positioned by detecting each alignment mark and is finally soldered with AuSn solder. The single-mode optical fiber is self-aligned on a Si V-groove.

New module packaging techniques are also used to reduce assembly cost and module size. A miniature optical coupling unit on a Si substrate enables us to develop a compact and low-cost ceramic package. A simple receptacle structure is designed at the fiber output port, using a glass ferrule with a short length of fiber and other mechanical parts. The optical coupling unit is passively assembled without monitoring output power, and the ferrule can be treated as a usual part without the need to handle a long fiber in the assembly process. This structure is very much suitable for automatic assembly because of the easy handling of the parts — high productivity and compact size can be achieved. In this chapter, the key techniques are described along with the SMT (surface mountable) module design [3], and a hybrid integrated wave guide module [4] is also introduced as one of the applications.

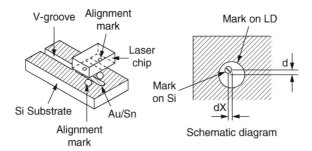

FIGURE 9.1 Schematic diagram of the developed passive alignment technique.

9.2 KEY TECHNIQUES

9.2.1 VISUAL ALIGNMENT

Figure 9.1 shows a schematic diagram of the developed passive alignment technique for positioning a LD on the Si substrate [2,5]. Figure 9.2 shows the fabrication process of the optical Si bench. The positioning marks and patterns of the V-groove are made by dry-etching gold. Relative accuracy between the marks and the V-groove depends on the patterning mask accuracy of 0.5 μm. The AuSn solder is made of a gold/tin multilayer on the substrate. The gold and tin are subsequently evaporated, and the gold is finally covered on the top layer to prevent oxidation of the tin. The optical fiber is aligned on the Si V-groove. A rectangular groove across the V-groove acts as a fiber stopper. The LD is mounted on the Si substrate by detecting the alignment marks on the LD bottom surface and the Si substrate, and high-accuracy LD mounting is attained, using a newly developed automatic mounter. Figure 9.3 shows a block diagram and a photograph of a newly developed LD mounter. It consists of air slide stage and stage movable

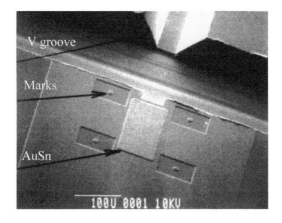

FIGURE 9.2 Scanning electron microscopy (SEM) photograph of the optical unit on the Si substrate.

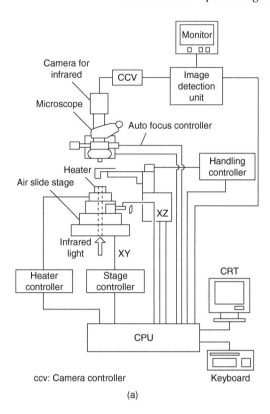

(a)

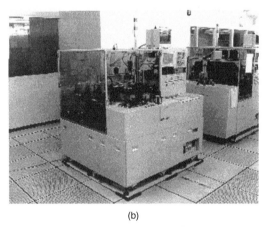

(b)

FIGURE 9.3 Automatic LD mounter. (a) Block diagram, (b) photograph of LD mounter. Courtesy of NEC corporation.

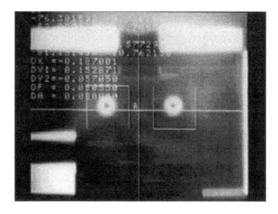

FIGURE 9.4 Observed alignment marks by IR microscope.

in X/Y directions and a heater located on the slide stage, with the sample placed on this heater. The sample is illuminated from the backside, using infrared light. The sample can be observed through the microscope-IR camera, and the images can be obtained on the monitor through the image detection unit. The X/Y movement is controlled through the stage controller-CPU and can be accessed through the keyboard. A heater controller provides heat to the slide stage. Information from the CPU can be fed back to the slide stage, handling the controller, the auto focus controller of the microscope, and finally, the image detection unit.

A pair of marks placed face to face can be observed at the same time through the substrate by infrared light, as shown in Figure 9.4. Accurate detection of marks and precise control of the adjusting stage are required in this technique. To get correct positioning information, we detected the center of gravity of each mark. The center of gravity is calculated from the summation of surface factors (dS) as shown in Figure 9.5. Each surface factor has area information, and the detection error appears from the infinite summation of the surface factors near by the edge area of the mark. The detecting error is decreased by summation of the surface factors in a particular area. Figure 9.6 shows the result of an actual detecting error obtained from the center of gravity. The x-axis denotes the radius of each mark, and the y-axis denotes the detection error. It can be observed that the accuracy increases (i.e., detection error decreases) as resolving power is decreased and as the radius of the mark decreases. We chose a minimum mark radius of 5 μm and a resolving power of an objective lens of 1.5 μm. An actual detecting error of 0.2 μm is achieved [6]. After positioning, the LD is soldered, using AuSn solder. The correct amount of AuSn solder is previously applied on the Si substrate. Figure 9.7 shows the mounting accuracy of the LD. An average mounting accuracy of 0.23 μm with a standard deviation of 0.14 μm is achieved. Figure 9.8 shows the scanning electron microscope (SEM) micrograph of the optical unit on Si substrate. The fiber end face is suitably positioned along the fiber axis by touching its end face to the fiber stopper with an accuracy of within 3 μm accuracy. Misalignment factors such

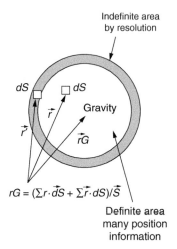

FIGURE 9.5 Calculation of the center of gravity.

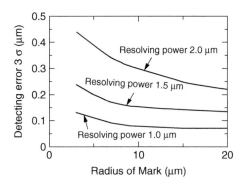

FIGURE 9.6 Deviation of the detected position.

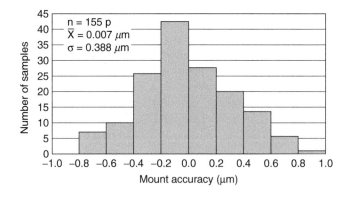

FIGURE 9.7 Mounting accuracy.

FIGURE 9.8 SEM micrograph of the optical coupling unit.

as Si V-groove and solder thickness are considered, and the standard deviation of total mounting accuracy is estimated. The optical alignment tolerance is 2 μm in a direction perpendicular to the optical axis, whereas that along the optical axis is 10 μm for three times the standard deviation.

In an alignment, the offset function is applied in the machine to eliminate the misalignment of the LD cleaving error and the LD mark patterning error. The distance between the LD facet and the mark is detected, and the facet position is adjusted to a suitable position for LD mounting. The LD mark position depends on the patterning process of LD fabrication, and this error is of almost the same order in the same wafer. LD mark position error is previously measured by sampling of LD. The LD patterning error is adjusted by applying an offset of position error. In this technique, misalignment of the optical axis between the LD and the fiber can be eliminated within 5 μm along the optical axis and less than 2 μm orthogonal to the optical axis.

9.2.2 Fabrication Process of Silicon Substrate

On a silicon substrate, the positioning marks and the Si V-groove used for fiber alignment are fabricated on the Si substrate, with standard photolithographic and anisotropic etching techniques. Figure 9.9 shows the fabrication process of the optical coupling unit bench on the Si substrate. At first, the Si substrate, which is covered by thermal oxidized SiO_2, is metalized by Cr/Pt/Au. The positioning marks and the pattern of the V-groove are made using a dry-etching process at the same time. The relative accuracy between the marks and the V-groove depends only on the patterning mask accuracy of 0.5 μm. After the SiO_2, the V-groove area is removed by hydrofluoric acid (HF) etching, and the AuSn solder is applied by vapor phase deposition. The AuSn solder consists of Au/Sn multilayers alternately evaporated, and an Au top layer finally covering the top prevents oxidation of Sn. Finally, the V-groove is etched by KOH. The vertical accuracy perpendicular

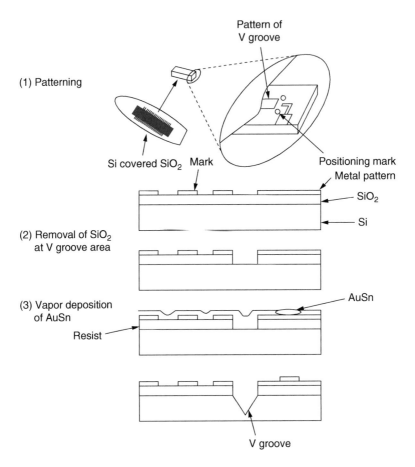

FIGURE 9.9 Fabrication process of the Si substrate.

to the Si substrate surface depends on the side etching of the V-groove and the thickness of the AuSn layer. The standard deviation width of the Si V-groove is controlled to about 0.6 μm, and the standard deviation of the AuSn thickness after soldering is 0.15 μm.

9.2.3 OPTICAL COUPLING DESIGN

Normally output power over 1 mW is required in long-distance transmission or subscriber loops in which a passive splitter in the fiber line is used. To achieve low coupling loss of less than 5 dB by passive alignment, a hemispherical lens fabricated by etching the fiber facet was attempted. Fabrication of this lens with fixed radius was easily reproducible. Figure 9.10 shows the fabricated microlens.

In general, optical coupling loss and alignment tolerance are trade-offs. Alignment tolerances have to be balanced with the actual positioning accuracy. The optical coupling characteristics (coupling loss and alignment tolerance) are varied

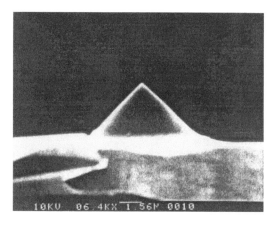

FIGURE 9.10 Photograph of the fabricated microlens.

by changing the lens parameters. To accomplish passive alignment by using alignment marks, the tolerance along the optical axis had to be relaxed because there was a maximum 10-μm misalignment deviation for this direction. Next, the alignment tolerance of 2 μm in a direction orthogonal to the optical axis was modified.

Relaxation of alignment tolerance is done by filling the gap between the LD and the fiber with low-refractive index materials. A change of refractive index changes the lens parameters (focal length and spot size at the beam waist) of the hemispherical lens, and the optical coupling characteristics (coupling loss and alignment tolerance) are varied by changing lens parameters. Figure 9.11a and Figure 9.11b show the approximate result for optical characteristics. The loss variation along the optical axis becomes gradual, and the alignment tolerance is relaxed according to the increase of the refractive index, while the coupling loss increases.

9.2.4 Lens Design

A hemispherical lens forms at the fiber facet through an HF etching process. The lens is formed because of the different rates by which the HF etchant etches the fiber core and the cladding. The spot size of the fiber can be controlled from 1.5 to 3.0 μm by varying the etching time and HF concentration. Figure 9.12 shows the etching time dependence of the fiber spot size.

Under reproducible conditions for achieving stable spot size, we can choose a size of 1.6 or 2.0 μm by changing etching conditions. In previous work, we chose per fluoride liquid with a refractive index of 1.3 requiring a spot size for the hemispherical lens of roughly 2 μm. However, sealing a per fluoride liquid usually is troublesome in module packaging. Because suppression of the liquid thermal expansion must be considered. Application of a resin solves these problems. We selected a Si resin as a low-refractive index material. Si resin, which has a refractive index of 1.39 (the lowest in all the Si resins), is selected. In consideration for the spot size

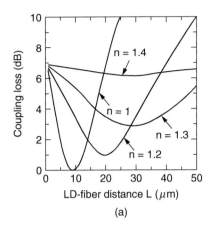

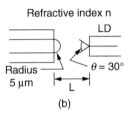

FIGURE 9.11 Optical coupling loss.

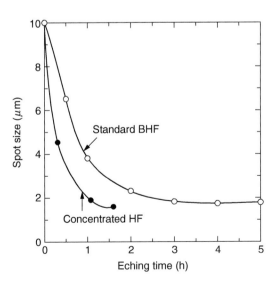

FIGURE 9.12 Spot size of the fiber with microlens.

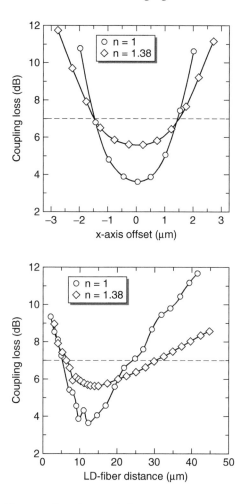

FIGURE 9.13 LD-fiber coupling characteristics.

in the liquid, a small spot size must be selected because of its slightly higher refractive index. We selected the fiber spot size of 1.6 μm to obtain a LD-fiber coupling loss of 5 dB.

Figure 9.13 shows experimental coupling characteristics for the LD to fiber. The alignment tolerance along the optical axis is around 12 μm wide, with 2 μm along the direction orthogonal to the optical axis.

9.2.5 MODULE DESIGN

New module packaging techniques that exhibit the advantages of passive alignment must be developed. In the conventional assembly technique, an optical pigtail cord for monitoring output power is necessary. In passive alignment, a miniature optical coupling unit can be made on a Si substrate without monitoring the optical

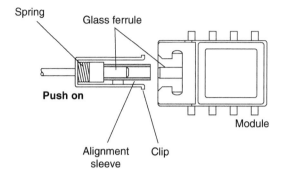

FIGURE 9.14 Receptacle structure.

output power. NEC recently designed a simple receptacle structure at the fiber output port. The receptacle structure is shown in Figure 9.14. This receptacle consists of a glass ferrule with a short length fiber and hooking parts. The connector pigtail with a clip and a ferrule in a sleeve is easily mated to the receptacle by using the clip-hooking parts. The two ferrule end faces (receptacle portion and connector cord) physically contact each other in the sleeve. Figure 9.15a and

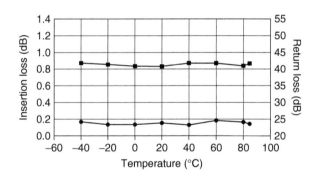

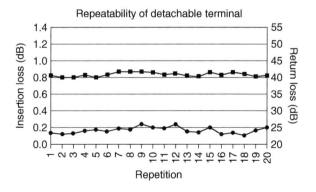

FIGURE 9.15 Typical connection characteristics.

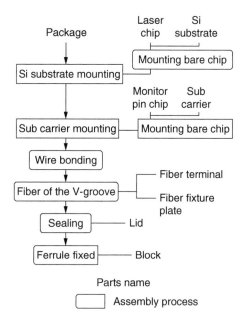

FIGURE 9.16 Module assembly chart.

Figure 9.15b show typical connection characteristics of the receptacle. Low insertion loss and high return loss are achieved, because of the high mechanical accuracy of the glass ferrule and fine polishing of the ferrule end face.

This structure has the advantages of both module assembly in manufacture and module mounting on the circuit board, as described in the following.

9.2.5.1 Advantage in Manufacturing

An optical pigtail cord has a problem in an automated module assembly line because of the handling difficulties. The optical pigtail cord must be rolled to make it compact and easy to handle. This process is timely and requires a large space. Hence, the conventional process may not be very suitable for mass production of a very low cost module. In this structure, handling problems can be avoided using small-ferrule fibers. Figure 9.16 summarizes the module assembly chart. It consists of individual subassemblies of the LD mount, PD mount, fiber mount, and final sealing. This module is tested after each subassembly, and completed with an auto power control (APC) test. We have succeeded in realizing an automated assembly line. Automatic assembly machines performing such tasks as Si substrate mounting and ferrule mounting have also recently been developed. The operating cycle time of all assembly machines is set within 60 sec, and all parts including the ferrules are supplied by trays. Figure 9.17 shows the view of the developed assembly line. The modules with different types of the connector models (FC, SC, or ST type) are assembled, using identical assembly lines.

FIGURE 9.17 Photograph of the assembly line. Courtesy of NEC corporation.

FIGURE 9.18 Mounting of the module on a circuit board. Courtesy of NEC corporation.

9.2.5.2 Advantage in Mounting on a Circuit Board

The connector pigtail cord is attached after mounting the module on a circuit board. Without the connector pigtail cord, the module can withstand reflow soldering temperatures — a feature not present in conventional modules. This feature leads us to automatic reflow soldering for mounting of the module, similar to ordinary electrical component soldering. Figure 9.18 shows one example of a module mounting on a circuit board.

9.3 APPLICATION

Many applications of optical modules can be envisaged using the developed passive alignment and packaging technique mentioned above. The various concepts of the applications and technical trends are indicated in a three-dimensional map. The three axes correspond to performance such as transmission speed or distance, function integration, and number of channels, respectively, as shown in Figure 9.19.

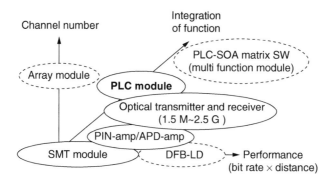

FIGURE 9.19 Three-dimensional map of application.

Low-power Fabry-Perot-LD and positive intrinsic N-type detector (PIN) modules with many basic functions and simple optical coupling structure are positioned at the origin. In the performance direction, hemispherical lensed fiber [7] or a spot size converted laser [8] is necessary to obtain higher output power. In this direction, improvement in the laser chip performance is also an important factor. For example, auto power control (APC) (partially corrugated waveguide)-laser [9] can improve modulation characteristics for optical feedback. This improvement is effective in attaining an isolator-free laser module with single longitudinal mode. In the direction of the function, the hybrid integration of an LD and PD as active chips with the PLC as a passive component are important techniques. To realize low-cost subscriber systems, the electrical circuits of a preamplifier or receiver integrated circuits (ICs) are hybridly integrated. Hybrid integration of PLC, LD, and PD enables us to design various types of functional modules such as bidirectional modules or WDM modules, using a similar design rule. Along the channel number direction, array modules are lined up to provide larger throughput for optical interconnect applications [10]. In the future, applications will extend to highly integrated functional modules, which are key components in all optical network systems. In the following section, typical modules in each application are introduced from the assembly point of view.

9.3.1 SMT LD MODULE

Figure 9.20 shows the structure of a simple mode transmitter (SMT) LD module. The SMT LD module is one of the most basic applications for using the low-cost packaging technique. A LD is passively mounted on a Si substrate, and a fiber is self-aligned to the LD by placing it on a Si V-groove. This module is designed to make use of the idea of reflow soldering on an electric circuit board. A compact ceramic package measuring $12 \times 7.6 \times 3$ mm houses the optical unit. In particular, its flat shape with a 3-mm height meets all requirements for high-density mounting on both sides of the electric circuit board. Usually, in telecommunication applications, very high receiving power (e.g., less than 34 dBm at 155 Mbps), is required. To achieve very high receiving power, electrical noise from the outside must be cut by the package. In addition, parasitic capacitance must be eliminated. A ceramic package is

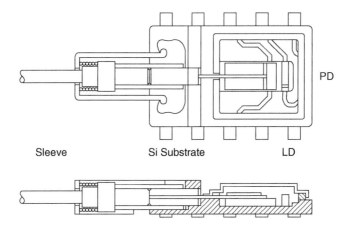

FIGURE 9.20 Structure of the SMT module.

effective in obtaining fine electrical sealing and low parasitic capacitance. A SMT PIN module similar to the SMT LD module and the PIN-Amp module was also newly designed, using the Si V-groove and planar type PIN on ceramic carrier. The Pre-amp IC is easily built into the ceramic package. The parasitic capacitance excluding the PIN becomes even smaller, less than 0.1 pF in this module. Figure 9.21 shows the view of the PIN-Amp module designed for 150 Mbit/sec burst systems. A minimum received power of −35 dBm is achieved [11]. In the near future, this type of application will extend to various modules, including the APD module or the APD-Amp module.

9.3.2 PLC MODULE

A bidirectional transmitter/receiver optical module is a basic component in a passive double star network. Attractive hybrid integrated PLC modules have been developed [12,13].

FIGURE 9.21 Photograph of the PIN-Amp module.

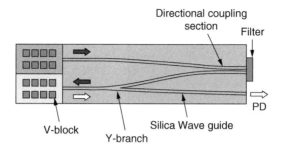

FIGURE 9.22 PLC transmission loss.

9.3.3 PLC

The PLC is fabricated on the Si substrate using TEOS/O3 APC vapor deposition-process. Figure 9.22 shows one example of a waveguide pattern used in the bidirectional Y-branch module. The waveguide consists of a Y-branch for a 3-dB coupler and a reflective directional coupler to enable the PD port to be placed on the side opposite the LD port. The 1.3-μm output signal from the LD passes through the waveguide and goes to the LINE port through the Y branch after being reflected at the waveguide facet by a filter. The 1.3-μm input signal from the LINE port then goes to the photodiode through the Y branch. For the purpose of fiber alignment, V-grooves are fabricated on the PLC. A groove length of 3 mm was selected to ensure sufficient strength in the fiber mount. These grooves consist of a silicon block fiber guide (SBF) [14]. This SBF is novel, having multiple Si mesas. The tolerance of the etching mask alignment can be much enhanced, and the corner of each Si mesa can support SMF, even if the etching mask is not completely parallel to the crystal orientation. On the other side of SBF, the LD-fiber coupling unit is used to couple the LD, and the fiber is attached to the mesa-blocks. The size of PLC is 3 mm wide and 12 mm long.

9.3.4 PLC Module

Figure 9.23 shows the structure of a bidirectional PLC module with a size 36 × 12.5 × 3 mm. Hybrid integration of optical elements on the silica waveguide and alignment free assembly is used for cost reduction. The packaging design is based on the SMT module. Optical elements (PLC, LD-fiber coupling unit, and PIN on chip career) are built in a ceramic package. The receptacle structure is formed at the fiber output port.

The PLC module is mainly made up of three subassembled optical components: a LD-fiber coupling unit fabricated on a Si substrate, a PLC unit on which an LD-fiber coupling unit and two kinds of optical fibers are mounted, and a ceramic PKG in which a bare receiver-amplifier and photodiodes are mounted.

These three components can be easily assembled into the module without optical adjustment.

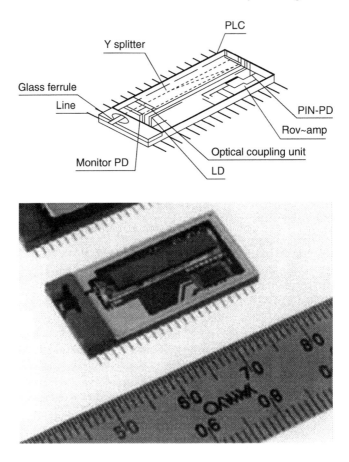

FIGURE 9.23 Structure of bidirectional module. (a) Structure of PLC module, (b) photograph of PLC module.

9.3.5 LD-Fiber Coupling and Mounting on PLC

The LD-fiber optical coupling unit is formed on a Si substrate, using the same method as the SMT LD module. The fiber mounted on the Si substrate is sawed off at the end of the Si substrate. When mounting on PLC, the LD-fiber coupling unit is turned over and mounted in the V-groove formed on the PLC. In this way, the LD and the waveguide are connected through the short-length fiber. The LD-fiber coupling unit has the same function as a spot size converted laser (Chapter 10).

For LD-PLC optical coupling, a subassembled optical coupling unit with an Si substrate is adopted. This structure is similar to SMT LD module in optical coupling structure. A low-threshold current LD with a short-length cavity is chosen for bias-free drive. This LD-fiber optical coupling unit has the same function as that of the spot size converted laser. Alignment free assembly of optical elements (LD-fiber coupling unit and output port fiber) can be obtained.

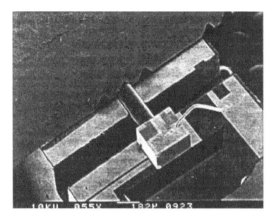

FIGURE 9.24 The LD-PLC coupling geometry.

The LD-fiber optical coupling unit is self aligned on the guiding of the fiber, as shown in Figure 9.24. The coupling loss of fiber-waveguide is estimated at roughly 0.5 dB.

9.3.6 PD AND LSI MOUNTING

The PLC unit is mounted in a ceramic package in which a receiver-amplifier and photodiodes are built with the use of mounter. The 1.3-μm receiver port on the waveguide and a photodiode whose detecting area diameter is 120 μm are mechanically aligned in the ceramic package without optical adjusting. The thin plate is inserted between the PLC and ceramic package for the purposes of height alignment. The PLC unit is placed by detecting the distance between the photodiode and the side edge of the PLC. Relative positioning accuracy between the PLC and the photodiode is shown in Figure 9.25. Standard deviation of 10 μm in the direction of waveguide is obtained. PLC-to-photodiode coupling loss is estimated to be less than 0.5 dB.

9.4 CHARACTERISTICS

The main optical characteristics are summarized in Table 9.1. Figure 9.26 shows the typical light-current (L-I) characteristics. Low-threshold current laser of 6.6 mA at 25°C can be driven bias-free.

Recently, research and development of spot size converter (SSC)-LD has progressed and has almost reached practical use level. The LD-fiber coupling unit can be replaced by SSC-LD in the near future. SSC-LD is very attractive for further cost reduction, because it decreases the number of parts and the size of a PLC chip. A trial has been made of direct mounting of a SSC-LD on the PLC chip, using the same mounting techniques as the SMT module. Figure 9.27 shows the structure of the mounting platform. The same mount accuracy was obtained as with the SMT module. As mentioned above, the basic structure of the optical passive alignment

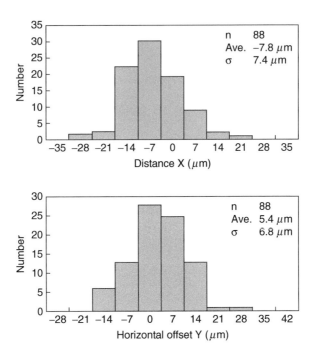

FIGURE 9.25 Relative mounting accuracy of PLC and PD.

on the substrate is common to both the surface mount modules and the PLC modules. In surface mount modules, the optical units are the LD and a fiber while the substrate is the Si substrate. In contrast, the PLC module optical units are the LD, the LD-fiber coupling unit, or the SSC-LD (in the near future) and the substrate is the PLC. Therefore, this standardization of the structure and assembly process enables us to realize the common assembly machines and line in all these modules. Various types of optical passive circuits can be designed on a PLC substrate. This hybrid integrated module technology can be used in other functional modules such as wavelength division multiplexing (WDM) modules or array switching modules in the similar process and assembly line.

TABLE 9.1
Module Optical Performance

Parameter	Characteristics
Output power	0 dBm
Modulation current	< 35 mA
Threshold current @25°C	4.5 mA
Threshold current @85°C	9.0 mA
Responsivity	0.3 W/A

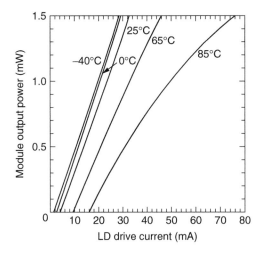

FIGURE 9.26 Typical L-I characteristics.

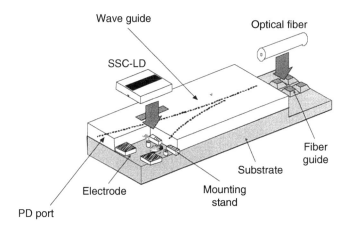

FIGURE 9.27 Structure of the mounting platform.

9.5 HIGH-PERFORMANCE AND LOW-COST PLASTIC PACKAGED OPTICAL MODULE AS AN APPLICATION OF THE PASSIVE OPTICAL ALIGNMENT METHOD

Cost reduction and mass production of optical modules while maintaining high-speed bidirectional digital fiber communication performance are essential requirements for the realization of optical access network systems. This section reviews plastic optical module technologies applying electronic ICs. Included will be passive surface mountings on Si submounts for bare optoelectronic components [15] and nonhermetic plastic packaging [16–19]. These technologies lead low-cost solutions through fully automated mass production of fiber communication modules, retaining high performance and reliability.

9.6 PLASTIC TRANSMITTER MODULE

The fabricated plastic transmitter modules are shown in Figure 9.28: the preinjection mold case type (on right) and the transfer mold type (on left). Both are made with 8-pin dual in-line packages similar to the ICs, shown in the middle for reference. The sizes are $6.6 \times 3.0 \times 11.5$ mm and $6.3 \times 3.0 \times 9.6$ mm, respectively. The main specifications of these modules are summarized in Table 9.2.

9.6.1 STRUCTURE AND FABRICATION PROCESS

Inside these transmitters there are integrated subassemblies mounted on a Si V-groove substrate, as shown in Figure 9.29, which consists of the conventional ridge-type Fabry-Perot LD, a photodetector for the LD output monitor, and a butt-jointed cleaved-facet single-mode optical fiber (SMF). LDs and PDs [20] are soldered on the Si substrate metallic electrodes, with the active regions closer to the Si substrate surface with the LD junction down to obtain higher accuracy in the vertical direction. The distance between the LD facet and the fiber facet is 40 μm to obtain –11 dB optical coupling, and that between the rear facet of the LD and the monitor PD is 30 μm. The LD facet is tilted 3° to reduce the return loss reflected back to the fiber.

Preinjection mold case type
6.6 (W) × 3 (H) × 11.5 (L) mm

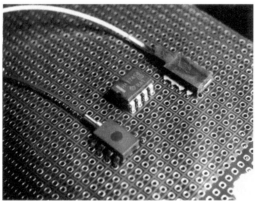

Transfer mold type
6.3 (W) × 3 (H) × 9.6 (L) mm

FIGURE 9.28 Plastic MiniDIL transmitter modules fabricated by a plastic packaging process commonly used in the field of ICs. The upper module is of the preinjection mold case type, and the lower module is made by transfer molding similar to the dynamic random access memories (DRAMs) shown in the middle.

TABLE 9.2
Specifications of the Plastic MiniDIL Transmitter Module that Meets OC 3 (STM 1) and OC 12 (STM 4)

Items	Unit	@25°C
Fiber output	mW	> 0.4
Threshold current	mA	< 20
Slope efficiency	W/A	0.02
Rise/fall time	ns	0.5
Monitor current	μA	> 200
PD dark current	nA	150 @ 5V
Operating temperature	°C	−40 ~ +85
Mean time to failure (MTTF) (85°C)	Hours	1.0E5

Butt joint optical coupling tolerances of LDs to the cleaved fiber are shown in Figure 9.30. Figure 9.30a is for the fiber direction dependency, which indicates the distance Z to be 40 μm for −11-dB optical coupling. Figure 9.30b is for the horizontal direction dependency. The −0.5-dB excess loss tolerance is ±1.5 dB for the −11-dB optical coupling.

To realize this high positioning accuracy, infrared visual microscopic index alignment techniques combined with a silicon V-groove plate for the fiber fixing

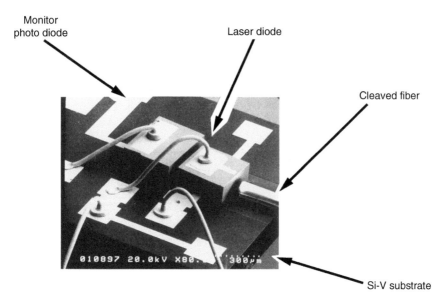

FIGURE 9.29 Subassembly inside the plastic MiniDIL transmitter package which consists of a conventional ridge-type laser diode, a monitor photodiode, a cleaved facet fiber, and a silicon substrate with a V-groove.

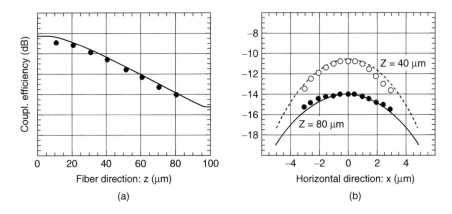

FIGURE 9.30 Butt-joint optical coupling tolerances between the laser and fiber in the (a) fiber direction and the (b) horizontal direction.

are introduced. An anisotropic etching process fabricates the silicon V-groove. The accuracy of the Si V-groove width is a critical issue for LD/fiber optical coupling. Figure 9.31 is a statistical result of the V-groove width distribution. Three-sigma placement accuracy was 1.05 μm converted to the fiber height for this case. The sample number was 128, which was fabricated by a conventional silicon aniso-tropic etching process, which consists of photolithography and dicing.

Infrared index alignment is shown in Figure 9.32. For this alignment, index marks are etched on the bottom surface of the LD and on the surface of the substrate. The marks are illuminated by infrared light from the rear side of the substrate.

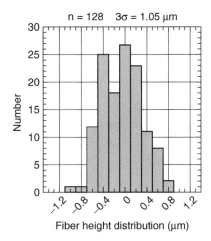

FIGURE 9.31 Silicon V-groove width distribution in terms of fiber height. Three σ was given to be 1.05 μm for the 128 samples. This is a sufficient margin for LD/fiber alignment.

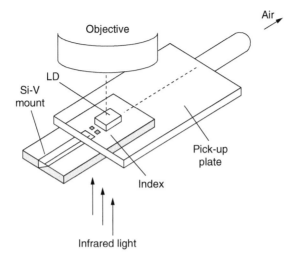

FIGURE 9.32 Schematic showing the principle of the infrared die bonding. Positioning marks fabricated on the bottom surface of the LD/PD and on the Si substrate are illuminated by infrared light. The marks are imaged on a CCD camera through the infrared microscope. The center of each mark and the center differences between the LD/PD and the substrate are calculated. A micrometer moves the substrate until the difference becomes zero and the components are soldered after tact bonding.

They are imaged on an infrared video camera and input to memory, and calculations are made to obtain the center of each mark. The substrate is moved by the micrometer until the marks relative to the position difference become zero. The diffraction-limited resolution of the infrared microscope is more than about 2 µm, which is the wavelength divided by the objective lens numerical aperture, 0.5. However, the center of the marks calculated with the electronic image processing gives better than 0.5 µm accuracy. A statistical distribution for the die bonding with 31 samples shown in Figure 9.33 indicates an accuracy better than ±5 µm, including soldering. These positioning experimental results support the theoretical predictions (indicated as solid and dotted lines) that the butt-joint coupling excess loss is less than –0.5 dB, referring to Figure 9.30b.

9.6.2 PERFORMANCES

The basic performances for the plastic transmitter module are shown in Figure 9.34 to Figure 9.37. Figure 9.34 shows typical temperature characteristics from –40 to 85°C. Using a conventional low-cost ridge-type LD, more than 0.4 mW of fiber output was obtained, with a maximum driving current of 75 mA. Figure 9.35 is a measurement for the return loss of the plastic transmitter modules. In the fabricated plastic transmitter module, the silicon V-groove substrate subassembly consists of the LD and the cleaved facet fiber, and the waveguide-type monitor photo detector

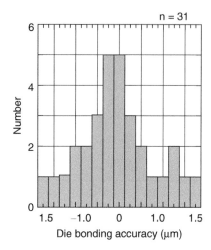

FIGURE 9.33 Die-bonding positioning distribution when using the infrared die bonder, shown in Figure 9.5. Accuracy of ±1.5 μm with 40-μm fiber distance gives −11 dB optical coupling (Figure 9.30a).

is encapsulated with transparent silicone gel to avoid water condensation, preventing corrosion in the nonhermetic plastic packaging. The Fresnel reflection on the laser diode facet is decreased because of the reduction of the refractive index difference between the facet of the laser diode and the silicone gel. In addition, the

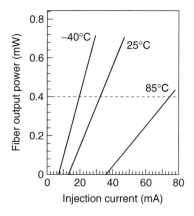

FIGURE 9.34 Typical temperature characteristics of the fabricated plastic MiniDIL transmitter. Full temperature operation was achieved from −40°C to 85°C with a maximum injection current of less than 75 mA.

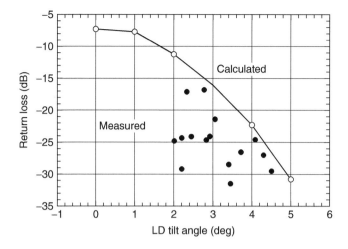

FIGURE 9.35 Measured and calculated return loss of the fabricated plastic MiniDIL was less than –15 dB with the LD tilted 3° from the optical axis.

tilted laser diode facet gives lower power reflected back to the facet of the fiber. As a result, the return loss can be reduced to less than –15 dB.

Figure 9.36 represents a frequency response of the plastic module working up to 622 Mbps, with 3 dB down frequency. The bit error rate performance is shown in Figure 9.37. Similar curves were obtained for 10 km transmission compared with back-to-back operation.

9.6.3 HIGHER-POWER VERSION

A shorter distance between the laser diode and the cleaved facet fiber gives higher fiber output even with butt-joint optical coupling. Unfortunately, it is not a realistic approach because the closer coupling is too sensitive to the fiber laser misalignment. A coupling lens such as an aspheric lens gives much higher efficiency because of better numerical aperture matching with the LD and fiber, but the lens loses its power when it is surrounded by the silicone gel because of the reduction of the refractive index difference. A solution is to introduce large spot laser diodes [21] (Chapter 10). The structure of the laser consists of an active-gain region and a spot-expanding region. The thickness of each layer is tapered in the vertical direction to obtain smaller effective refractive indices, which results in larger spot size. In the horizontal direction, stripe width is fabricated to increase gradually to have larger optical spot size. Coupling loss measurements for large spot LD with the cleaved facet optical fiber are reported to be –2.9 dB with an alignment tolerance of ±1.5 μm.

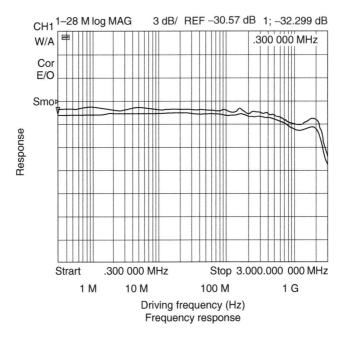

FIGURE 9.36 Frequency response of the fabricated plastic MiniDIL transmitter. Operation up to 622 Mbps was confirmed.

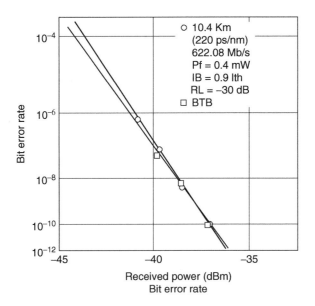

FIGURE 9.37 Bit error rate of the plastic MiniDIL transmitter module. There is no significant difference between the 10.4-km transfer signal and the back-to-back case.

9.7 PLASTIC PACKAGING

Figure 9.38 shows a historical road map of IC packaging. It started with metallic in the sixties and evolved to plastic via ceramic in the seventies. There are mainly two methods for the plastic packaging: the preinjection mold case type and the transfer mold type. In the former method, silicon chips are mounted inside the case and filled with resin. This method is mainly applied to hybrid ICs and relatively larger packages. The latter is applied to millions of mass produced packages like DRAMs because of the parallel molding capability, using multicavity processes. In this method, the thermal expansion ratio of the resin is designed and fabricated with filler to have a similar ratio to that of the silicon chip to avoid thermal distortion resulting from the stress.

9.7.1 PLASTIC SEALING

Sealing against moisture is one of the most important issues in plastic packaging to ensure long-term reliability. Figure 9.39 is the package model used for the moisture-sensing experiment that tests several resin adhesives [22]. A humidity sensor and thermometer are installed inside the metallic package. Several kinds of adhesives: silicon gel, acrylic, and epoxy, are applied between a silicon plate and the pipe of the package model and tested. Silicon gel is quite transparent, and acrylic is in between. Even with epoxy-sealed samples the humidity inside reached the condensation point at room temperature after about 1500 h, at 85°C, 85% relative humidity.

Two methods were proposed on the basis of these experiments. One is to seal the package between the lid and the substrate with an epoxy adhesive. In this case, even though the leakage path is long and narrow, moisture will accumulate, and hence condensation will take place. Therefore, LD/PD passivation must be

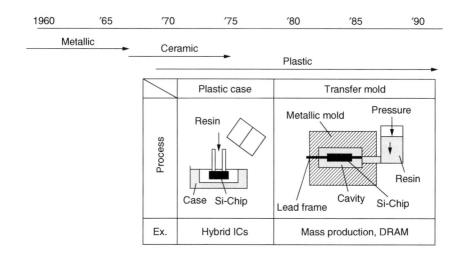

FIGURE 9.38 Historical road map of the IC packaging evolving from metallic to ceramic then plastic molding.

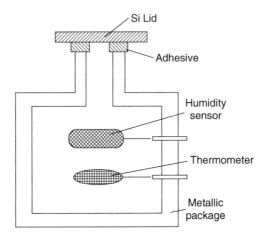

FIGURE 9.39 A metallic package with a thermometer and a humidity sensor inside was used to measure the moisture transparency of several adhesives (epoxy, acrylic resin, and silicone gel).

completed [23,24]. Another method is to encapsulate the optoelectronic chips with silicone gel. The silicone gel exhibits intimate contact to the surface of the materials, and no void is generated. Thus, no condensation takes place, preventing the ionic chemical reaction that results in corrosion [20,25,26]. These silicon gel encapsulation techniques are widely used in hybrid ICs, for example, in air flow sensors in an automobile engine compartment.

9.7.2 THERMAL RESISTANCE

The thermal resistance of the plastic package is an important design consideration because of the low thermal conductivity of the plastic relative to that of metal. Figure 9.40 shows the measured thermal resistance of the high- (A) and low- (B)

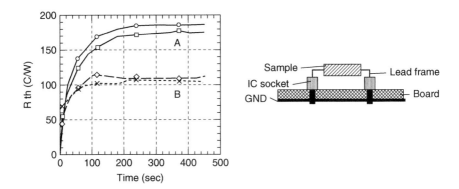

FIGURE 9.40 Measured thermal resistance of plastic MiniDIL transmitter packages with a lead frame made of metallic materials A and B.

conductivity metallic lead frame [27]. The metal B shows lower thermal resistance and indicates that the heat transfers more easily through the metallic lead frame than through the plastic resin. The thermal resistance of the plastic package using lead frame of metal B is close to that of conventional ceramic hermetic packages.

9.8 RELIABILITY

The reliability of the plastic packaging is one of the most important issues from the aspect of the degradation of optoelectronic components and fiber-fixing mechanical stability. Figure 9.41a shows the results of reliability testing under automatic constant current (ACC) operation of the plastic optical transmitter modules at 85°C and 85% relative humidity (RH). The monitored fiber output power deviation was less than ±0.5 dB during more than 2000 h of operation for both the plastic case type and the transfer mold type. Figure 9.41b shows the results of the temperature cycling from −40 to +85°C each for more than 3000 h. These results demonstrate that the Si V-groove fiber-fixing method is reliable for optoelectronic plane mountings.

The established testing standard for optical modules is the well-known Bellcore or Telecordia standard. However, this standard is limited to hermetically sealed packages, and hence a new standard for nonhermetic packages is necessary. We propose that the new standard should be based on the biased testing such as the findings described in this chapter. In other words, the new standard can be the testing

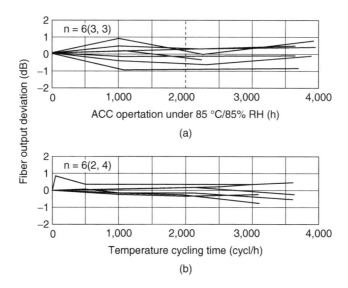

FIGURE 9.41 Reliability testing results for (a) ACC operation at 85°C and 85% RH condition, and (b) temperature cycling testing. Fiber output deviation was less than ±0.5 dB during more than 2000 h of automatic constant current operation and more than 3000 h of heat cycling for both the preinjection mold plastic case type and the transfer mold type.

condition for the plastic-sealed electronic ICs used for the operation of the hermetically sealed optical modules because those ICs are operated on the same circuit board as the hermetically sealed optoelectronic components that satisfy Bellcore standard.

9.9 CONCLUSIONS

Plastic MiniDIL transmitter modules were proposed and fabricated as low-cost solutions. Their operating performances were confirmed to satisfy the OC 3 (STM 1) and OC 12 (STM 4) specifications. The reliability testing results demonstrate that they work longer than 2000 h under biased 85°C, 85% RH conditions. These results indicate that the plastic packaging technologies used in the field of ICs can also be applied to the optical modules for the fiber-optic communication systems with high reliability. The choice of the preinjection mold case type and transfer mold type comes from the size of the production. The preinjection mold case type can be applied to moderate volume production such as hybrid ICs, and the transfer mold process can be used for mass production, like DRAMs.

In addition to Bellcore's hermetic testing standards, nonhermetic standards need to be established for the plastic module market to be successful. These plastic packaging technologies can be extended to other cost-driven optical modules like detachable-type MiniDIL [29] array-type modules, and the modules for the data communication systems that require more cost reduction.

REFERENCES

1. J. Yoshida et al., 7th International Workshop on Optical Access Networks, 6.1-1, 1995.
2. Sigeta Ishikawa, "A Passive Alignment Technique Using Au/Sn/Au Multilayer Solder for LD to Fiber/Waveguide Coupling" in Technical Digest Series of the Integrated Photonics Research, Vol. 3, pp. 365–366, 1994.
3. K. Kurata, "A Surface Mount Type Single-Mode Laser Module Using Passive Alignment" in Proceedings of the Conference 6th International Workshop on Optical Access Networks, S 3.4–1, 1994.
4. K. Kurata et al., "A Hybrid Integrated Bi-Directional Transmitter/Receiver Optical Module Based On Silica Waveguide Using Alignment-Free Hybrid Assembly Technique" in Technical Digest of OFC'96, WL3, pp. 168–169.
5. M.S. Cohen et al., "Passive Laser-Fiber Alignment by Index Method," Transactions Photonics Technology Letters, Vol. 3, No. 11, pp. 985–987, November 1991.
6. K. Kurata et al., "A Surface Mount Type Single Mode Laser Module Using Passive Alignment" in Proceedings of the 45th ECTC, pp. 759–765, May 1995.
7. H. Honmou et al., "Optical Coupling of Laser Diode Array to Single-Mode-Fiber Array with Heat-Treated Hemispherical Micro Lens" Electronics Letters, Vol. 31, No. 10, pp. 793–794.
8. Y. Sakai et al., "Improved FFP of a 1.3-/spl mu/m Spot-Size Converted Laser for Highly Efficient Coupling to Optical Fiber" in Technical Digest of OFC '96, WL9, pp. 175–176.

9. Y. Haung et al., "External Optical Feedback Resistant Characteristics in Partially Corrugated-Waveguide Laser Diodes" in Electron Letters, Vol. 32, No. 11, pp. 1008–1009, 1996.

10. I. Hatakeyama et al., "A Low Crosstalk 8-channel Array Optical Transmitter and Receiver for Star Topology Access Networks" in Technical Digest of OFC'96, TuM5, pp. 66–67.

11. H. Yanagisawa et al., "Wide Dynamic Range 156 Mbit/sec Burst Mode Optical Receiver Module" in Proceedings of the 1995 Electronics Society Conference of IEICE, 1995.

12. N. Uchida et al., "Passively Aligned Hybrid WDM Module Integrated with a Spot-Size Converted Laser Diode and Waveguide Photodiode on a PLC Platform for Fiber-to-the-Home" OFC '96, PD15.

13. Y. Yamada et al., "Silica-on-Terraced Platform for Optical Hybrid Integration" in Technical Digest of OEC'94, pp. 326–327, 1994.

14. N. Kitamura et al., "Silica-Based Optical Waveguide Devices with Novel Fiber Guide Structure for Alignment-Free Fiber Coupling" in Proceedings of IPR'96, Ith B2 pp. 608–611, 1996.

15. S. Turley et al., "High Stability Low-Cost Optical Module Packaging for the Access Market," Technical Digest, OECC '96 (Makuhari Messe), 18D3-4, pp. 408–409, 1996.

16. J.V. Collins et al., "New Technology Developments Make Passive Laser/Fiber Alignment a Reality," Proceedings SPIE, Vol. 2610, pp. 108–116, 1995.

17. M. Fukuda et al., "Pigtail Type Laser Modules Entirely Molded in Plastic," Technical Digest ECOC '95 (Brussels), We.A.3.5, p. 549, 1995.

18. K. Tatsuno et al., "Low-Cost Plastic Receptacle Type Transmitter Module for the Optical Subscriber Access Network Systems," Technical Digest, OECC'96 (Makuhari Messe), 18P20, pp. 460–461, 1996.

19. K. Tatsuno et al., "High-Performance and Low-Cost Plastic Optical Modules for the Access Network Application," Technical Digest, OFC'97 (Dallas), WB3, pp.111–112, 1997.

20. H. Nakamura, "Highly Reliable Operation of InGaAlAs Mesa-Waveguide Photodiodes in Humid Ambient," Material Research Symposium Proceeding, Vol. 531, pp. 317–325, 1998.

21. M. Aoki et al., "Reliable Wide Temperature Range Operation of 1.3 μm Beam Expander Integrated Laser Diode for Passively Aligned Optical Modules," IEEE Journal of Selected Topics in Quantum Electronics, Vol. 3, No. 6, pp. 1405–1412, 1997.

22. T. Ishikawa, M. Shimaoka, and K. Fukuda, "Estimation of Water-Resistant Adhesion for Optical Module," Proceedings of the IEICE (in Japanese) C-220, p. 220, 1996.

23. J.W. Osenbach et al., "Temperature-Humidity-Bias Behavior and Acceleration Factors for Nonhermetic Uncooled InP-Based Lasers," Journal of Lightwave Technology, Vol. 15, No. 5, pp. 861–873, 1997.

24. J.W. Osenbach and T.L. Evanosky, "Temperature-Humidity-Bias Behavior and Acceleration Model for InP Planar PIN Photodiodes," Journal of Lightwave Technology, Vol. 14, No. 8, pp. 1868–1881, 1996.

25. K. Otsuka et al., "The Mechanisms of That Provide Corrosion Protection for Silicone Gel Encapsulated Chips," IEEE Transactions on Component, Hybrids, and Manufacturing Technology, Vol. CHMT-12, No. 4, pp. 666–671, 1987.

26. J.W. Balde, "The Effectiveness of Silicone Gels for Corrosion Prevention of Silicon Circuits: The Final Report of the IEEE Computer Society Computer Packaging Committee Special Task Force," IEEE Transactions on Component, Hybrids, and Manufacturing Technology, Vol. CHMT-14, No. 2, pp. 352–365, 1991.

27. K. Fukuda et al., "Evaluation of Thermal Resistance in Plastic Optical Modules," Proceedings of the IEICE (in Japanese) C-3-69, p. 254, 1997.

28. K. Tatsuno et al., "High Performance and Low-Cost Plastic Optical Modules for Access Network System Applications," IEEE Journal of Lightwave Technology, Vol. 17, pp. 1211–1216, Issue 7, 1999.

29. K. Tatsuno et al., "Fiber Pigtail-Detachable Plastic MiniDIL Transmitter Model Optical Connector," IEEE Journal of Lightwave Technology Vol. 21, (4) pp. 1066–1070, 2003.

Section 3

Utilities for Passive Alignment

10 Large Spot Devices, Mode Transformers, and Optics

John V. Collins

CONTENTS

10.1 INTRODUCTION

The semiconductor chip-to-fiber alignment process is the main contributor toward the cost of a pigtailed device. The alignment process should give good stable coupling of the laser output into the fiber, and for monomode devices, this usually means an active alignment process with submicron positional accuracy. In simple devices, such as a laser transmitter, the semiconductor chip itself is not

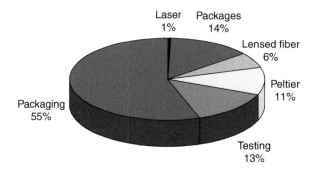

FIGURE 10.1 Proportionate cost of laser transmitter.

a major factor in the cost of the complete pigtailed package. This is shown in Figure 10.1, where the cost of producing a laser transmitter has been analyzed, it is obvious that the cost of packaging dominates the device price, and the chip cost is a very small fraction of the overall total. Developments of the semiconductor chip structure can help make the fiber alignment process less dependent on positional tolerances, and so give potentially lower packaging costs. This chapter examines a number of developments in laser device processing, which, together with novel package designs described in previous chapters, open up opportunities for reduced component manufacturing costs.

Conventional lasers emit their radiation from a mode spot around 1×1.5 μm in size that does not match well to the mode spot size of a cleaved single-mode fiber (~8 μm) as shown in Figure 10.2. In general, some form of optics is used between the laser and the fiber to match these different mode spot sizes. A variety are shown in Figure 10.3. These optical elements usually increase the optical coupling from the laser into the fiber from around 10% with no optics ("butt coupled" to a cleaved fiber end) to over 50%; however, this increase in coupling is achieved at the expense of tightening the positional tolerances between the laser and the coupling element. The simplest and most widely used optical element is the lens formed on the fiber end. Figure 10.4 shows a variety of lensed fiber ends.

Figure 10.5 shows how the coupling efficiency between a laser and lensed fiber varies with the offset of the fiber from its optimum coupling position.

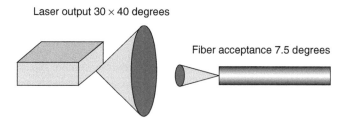

FIGURE 10.2 Schematic of laser and fiber optical modes.

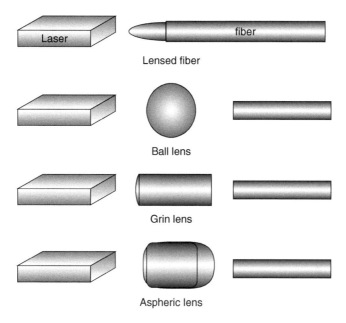

FIGURE 10.3 Optical coupling elements.

Movements of less than 1 μm result in a loss of coupling efficiency of 3 dB. It is these tight alignment tolerances that make the coupling problem so acute and expensive.

Several laboratories have developed lasers with the mode-matching element fabricated on the laser chip itself (denoted as large spot size lasers), removing

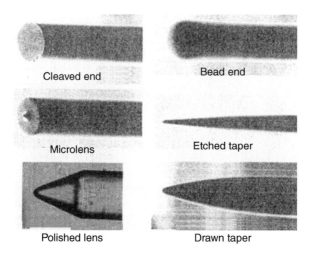

FIGURE 10.4 Types of lensed fiber.

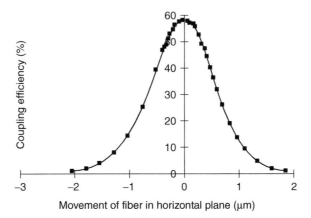

FIGURE 10.5 Laser-to-fiber coupling tolerance.

the need for the separate alignment of an optical element but still yielding good coupling efficiency to a cleaved optical fiber. There are several methods for fabricating the mode-matching element on the laser. These will be discussed, comparing the ease of fabrication with coupling characteristics and the relaxation of position alignment tolerance.

10.2 LARGE SPOT LASERS

Figure 10.6 shows a typical conventional semiconductor laser structure. The small size of the active region combined with the refractive index step down to the cladding layers cause a highly divergent output beam pattern to the light. This pattern is usually elliptical, which further complicates the coupling (of the laser) to the circular mode of the optical fiber.

There are several possible methods to expand the laser spot size. The laser active region itself may be modified, or a separate region may be fabricated beyond the laser active area in which the mode is allowed to expand. These expansions will serve to better match the optical modes between the laser and the fiber.

10.2.1 MODIFIED ACTIVE REGIONS

10.2.1.1 Dilute Guides

Lasers may be fabricated in which the guiding of the optical mode is weak. The optical mode is thus larger than in the tightly confined devices, and hence the output radiation is less divergent, leading to better coupling into the optical fiber. This works quite well in GaAs lasers. However, in InP lasers, because of the weaker guiding, these devices usually have higher threshold currents, and their temperature performance is poor. These devices achieve their increased coupling performance at the expense of other parameters and can struggle to meet the criteria for stable optical telecommunication system operation.

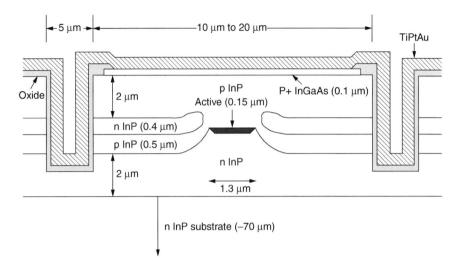

FIGURE 10.6 Typical buried heterostructure laser layer structure.

10.2.1.2 Lateral Tapers

Figure 10.7 [1] shows a number of methods of changing the width of the active guide in a laser diode. As the waveguide width tapers down, the optical mode expands. This is quite easy to achieve in structures such as those in Figure 10.7a. There is a little change to the spot size until the active stripe width is very narrow. The index of the guided mode is similar to the substrate, and hence there can be mode leakage and high losses. However, various authors have shown this approach can give an increase in spot size [25–29] and quite acceptable coupling efficiencies to cleaved optical fibers (~ 3 dB). Structures 10.7c, d, e are compound guides where the active stripe is tapered and the optical mode transfers into an underlying guide. In the ridge waveguide structures, it is necessary to make a tight guide cutoff.

Lealman et al. at BT laboratories [2] have produced a buried heterostructure (BH) device in which a passive guide is grown beneath the laser active layer and the laser active region is tapered in the horizontal plane to force the optical mode to expand in both the horizontal and vertical directions shown in Figure 10.8. This allows the optical mode to couple into the underlying passive guide where the mode is free to expand. The taper length and the length of the passive region are chosen so that the laser mode at the facet matches better to the mode size of the fiber.

The optical mode size of the laser has thus been closely matched to that of an optical fiber, resulting in coupling losses as low as −1.8 dB. Figure 10.9 shows the coupling between a large spot size laser and a cleaved optical fiber compared to an ordinary laser and a polished lensed fiber. It can be seen that the coupling efficiencies are similar, but the tolerance to movement from the optimum position was increased when the large spot laser is used.

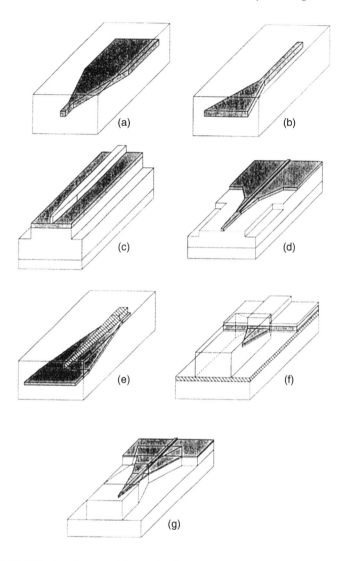

FIGURE 10.7 Various large spot laser structures.

The following section describes this work in detail. Increasing the spot size of the laser required both the passive guide dimensions and taper length to be altered. The dimensions of the passive guide were determined by the need to closely match the near field laser spot size to that of the cleaved fiber. A Gaussian profile with a 9-μm 1/e diameter was used to simulate the electric field of the guided mode in the single-mode fiber. The laser spot size was then calculated, using a variable grid finite difference program. A 1.1-μm wavelength composition quaternary guide, 7-μm wide, and 0.07-μm thick, gave a close match to the fiber-guided mode in the direction parallel to the active junction. This is the x direction

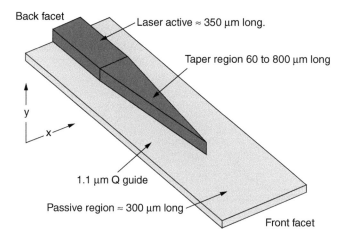

FIGURE 10.8 Schematic of laser active region.

in Figure 10.8. The match between the fiber and laser is not so close as to be perpendicular to the active layer. The perpendicular is the *y* direction in Figure 10.10. To estimate the coupling loss, an overlap integral was determined between the fiber and device fields for both profiles. The product of these overlaps yielded a total coupling loss of 0.7 dB for the device. Calculations of field overlap between the passive guide and the tapered active layer of the laser showed that an active width of ≤ 0.2 µm was required at the end of the taper if the coupling loss between

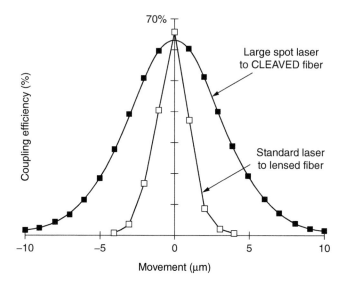

FIGURE 10.9 Laser-to-fiber coupling results.

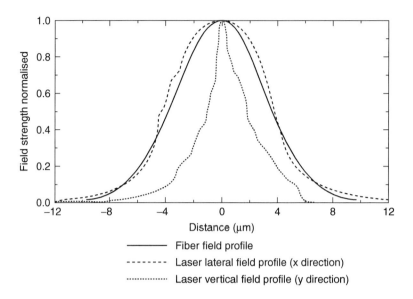

FIGURE 10.10 Calculated field profiles for 10-μm core single-mode fiber and the large spot lasers.

these two sections was to be kept below 1 dB [3]. Calculation of the critical taper length for this structure and a minimum active width [3,5] yielded a minimum taper length of ≈ 440 μm.

10.2.1.2.1 Growth and Fabrication

The planar wafer design used in this device was similar to that of [4], employing an eight-well strained multi-quantum-well (MQW) active. The planar was epitaxially grown on an n doped InP substrate by atmospheric pressure metal organic vapor phase epitaxy (MOVPE) and consisted of a 3 μm thick n doped InP buffer layer, a 0.07 μm thick n doped 1.1 μm wavelength quaternary guide, a 0.2 μm thick n doped InP spacer layer, an undoped 8 well strained MQW active layer, and 0.2 μm thick p doped InP cap. The MQW active layer consisted of 8 wells of $In_{0.84}Ga_{0.16}As_{0.68}P_{0.32}$, which had a mismatch of +1% with respect to InP, and a nominal well thickness of 70 Å. The 140 Å thick 1.3 μm wavelength quaternary barriers were strained −0.5% relative to the substrate to compensate for the strain in the wells. The MQW was surrounded by a 100 Å thick SCH with the same composition and strain as the barriers. This layer structure was fabricated into BH lasers (as reported in [3]), and the thickness of the second burying regrowth was 5 μm to accommodate the increased laser spot size.

10.2.1.2.2 Results

Devices were bonded junction side up for continous wave (CW) assessment. Linear and three-section tapers with taper lengths of 60, 180, 400, and 800 μm

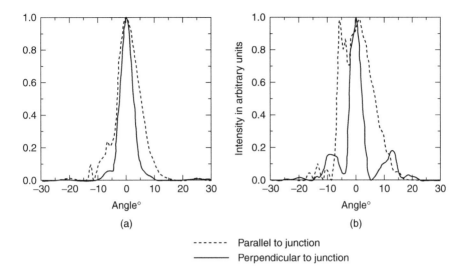

------- Parallel to junction
——— Perpendicular to junction

FIGURE 10.11 Farfield angles for large spot lasers with (a) a 400-µm taper and (b) a 60-µm long taper.

were employed, as well as devices cut with 300-µm-long untapered active and passive sections. In a linear tapered device, the active width decreases uniformly along the taper section, whereas in a three-section tapered device, three linear gradients are used to more closely follow the calculated adiabatic taper profile. Farfield measurements at 50 mA and 20°C showed a reduction from ≈ 27° to ≈ 5.5° by 10° at the full width at half maximum point (FWHM) for the tapered devices with 400- and 800-µm tapers (Figure 10.11a).

These compare to Farfield angles, based on the calculated nearfield of 7.1° by 10.7°. The FWHM of single-mode fiber is ≈ 7.5 µm, indicating that the laser spot is slightly larger than intended in the x direction. Although the shorter tapers had similar farfield angles, there was evidence of higher-order modes, indicating that these taper lengths were insufficient for this larger spot size (Figure 10.11b). No significant difference was observed in farfield angle between the linear and three-section tapers. Coupling measurements carried out to 10 µm of core-cleaved fiber showed Fabry-Perot ripple due to reflections from the cleaved fiber facet. Fiber was butt coupled to the chip and then moved out along the z direction axis to obtain the peak and trough of the ripple. Typical ripple observed was ≈ 2.2 dB. The coupling loss was calculated from the mean of these two values in mW. The results obtained are shown in Figure 10.12, with a minimum coupling loss of ≈ 1.8 dB being obtained for a device with an 800-µm taper. At the time of printing, this is the lowest-ever coupling loss obtained between a laser and cleaved 10-µm core fiber. The increased coupling loss observed for the shorter (60 and 180 µm) tapers confirms that these taper lengths are indeed too short and indicates that the approximation of the calculation method proposed in reference [3] is reasonable.

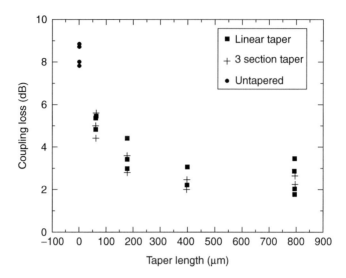

FIGURE 10.12 Coupling loss measured to 10-μm core-cleaved fiber as a function of taper length measured at 50 mA drive current and at 20°C.

Although the larger-than-desired spot size in the lateral direction resulted in the coupling loss being ≈ 1 dB larger than the predicted 0.7 dB, it also results in an increased lateral alignment tolerance of 5.5 μm (3 dB), compared to 3.5 μm in the vertical direction. This increased lateral alignment tolerance further simplifies packaging and allows low-cost techniques such as pick-and-place assembly. The ripple observed with these measurements does, however, indicate that if cleaved fiber is to be used with these devices, then they will need to be AR facet coated, cleaved at an angle, or glued to the laser facet with glue index matched to the fiber. Further reduction in coupling loss may be achievable by more closely matching the guided mode of the laser in the direction perpendicular to the active junction to that of the fiber.

This work demonstrates a large spot size laser with a spot size closely matched to that of 10-μm core-cleaved fiber. A coupling loss of 1.8 dB has been observed for a device with an 800-μm-long taper. This represents the lowest reported coupling loss for a semiconductor laser butt coupled to 10-μm core-cleaved fiber. A 3-dB alignment tolerance of 5.5 μm was obtained in the direction parallel to the active junction. The fabrication processes described above are typical for laterally tapered devices. The lateral tapering is generally achieved by photolithographic masking and wet or dry etching. The end of the taper is usually the most critical point of the processing as it needs to be either as fine as possible or, as in the example above, designed to be cut off at a given small value. Usually the wafer has to be returned to the growth chamber after the laterally tapered active regions have been defined for the cladding layers to be laid down.

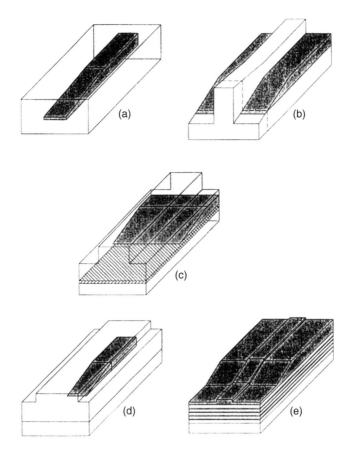

FIGURE 10.13 Schematics of vertical tapers.

10.2.1.3 Vertical Tapers

Figure 10.13 [1] illustrates several of the approaches taken to change the thickness of the guiding layers along the device length. The decrease in the thickness again allows the optical mode to expand. In the ridge waveguides (Figure 10.13c, d) the tight mode is cut off, forcing the mode to transfer to an expanded guide structure.

10.3 FABRICATION TECHNIQUES OF VERTICALLY TAPERED LASERS

Vertically tapered layers can be fabricated by either etching or growth techniques.

10.3.1 Etching

The simplest technique is to dip the sample into an etch in a controlled way [6]. However, this usually produces long, shallow tapers and cannot be used to process

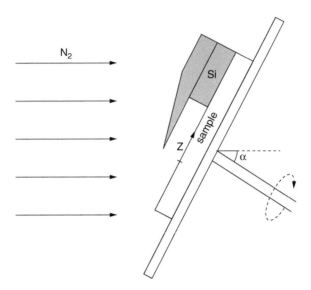

FIGURE 10.14 Ion beam shadow etching [12].

full wafers. Other etching methods include dynamic etch masking [7], diffusion limited etching [8], or stepped etching [9].

Dynamic etch masks gradually dissolve in the etch material, exposing more semiconductor to the etching. Diffusion-limited etching is a self-limiting etch through a mask, and stepped etching involves subsequent masking and etching stages. All of these methods have significant drawbacks. The stepped etching is very time consuming and the others are not very reproducible.

Dry etching can be better controlled and shadow etching techniques have been developed [10,11]. Basically, tapering can be achieved by etching at an angle through a mask held above the sample surface (Figure 10.14). Uniformity across a full wafer is often hard to obtain, and the etching techniques tend to be interesting laboratory experiments that are not controllable enough to consider for laser manufacture.

10.3.2 GROWTH

Growth methods of producing tapered structures are usually more controllable and can be applied evenly over entire wafers, making them more suitable as manufacturing techniques.

Thickness and compositional changes can be obtained during growth by applying a temperature gradient across a sample. By using a profiled mounting plate [13] (Figure 10.15) in a growth chamber, periodic variations of temperature can be induced in the wafer, and tapers can be fabricated. In this case, the compositional change (and hence refractive index change) results in the majority of the change in the optical mode shape.

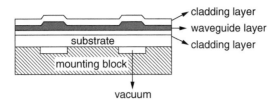

FIGURE 10.15 Temperature-induced tapers.

Selective area growth (SAG) [14] is another method of producing tapers. If areas of the wafer are masked with silicon oxide (Figure 10.16) [15], a growth-rate enhancement can be achieved by limiting the mask window and changing the reactor pressure. Shadow masked growth (SMG) [16] is a similar technique, in which layer thickness variations can be obtained by growing through a mechanical or monocrystalline mask. Figure 10.17 shows the regime for growth through a mechanical mask [17]. Extra processing steps are required to define and remove the monocrystalline mask. In both SAG and SMG, the mode tapering is attained by a combination of thickness and compositional changes. SAG tends to be employed in low-pressure MOVPE systems, whereas SMG is usually atmospheric pressure growth.

Kobayashi et al. [24] describe the results they obtain with a tapered-thickness waveguide MQW laser grown by the SAG technique. Both the gain and tapered-thickness MQW waveguide regions are simultaneously grown on an n-doped, InP-patterned substrate. Openings between 10 to 20 μm wide were formed in a SiO_2 mask to deposit the active layers while the mask was completely removed

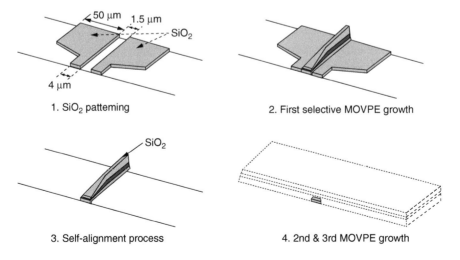

FIGURE 10.16 Selective area growth [15].

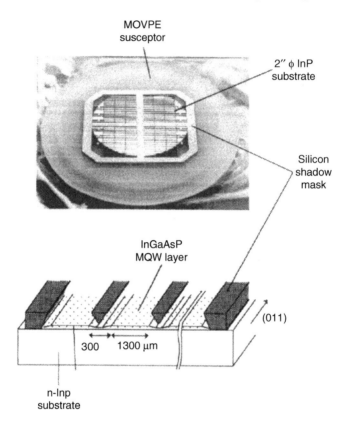

FIGURE 10.17 Shadow mask growth [17].

where the waveguides were deposited. SAG employing trimethylindium, triethyl-garium, arsine, and phosphine as the source materials at 0.1 atmospheric pressure and 600°C gave the thickness and photoluminescence wavelengths shown in Figure 10.18.

The lasers yielded thresholds of 6.5 and 22.2 mA at 25 and 85°C, respectively, and lased up to 120°C. The full-width, half-maximum angles of their emitted radiation were 8 and 9° in the lateral and vertical directions. The coupling loss into a single-mode optical fiber was −2.2 dB with a tolerance of ±2 μm for 1-dB degradation in coupling.

10.4 COMBINED TAPERS

In combined tapers, the lateral and vertical dimensions of the guiding layers are changed. SAG and SMG can be used to taper the layers in both lateral and vertical directions simultaneously to optimize the mode expansion. Good results can be obtained [14,18]; however, the growth and fabrication are complex, and this approach does not appear to have been adopted for production.

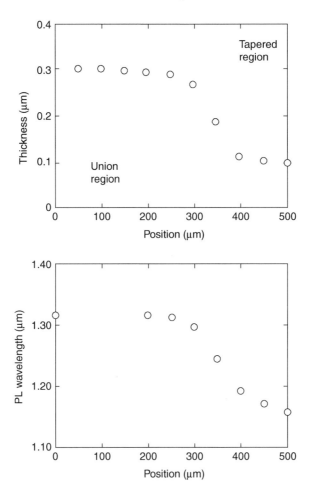

FIGURE 10.18 Thickness and compositional changes from selective area growth [24].

10.5 SEPARATE BEAM EXPANSION REGIONS

Selective area growth techniques can be used to define separate laser active and mode expansion regions. If the laser layers are grown and then masked, the regions outside the mask can be removed and a taper layer structure grown [19]. The growth rate is enhanced near the dielectric mask, and hence the regrown layers will be tapered. Obviously, this involves at least one regrowth. In fact, the process usually involves two regrowths, as the cladding layers are often grown separately because the butt joint of the first regrowth is a critical step and requires much growth optimization. This method does, however, allow the separate design of the device active and taper layers. Table 10.1 [1] shows a review of the work of a number of laboratories around the world.

TABLE 10.1
Selection of Work on Expanded Beam Lasers

Group	Technology	λnm	Type	Ith (mA)	QE (%)	CL (dB)	Alignment Tolerances (μm)	Remarks	Reference
Laterally tapered buried waveguide									
NTT	Direct e-beam writing + RIE	1550	DH, DBR, BH	9	21	2.8	2 × 2	FF 12 × 12	[25]
NTT	Stepper litho + RIE	1300	DH, P, BH	4.7	57	2.3	2.8 × 3.2	HR coat, high temperature 85°C	[26]
AAR	Chemical undercut of SiO$_2$ mask + RIE	1480	MQW, BRS	25	52	5.2		HR coating FF 13 × 15	[27]
Matsushita	Lateral taper over whole cavity	1300	MQW, PBH	6.9	6.5	4	1.7 × 1.7	HR coating FF 13 × 13	[28]
NEC	Chemical etching	1300	MQW, PBH	3	63	6.2	2	HR coating FF 14 × 16	[29]
Single lateral taper from ridge to FM waveguide									
Sandia NL	Direct e-beam + RIBE	980	GRIN SCH ridge	110	19.8	0.9		FF 5.6 × 7.4	[30]
Hitachi	Wet etching	1300	MQW ridge	25	62	4.8	2.4 × 1.5	HR coatings FF 5.4 × 19	[31]
Hitachi	Wet etching	980	DQW ridge	19	63			FF 10 × 15 high power 390 mW	[32]

Lateral taper plus buried waveguide

DT	e-beam+RIE+regrowth	1530	DFB MQW	40	14.8	1.85	2.35 × 2.15		[33]
NTT	SAG	1300	DH, BH	21.4	26.2	2.4	2 × 2	Two-inch wafer	[34]
AT&T	Knife edge litho + selective etching + regrowth	1550	MQW PBH	12	52	3.5	3.1 × 2.6	FF 5 × 7	[35]
BTL	Wet etching	1300		22	51	3.5	2.8 × 2.5	FF 6 × 8	
		1550	MQW PBH			1.8	5.5 × 3.5 (3 dB)	FF 5.5 × 10	[2]
CSELT	Two litho + RIE	1550	MQW BR	30	21	2.2	2.3 × 2.15	FF 6.6 × 11	[36]

Vertical down tapered buried waveguide

NTT	SAG butt joint	1300	MQW PBH	7	33.3	1.06	2 × 2	HR coating FF 7 × 10	[37]
NTT	SAG butt joint 2 inch wafer	1300	MQW PBH	5.6		0.94	2 × 2	Two HR coats 50 mW high temperature 134°C	[38]
AT&T	Stepped etching	1500	MQW PBH	28	17	5	3 × 2	FF 10 × 15	[39]
Hitachi	Si shadow masked growth	1300	MQW PBH	16	56	3.5	2 × 2	FF 12 × 13 high temperature 85°C	[40]
Fujitsu	Selective area growth	1300	MQW PBH	6.5	40	3.8	2 × 2	HR coating FF 9 × 11	[41]

(*continued*)

TABLE 10.1 (Continued)
Selection of Work on Expanded Beam Lasers

Group	Technology	λnm	Type	Ith (mA)	QE (%)	CL (dB)	Alignment Tolerances (μm)	Remarks	Reference
University of Gent	Shadow masked growth	1550	MQW PBH	8.2	31.5	3.3	2.1 × 1.7	FF 12 × 16	[42]
Vertical down tapered ridge waveguide									
ETHZ	Diffusion limited etch + wet etch (including lateral taper)	1300	DH ridge	55	8.5	3	1.8 × 1.8		[43]
Mitsubishi	LPE growth on ridge	780	BH BTRS	60	42			AR-HR coating FF 9 × 10	[44]
University of Gent	Shadow masked growth	980	SQW ridge	11.2	36			FF 15 × 29	[45]
Vertically down tapered and laterally up tapered waveguide									
AT&T	Stepped etch (vertical), wet etch (lateral)	1500	MQW DBR PBH	52	30	4.2	3.8 × 2.1	HR coating FF 12 × 12	[39]

10.6 PASSIVE ALIGNMENT

If the laser and fiber could be assembled by a "passive" method, the cost would be considerably reduced. To achieve passive alignment, the laser has to be bonded to a carrier so that the position of its emitting area is accurately determined, and the fiber has to be fixed to this carrier so that the light emitted by the laser is captured by the fiber core.

There are several methods to achieve this, most of which have been described in previous chapters (e.g., solder-bump flip-chip bonding and infrared aligned flip-chip bonding). This section describes how a laser can be dimensioned extremely accurately and how a micromachined carrier, known as a silicon optical bench, can be used for "passive alignment" of a fiber to a laser.

10.7 PRECISION CLEAVING OF LARGE SPOT LASERS

For passive alignment to be feasible, it is vital to know exactly where the light-emitting area of the laser is with respect to some feature on the laser. The normal way of separating lasers into individual die from the growth wafer is to use a diamond to initiate a crack along a crystallographic plane in the wafer and then to flex the wafer in this direction to propagate the crack right through the wafer. The positioning of the diamond and the exact position from which the crack starts could only be controlled to approximately ±10 μm, which is obviously nowhere near accurate enough. Therefore, a novel technique that can define its dimensions to submicron accuracy has been developed to cleave up the laser die [20]. This is achieved by defining a cleave channel at the same time that the active and passive areas are produced.

After the growth of the passive laser wafer is completed and the passive layer is etched to define the passive guide beneath the laser active region, a cleave channel is opened by etching away the passive layer between adjacent lasers at the same time (using the same photoresist mask). The laser wafer is returned to the epitaxial growth kit, and the remaining layers are grown over the etched passive layer. The laser fabrication is completed, including the metallization layers, and then finally a window is etched between the individual lasers, down to the gap in the passive layer. A selective etch is then used to etch a deep channel in the laser wafer through the gap in the passive layer, the edge of which is not affected by the etch. These deep channels are then used to propagate cleave planes in the laser wafer by flexing the wafer (Figure 10.19).

The unetched corner of the passive layer can then be used as a reference edge for bonding the laser (i.e., it is this edge that is pushed up to the alignment stop on the silicon optical bench). The distance of this edge to the laser-emitting area was defined in one photolithographic step and is therefore accurate to less than 0.5 μm. A scanning electron micrograph of the corner of a precision-cleaved laser diode is shown in Figure 10.20.

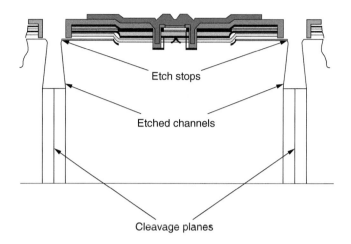

FIGURE 10.19 Schematic of precision cleaving.

FIGURE 10.20 Scanning electron micrograph of the corner of a precision-cleaved laser.

10.8 SILICON OPTICAL BENCH

This section describes some work from a paper presented at the OFC '95 conference [20] and an associated *Electronics Letter* [21], which detailed how greater than 50% coupling between a laser diode and a cleaved-ended single-mode optical fiber was achieved, using passive alignment techniques.

A precision-cleaved, large spot size laser was mounted on a silicon optical bench that has an alignment stop against which to bond the laser, and a V-groove in which to lay the cleaved fiber (Figure 10.21). An oxide alignment stop needed to be tall enough to be an effective stop for the laser to be bonded to (Figure 10.22).

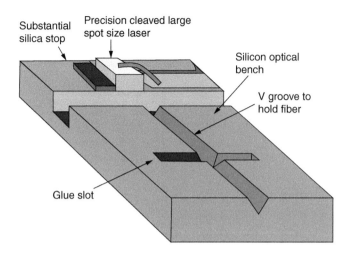

FIGURE 10.21 Schematic of silicon laser optical bench.

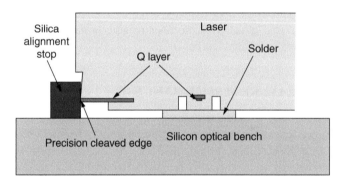

FIGURE 10.22 Schematic of alignment of laser on optical bench.

The height of the alignment stop was dictated by solder height, and it was found that 3 μm of solder gave a good void free bond. The thicknesses of the epitaxial layers and metal layers of the laser set the distance from the laser surface to the emitting area (approximately 5.5 μm). This vertical distance also dictated the width of the fiber V-groove, as the fiber core center needed to be at the same height as the laser emitting area.

10.9 LASER DIE BONDING

A precision-cleaved large spot laser was bonded to the bench using a conventional laser die bonder. The laser was held on the end of a vacuum pickup tool. It was viewed through a stereo microscope and was pushed against the alignment upstand.

FIGURE 10.23 Assembled silicon optical bench.

It was held in place while the solder was melted. The heat to melt the solder was applied by passing current through a carbon strip heater; when the current was turned off, the solder solidified and the laser was released from the pickup.

The fiber was then laid in the V-groove, and a quick-setting epoxy was applied. The epoxy was fed under the fiber so that as it set, it pulled the fiber into the groove. A silicon lid was also glued over the fiber to give protection and robustness. Figure 10.23 shows an assembled bench.

10.10 TEMPERATURE PERFORMANCE

Diode lasers are very temperature sensitive. As the temperature increases, the light output decreases; the lasers are therefore normally bonded onto diamonds to give the best possible heat sinking. Silicon is a reasonable heat sink, but its thermal conducitivity is considerably less than diamond (approximately 10% of diamond's conductivity), and there was a concern that the lasers would perform much worse on the silicon benches than on ordinary laser packages. The threshold currents of lasers bonded to a diamond heat sink and to a silicon optical bench were compared (Figure 10.24). It was found that the laser on the bench was slightly better than the laser on the diamond. This was probably because diamond is expensive and therefore small heatsinks are used, whereas the silicon bench, although of poorer conductivity, is physically much larger and thus is effectively a better heat sink.

10.11 COUPLING RESULTS USING LARGE
SPOT LASERS

Using the totally passive method described previously, coupling efficiencies as high as 55% have been obtained [20]. These are among the best reported results in the world to date for passively aligned devices. These coupling efficiencies were determined by measuring the amount of power in the optical fiber at a given

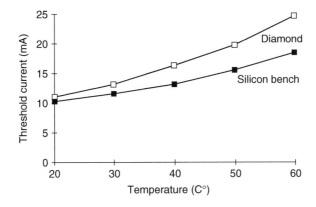

FIGURE 10.24 Comparison of diamond and silicon bench as heat sinks.

laser drive current (50 mA) and comparing this value to the average value of the total output from large spot size lasers bonded to conventional headers at the same drive current measured with a large area photodetector.

Electron micrographs of lasers bonded on to the optical benches are shown in Figure 10.25, and a histogram of the coupling results are shown in Figure 10.26. It can be seen that the results on the second wafer were much better. Two contributing

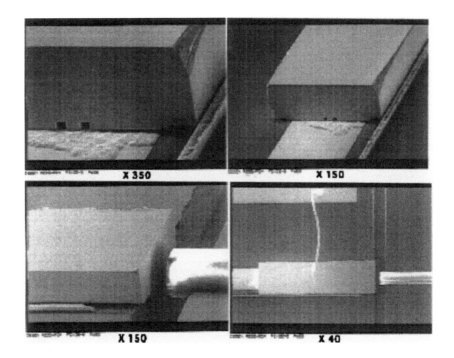

FIGURE 10.25 Scanning electron micrographs of laser on silicon bench.

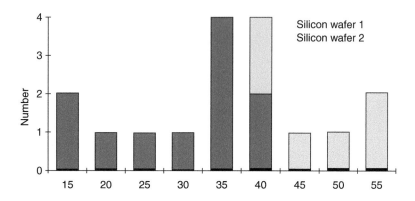

FIGURE 10.26 Coupling results of lasers on optical bench.

factors for this difference were that the solder thickness on the first wafer was half of the correct value, and some early devices were bonded at an angle to the silica stop. On later devices and on the second wafer, the laser was observed along the alignment edge during bonding, and the bonding was found to be easier and more accurate.

10.12 WIDER APPLICATIONS

So far, only the design and packaging of low-cost semiconductor lasers have been described. However, the technology developed has wider applicability than just to simple lasers. Future optical systems will involve large split passive optical networks and wavelength division multiplexing (WDM). The key components required if these networks are to become a reality are low-cost optical amplifiers and stable wavelength sources. Erbium fiber amplifiers are currently bulky, expensive components, but semiconductor optical amplifiers (SOA), which are much smaller, could be attractive if they could be realized at a very low cost with adequate performance. A very low cost SOA could dramatically affect the economics and topology of passive optical networks.

WDM systems require sources with well-defined wavelengths that remain stable over large operating temperature ranges. A potential wavelength stable source is the fiber grating laser (FGL). This component consists of a laser with a very low reflectivity facet coupled to a fiber in which a reflection grating has been written using etching to provide the feedback for lasing (Figure 10.27).

Because the feedback grating is in silica, its temperature coefficient of expansion and is about one order of magnitude less than indium phosphide, the wavelength temperature stability of an FGL is considerably better than a distributive feedback (DFB) laser. These devices could also be useful as upstream sources in PON systems in which the noise performance of amplifiers could be improved by narrowband optical filters if the wavelength of the sources is well defined.

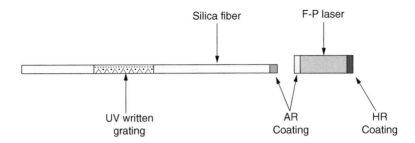

FIGURE 10.27 Schematic of fiber grating laser.

Both laser amplifiers and fiber grating lasers require very low levels of optical feedback into the semiconductor chip for stable operation. This is usually achieved by angling the device facet and applying a complex multilayer antireflection coating. The angled facet further complicates the normal packaging processes, and to date, devices have required very delicate hand assembly, resulting in low-yield, high-cost components. Using a large spot device, the angled facet has been shown to reduce the effective facet reflectivity [22], and the need for a complicated expensive coating has been removed. By using a modified passive alignment silicon submount, a very simple passively aligned amplifier and fiber grating laser have been produced. Figure 10.28 shows the fiber grating laser chip on its carrier. The alignment silica stop for the chip is fabricated at an angle with respect to the fiber V-groove to allow for the deflection of the laser chip emission by its angled facet.

The spectral output of a passively aligned fiber grating laser is shown in Figure 10.29. The side mode suppression ratio is seen to be greater than 30 dB.

The same technology has been applied to multiple laser amplifier arrays, as shown in Figure 10.30. Eight element amplifier arrays have been bonded to a silicon carrier, and silica-on-silicon waveguide splitter and delay devices have

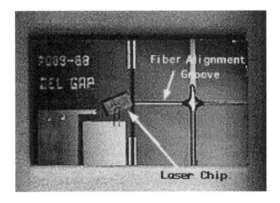

FIGURE 10.28 Laser chip on angled silicon bench.

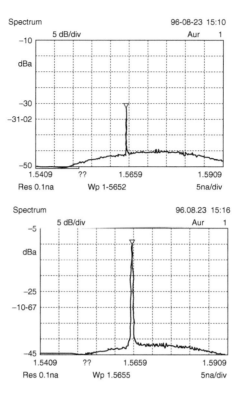

FIGURE 10.29 Spectral characteristics of passive fiber grating laser.

been passively aligned to the array to produce a "packet header pulse generator" which works at 100 Gbit/sec [23].

This array device (or any of similar complexity) would be virtually impossible to fabricate without the relaxed alignment tolerances of the large mode size

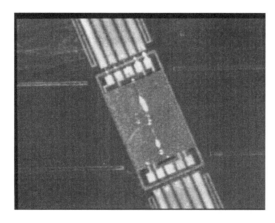

FIGURE 10.30 Laser amplifier array bonded on silicon carrier.

amplifiers and the simplified coupling obtained. Such devices offer up to 25 dB fiber-to-fiber gain with low polarization sensitivity.

10.13 SUMMARY

Matching the emission spot size of the semiconductor laser device to that of a single-mode fiber has been shown to be a key development on the path to low-cost packaging. The relaxed alignment tolerance that results has allowed the development of simple silicon submount-based passive alignment schemes for low-cost packaging. Examples of some of the methods of producing large spot lasers and techniques to realize passively aligned lasers, laser amplifiers, fiber grating lasers, and array devices have been discussed.

REFERENCES

1. I. Moerman, P. Van Daele, P. Demeester. "A review on fabrication technologies for the monolithic integration of tapers with III-V semiconductor devices." Journal of Selected Topics in Quantum Electronics, Vol. 3(6) Dec 1997 pp. 1308–1320.
2. I.F. Lealman, L.J. Rivers, M.J. Harlow, S.D. Perrin. "InGaAsP/InP tapered active layer multiquantum well laser with 1.8 dB coupling loss to cleaved singlemode fiber." Electronics Letters, Vol. 30 (20), 1994, pp. 1685–1687.
3. I.F. Lealman, L.J. Rivers, M.J. Harlow, S.D. Perrin, M.J. Robertson. "1.56 μm InGaAsP/InP tapered active layer multiple quantum well laser with improved coupling to cleaved single mode fiber," Electronics Letters, Vol. 30 (11), 1994, pp. 857–859.
4. I.F. Lealman, C.P. Seltzer, L.J. Rivers, M.J. Harlow, S.D. Perrin. "Low threshold current 1.6 μm InGaAsP/InP tapered active layer multiple quantum well laser with improved coupling to cleaved single mode fiber," Electronics Letters, Vol. 30 (12), 1994, pp. 973–975.
5. J.D. Love. "Application of a low-loss criterion to optical waveguides and devices," IEEE Proceedings — Optoelectronics, Vol. 136 (4), 1989, pp. 225–228.
6. T. Brenner, W. Hunziker, M. Smit, M. Bachmann, G. Guckos, H. Melchior. "Vertical InP/InGaAsP tapers for low-loss optical fiber-waveguide coupling." Electronics Letters, Vol. 28 (22), 1992, pp. 2040–2041.
7. M. Chien, U. Koren, T.L. Kock, B.I. Miller, M. Oron, M.G. Young, J.L. Demiguel. "Short cavity distributed Bragg reflector laser with an integrated tapered output waveguide." IEEE Photonics Technology Letters, Vol. 3 (5), 1991, pp. 418–420.
8. T. Brenner, H. Melchior. "Integrated optical modeshape adapters in InGaAsP/InP for efficient fiber to waveguide coupling." IEEE Photonics Technology Letters, Vol. 50 (9), 1993, pp. 1053–1056.
9. T.L. Koch, U. Koren, G. Eisentein, M.G. Young, M. Oron, C.R. Giles, B.I. Miller. "Tapered waveguide InGaAs/InGaAsP multiple-quantum well lasers." IEEE Photonics Technology Letters, Vol. 2 (2), 1990, pp. 88–90.
10. G. Muller, G. Wender, L. Stoll, H. Westermeier, D. Seeberger. "Fabrication techniques for vertically taered InP/InGaAsP spot-size transformers with very low loss." Proceedings of the European Conference on Integrated Optics ECIO '93, Neuchatel, Switzerland, p. 14.

11. B. Jacobs, R. Zengerle, K. Faltin, W. Weiershausen. "Vertically tapered spot size transformers by a simple masking technique." Electronics Letters, Vol. 31 (10), 1995, pp. 794–796.

12. G. Wenger, M. Schienle, J. Bellerman, B. Acklin, J. Muller, S. Eichinger, G. Muller. "Self-aligned packaging for an 8 × 8 InGaAsP/InP space switch." IEEE Journal of Selected Topics in Quantum Electronics, Vol. 3(6) Dec. 1997 pp. 1445–1456.

13. D.E. Bossi, W.D. Goodhue, M.C. Finn, K. Rauschenbach, J.W. Bales, R.H. Rediker. "Reduced-confinement antennas for GaAIAs Integrated optical waveguides." Applied Physics Letters, Vol. 56 (5), 1990, pp. 420–422.

14. R.J. Deri, C. Caneau, E. Colas L.M. Schiavone, N.C. Andreadakis, G.H. Song, E.C.M. Pennings. "Integrated optic mode-size tapers by selective organometallic chemical vapor deposition of InGaAsP/InP." Applied Physics Letters, Vol. 61 (9), 1992, pp. 953–954.

15. H. Yamazaki, K. Kudo, T. Sasaki, J. Sasaki, Y. Furushima, Y. Sakata, M. Itoh, M. Yamaguchi. "1.3 μm spot-size converter integrated laser diodes fabricated by narrow stripe selective MOVPE." IEEE Journal of Selected Topics in Quantum Electronics, Vol. 3(6) Dec. 1997 pp. 1392–1398.

16. G. Coudenys, I. Moerman, Z.Q. Yu, F. Vermaerke, P. Van Daele, P. Demeester. "Atmospheric pressure and low pressure shadow masked MOVPE growth of InGaAs(P)/InP and (In)GaAs/(Al)GaAs heterostructures and quantum wells." Journal of Electronic Materials, Vol. 23, 1994, pp. 227–234.

17. M. Aoki, M. Komori, H. Sato, T. Tsuchiya, A. Taike, M. Takahashi, K. Uomi. "Reliable wide temperature range operation of 1.3 μm beam expander laser diode for passively aligned optical modules." IEEE Journal of Selected Topics in Quantum Electronics, Vol. 3(6) Dec. 1997 pp. 1405–1412.

18. G. Wenger, L. Stoll, B. Weiss, M. Schienle, R. Muller-Nawrath, S. Eichinger, J. Muller, B. Acklin, G. Muller. "Design and fabrication of monolithic optical spot-size transformers (MOSTs) for highly efficient fiber-chip coupling." Journal of Lightwave Technology, Vol. 12 (10), 1994 pp. 1782–1790.

19. Y. Tohmori, Y. Suzaki, H. Fukano, O. Okamoto, Y. Sakai, O. Mitomi, S. Matsumoto, Y. Yamamoto, M. Fukuda, M. Wada, Y. Itaya, T. Sugic. "Spot-size converted 1.3 μm laser with butt-jointed selectively grown vertically tapered waveguide." Electronics Letters, Vol. 31 (3), 1994, pp. 1069–1070.

20. J.V. Collins, R.A. Payne, C.W. Ford, A. R. Thurlow, I.F. Lealman. "Technology developments for low-cost laser packaging." OFC '95 Conference Proceedings, San Diego, CA Feb. 26–Mar. 3, 1995.

21. J.V. Collins, I.F. Lealman, P.J. Fiddyment, C.A. Jones, R.G. Waller, L.J. Rivers, K. Cooper, S.D. Perrin, M.W. Nield, M.J. Harlow. "Passive alignment of a tapered laser with more than 50% coupling efficiency." Electronics Letters, Vol. 31 (9), 1995, pp. 730–731.

22. B. Mersali, H.J. Bruckner, M. Feuillade, S. Sainson, A. Augazzaden, A. Carenco, "Theoretical and experimental studies of a spot-size transformer with integrated waveguide for polarization insensitive optical amplifiers." Journal of Lightwave Technology, Vol. 13 (9), 1995, pp. 1865–1871.

23. D.C. Roger, J.V. Collins, C.W. Ford, P.J. Fiddyment, J. Lucek, M. Shabeer, G. Sherlock, D. Cotter, K. Smith, J.D. Burton, C.M. Peed, A.E. Kelly, P. McKee. "Demonstration of a programmable optical pulse pattern generator for 100 Gbit/sec networks." Electronics Letters, Vol. 31 (23), 1995, pp. 2001–2002.

24. H. Kobayashi, T. Yamamoto, M. Ekawa, T. Watanable, T. Ishikawa, T. Fujii, H. Soda. "Narrow-beam divergence 1.3 μm multiple-Quantum-Well laser diodes with monolithically integrated tapered thickness waveguide." IEEE Journal of Selected Topics in Quantum Electronics, Vol. 3 (6), 1997, pp. 1384–1391.

25. K. Kasaya, Y. Kondo, M. Okamoto, O. Mitomi, M. Naganuma. "Monolithically integrated DBR lasers with simple tapered waveguide for low loss fiber coupling." Electronics Letters, Vol. 29 (23), 1993, pp. 2067–2068.

26. H. Fukano, Y. Kadota, Y. Kondo, M. Ueki, Y. Sakai, K. Kasaya, K. Yokoyamam, Y. Tohmori. "1.3 μm large spot laser diodes with laterally tapered active layer." Electronics Letters, Vol. 31 (17), 1995, pp. 1439–1440.

27. P. Doussiere, P. Garadedian, C. Graver, E. Derouin, E. Gaumant-Gaorin, G. Michaud, R. Meilleur. "Tapered active stripe for 1.5 μm InGaAsP/InP strained multiple quantum well lasers with reduced beam divergence." Applied Physics Letters, Vol. 64 (5), 1994, pp. 539–541.

28. Y. Inaba, M. Kito, T. Nishikawa, M. Ishino, Y. Matsui. "Multiquantum well lasers with tapered active stripe for direct coupling to singlemode fiber." IEEE Photon Letters, Vol. 9, 1997, pp. 722–724.

29. A. Uda, K. Tsuruka, N. Suzuki, F. Fukushima, T. Nakamura, T. Torikai. "Spot size expanded high efficiency 1.3 μm MQW laser diodes with laterally tapered active stripe." Proceedings of the Indium Phosphide and Related Materials Conference, Hyannis, MA, May 1997, pp. 657–660, paper ThF2.

30. G.A. Vawter, R.E. Smith, H. Hou, J.R. Wendt. "Semiconductor laser with tapered rib adiabatic following fiber coupler for expanded output mode diameter." IEEE Photon Technology Letters, Vol. 9, 1997, pp. 427–427.

31. H. Sato, M. Aoki, M. Takahashi, M. Komori, K. Uomi, S. Tsuji. "1.3 μm beam expander integrated laser grown by single step MOVPE." Electronics Letters, Vol. 31 (15), 1995, pp. 1241–1242.

32. K. Shinoda, K. Kiramoto, M. Sagawa, T. Toyonaka, K. Uoni. "Circular beam high power operation of .98 μm InGaAs/InGaAsP lasers with a tapered waveguide spot size expander." Electronics Letters, Vol. 32 (12), 1996, pp. 1101–1102.

33. R. Zengerie, B. Hubner, C. Greus, H. Burkhard, H. Janning, E. Kuphal. "Monolithic integration of spot size transformer for highly efficient laser fiber coupling." Electronics Letters, Vol. 31 (14), 1995, pp. 1142–1145.

34. M. Wada, K. Kohtoku, K. Kawano, H. Okamoto, Y. Kadota, Y. Kondo, K. Kishi, S. Kondo, Y. Sakai, I. Kotaka, Y. Noguchi, K. Itaya. "High coupling efficiency laser diodes integrated with spot size converters fabricated on 2 inch substrates." Electronics Letters, Vol. 31 (24), 1995, pp. 2102–2103.

35. R. Ben-Micheal, U. Koren, B.I. Miller, M.G. Young, M. Chien, G. Raybon. "InP based multiple quantum well lasers with integrated tapered beam expander waveguide." IEEE Photon Technology Letters, Vol. 6, 1994, pp. 1412–1414.

36. R.Y. Fang, D. Bertoni, M. Meliga, I. Montrosset, G. Olivetti, R. Paoletti. "Low cost 1.55 μm InGaAsP-InP spot size converted (SSC) laser with conventional active layers." IEEE Photon Technology Letters, Vol. 9, 1997, pp. 1084–1086.

37. Y. Sakai, Y. Tohmori, Y. Susaki, Y. Kondo, O. Mitomi. "Improved FFP of 1.3 μm spot size converted laser for highly efficient coupling to optical fiber." Electronics Letters, Vol. 32 (15), 1993, pp. 1372–1374.

38. Y. Tohmori, Y. Susaki, H. Oohashi, Y. Sakai, Y. Kondo, H. Okamoto, Y. Kadota, O. Mitomi, Y. Itaya, T. Sugie. "High temperature operation with low loss coupling

to fiber for narrow beam 1.3 µm lasers with butt jointed selective grown spot size converters." Electronics Letters, Vol. 31 (21), 1995, pp. 1831–1839.

39. T.L. Koch, U. Koren, G. Esenstein, M.G. Young, M. Oron, C.R. Giles, B.I. Miller. "Tapered waveguide InGaAs/InGaAsP multiple quantum well lasers." IEEE Photon Technology Letters, Vol. 2, 1990, pp. 88–90.

40. M. Aoki, M. Komori, M. Suzuki, H. Sato, M. Takahashi, T. Ohtoshi, K. Uomi, S. Tsuji. "Wide temperature range operation of 1.3 µm beam expander integrated laser diodes grown by in-plane thickness control MOVPE using a silicon shadow mask." IEEE Photon Technology Letters, Vol. 8, 1996, pp. 479–481.

41. H. Soda, H. Kobayashi, T. Yamamoto, M. Ekawa, S. Yamazaki. "Tapered thickness waveguide InGaAsP/InP BH MQW Lasers." Proceedings of the 8th Annual Meeting, IEEE Lasers and Electro Optics Society, San Francisco, Oct. 30, 31, 1995, p. 13, paper EMGW2.3.

42. I. Moerman, M. D'Hondt, W. Vanderbauwhede, P. Van Daele, P. Demeester, W. Hunziker. "InGaAsP/InP strained MQW laser with integrated modesize converter using shadow masked growth technique." Electronics Letters, Vol. 31, 1995, pp. 452–454.

43. T. Brenner, R. Hess, H. Melchior. "Compact InGaAsP/InP laser diodes with integrated mode expander for efficient coupling to flat ended singlemode fibers." Electronics Letters, Vol. 31 (17), 1995, pp. 1443–1445.

44. T. Murakami, K, Ohtaki, H. Matsubara, T. Yamawaki, H. Saito, K. Isshiki, Y. Kokubo, A. Shima, H. Kumabe, W. Susaki. "A very narrow beam AlGaAs laser with a thin tapered thickness active layer (T3-laser)." IEEE Journal of Quantum Electronics, Vol. QE-23 (6), 1987, pp. 712–719.

45. G. Vermeire, F. Vermacke, I. Moerman, J. Haes, R. Baets, P. Van Daele, P. Demeester. "Monolithic integration of a SQW laser diode and a mode size converter using shadow masked MOVPE growth." Journal of Crystal Growth, Vol. 145, 1994, pp. 875–880.

11 Monte Carlo Analysis of Passive Alignment Methods

Randall B. Wilson

CONTENTS

The Monte Carlo method is a numerical algorithm widely used in the simulation of complex physical problems such as neutron scattering, carrier transport, and other problems typically involving the solution of integral equations or simulation of stochastic events [1–3]. Originally developed and used as a technique for numerically solving problems associated with war-time development in Los Alamos in the 1940s, the method has found a wide utility in the solution of problems that are stochastic in nature and in which random variables play a central role [4]. With the availability of faster computers and improved random-number generation algorithms, Monte Carlo simulation methods have become accessible for many new problems of interest, and because of the simplicity of the basic algorithm, this technique can be understood and used effectively by workers in many fields not traditionally used to using numerical simulation methods.

In the present context, the Monte Carlo method will primarily be used to estimate and predict the yield associated with manufacturing optoelectronic assemblies in which passive alignment methods are used for the optical alignment assembly step [5]. As described elsewhere in this monograph, passive alignment approaches are at the forefront of low-cost optoelectronic assembly and manufacturing methods and are expected to enable a new generation of fiber-optic-based telecommunication products competing in cost with more traditional, copper-based solutions [6–16]. Key to this revolution in cost competitiveness will be the ability of these new passive alignment methods to enable high-volume manufacturing with very high yield for products spanning the entire range of performance specifications. In spite of the large number of publications reporting on the application of passive alignment methods for low-cost design and manufacturing, relatively few articles have appeared addressing the use of Monte Carlo simulation [17–21] or other statistical methods [22] as a design and analysis tool. High-volume manufacturing and scalability of passive alignment methods has been previously discussed, and it is the purpose of this article to address the feasibility of achieving high yields using these methods and to ascertain the relationships and cost tradeoffs between yield and specified performance.

In Section 11.1 we define some terminology and concepts associated with a general manufacturing process that will be used in the rest of the article. In addition, several subsections offer some questions that can be answered using Monte Carlo analysis as the primary tool. Closing this section is an example illustrating how the method is used to calculate yields in manufacturing process simulation. Section 11.2 gives an overview of commercially available software that can be used to perform Monte Carlo analysis, and in Section 11.3, the basic method and approach is provided, along with a discussion of how to estimate the required input data for a simulation, choice of appropriate optical coupling models, and some complications of the method. In Section 11.4, a number of application examples are presented, with conclusions and suggestions for future research given in Section 11.5.

11.1 OVERVIEW AND INTRODUCTION TO MONTE CARLO ANALYSIS

11.1.1 THE MONTE CARLO METHOD

The Monte Carlo method was originated in 1949 by the mathematicians John von Neumann and Stanislav Ulam [4]. To illustrate how the method is used for yield modeling, we start with an easily visualized electrical example, based on one given previously by Sobol [2] and recently analyzed by Smith [28]. Consider an electrical circuit made up of a variety of individual electrical elements such as resistors (R_i), capacitors (C_i), transistors (Q_i), etc. The overall characteristics of this circuit will, in general, be functions of the values of each of the individual elements, and a given characteristic can be computed from the individual values (at least in principle):

$$U = f(R_1, R_2, \dots, C_1, C_2, \dots, Q_1, Q_2, \dots) = f(P_1, P_2, P_3, \dots P_k, \dots P_N) \quad (11.1)$$

For example, if U is the voltage across the load resistor in a resistor network, then using Ohms law, it is possible to solve the equations for the circuit and find an analytical solution for U in terms of R_1, R_2, and so forth.

Now let us say that one desires to create a manufacturing line to produce this circuit in large quantities and that the application in which the circuit will be used requires that the quantity U lie within a certain range or specification. It will be quickly found that the individual components purchased to build the circuit are not all identical, nor are their values necessarily exactly equal to the indicated values. If one measures the actual values for a large number of each component, they will be found to have a distribution. From the standpoint of our manufacturing line, the essential question is what will be the effect of these variations on the final parameter U and how many manufactured circuits will lie outside the required specification.

One approach to answering this question is to do a "worst case analysis" [28], usually by taking the worst-case values of each of the components and computing worst-case values of U from Equation (11.1). However, if the circuit is even moderately complex, this approach fails because of the interaction between components, making it unclear as to which set of component values are the worst. In addition, under normal conditions, it is very unlikely that a circuit will be built in which all of the component parameters are simultaneously at their worst-case values; therefore, a worst-case analysis is too pessimistic.

A more reasonable approach is to directly consider the random nature of the problem, treating the parameter values P_i and U as random variables while considering the function f as known and deterministic. In this case, it is possible to characterize the distributions of each of the parameter values P_i using the mean and variance of the individual distributions such that $\mathbf{MR}_1 = <R_1>$, and $\mathbf{DR}_1 = <[R_1 - \mathbf{MR}_1]^2>$, and so forth, and $\mathbf{MU} = <U>$ and $\mathbf{DU} = <[U - \mathbf{MU}]^2>$. What is required is knowledge of the distribution of U or at least the distribution characterizing parameters of U such as $\mathbf{MU}$ and $\mathbf{DU}$. Of course, once our production line is up and running, we can experimentally measure the variation in U as the circuits are fabricated, but it is far more desirable to have the knowledge of the variation in U in the design stage of the circuit, when the circuit can still be optimized. Nevertheless, it still remains problematic to determine the distribution of U in terms of the distributions of the individual circuit elements even when f is known. In fact, it is usually impossible to analytically determine the distribution of U from a knowledge of the elemental distributions if f is even moderately complex. Moreover, from the theory of random variables, it can easily be shown that in general:

$$\mathbf{MU} \neq f(\mathbf{MR}_1, \mathbf{MR}_2, \ldots, \mathbf{MC}_1, \mathbf{MC}_2, \ldots, \mathbf{MQ}_1, \mathbf{MQ}_2, \ldots) \qquad (11.2)$$

Therefore it is not even possible to directly determine the distributional characteristics of U (i.e., $\mathbf{MU}$ and $\mathbf{DU}$) from a knowledge of the means and variances of the circuit component parameters P_i and the function f.

The solution is to apply the Monte Carlo method to the problem. The basic idea is to simulate (on a computer) the assembly of a large number of circuits in a series of independent trials. In each trial, values for $R_1, R_2, \ldots, C_1, C_2, \ldots, Q_1, Q_2$ and so on are randomly generated consistent with the known distributions for each component. For each of these trial values, a value of U is computed from Equation (11.2) and stored. The process is repeated a large number of times until the distribution of U emerges. Once we have the distribution of U, the problem at hand is solved; now we can readily determine the proportion of circuits that fall outside the required specification and evaluate the viability of our design.

11.1.2 Manufacturing Flow Networks, Design Variables, and Yield

A simple generalized manufacturing flow process is shown in Figure 11.1. Such a process might represent the assembly of an optoelectronic product consisting of a pigtailed semiconductor diode laser and an ASIC driver chip. For example, the input components labeled a, b, and c might be the laser chip, optical fiber

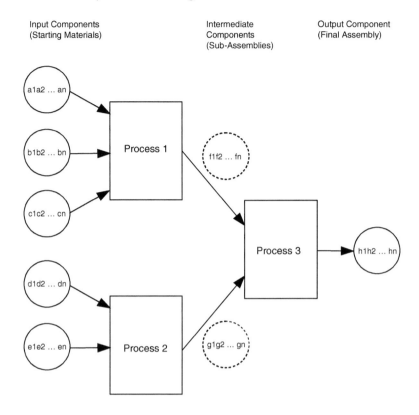

FIGURE 11.1 Generalized manufacturing process flow diagram.

stub, and silicon submount, respectively, while *d* and *e* could represent an ASIC driver chip and a ceramic submount. In this case we take process 1 to be a passive alignment assembly process used to produce a pigtailed laser intermediate component or subassembly (*f*), while process 2 is the submounting process for the ASIC chip, resulting in subassembly *g*. In the last step to this procedure, process 3 attaches the pigtailed laser subassembly *f* to the ASIC subassembly *g* to produce the final assembly *h*.

For each of the individual components or subassemblies we can identify and define a set of design variables, designated as a_1, a_2, a_n, b_1, b_2, b_n, and so forth in Figure 11.1, each of which affects the performance of the final assembly *h*, cither directly or through interaction with the other design variables. In a passive alignment assembly, examples of design variables might be V-groove width, fiber diameter, laser threshold, pedestal height, and so on.

Each unique component that is carried through the process depicted in Figure 11.1 has a distinct value for each of its associated design variables. In general, however, these values vary from component to component as the process is run. The variability in the design variables is a natural consequence of the fact that in any process or process sequence used to manufacture a part or subassembly, unintentional process variations inevitably lead to variations in the final characteristics of the parts. Such process variations can come from many different sources and the measurement and control of these variations is the subject of statistical process control and the reader is referred to the vast literature available on the subject [23]. These variations are sometimes called natural variations in the statistical process control literature. The primary aim here is to acknowledge that such variations always exist and to examine how such variations can limit the yield and affect the manufacturing cost associated with optoelectronic products using passive alignment techniques for assembly.

Associated with each design variable is a set of constraints called the design variable specifications, or simply design specifications. Usually these specifications consist of a pair of scalar values: the lower (lsl) and upper (usl) specification limits. The design variable specifications may be one-sided as well, either a single upper or lower limit. Design variable specifications will be designated as bold uppercased versions of the corresponding design variable. Thus $A_1 = \{lsl_1, usl_1\}$ is the design specification for the design variable a_1. A component part or subassembly in which the values of all the design variables fall inside of the corresponding design specification limits is in conformance and may contribute to the overall product yield. As large numbers of components are run through the process, the yield is defined as the fraction of parts with conforming design variables. We define scrap as any part in which one or more of the design variables is outside the range of the corresponding specification limits.

Because the exact values of the design variables vary in an unpredictable way from process run to process run and part to part, we can treat these as random variables, each with an associated probability density function (pdf) characterizing the likelihood that a design variable's value will fall in a particular range [24, 25]. Sometimes this will be referred to as the natural distribution function, or natural distribution. In general, the pdf associated with a given design variable is not known explicitly. However, measurements of the design variable over a large sample of parts allow the pdf to be estimated. Graphical techniques such as probability chart plotting can be used to estimate the form of the underlying pdf and extract the parameters of the distribution [26, 27]. More sophisticated distribution fitting techniques may also be used. Often a simple histogram and determination of the mean and standard deviation are adequate for distributions that are nearly gaussian. Some examples of histograms and the corresponding best-fit probability density functions are shown in Figure 11.2 for some typical optoelectronic component design variables. As demonstrated in this figure, it is not uncommon to find that key design variables have distinctly non-Gaussian-associated probability density functions.

The total integrated area of the pdf over the entire possible range of a design random variable is equal to unity. The area under the probability density function between two values of the random design variable is equal to the probability of the design variable taking on a value between the two limits, as indicated in Figure 11.3a. In other words, the pdf is useful in that one can determine the

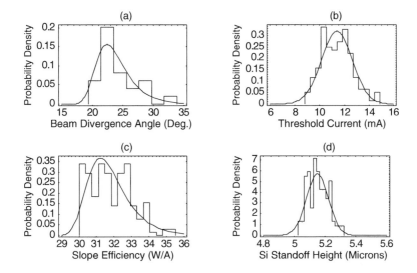

FIGURE 11.2 Best-fit probability density functions for typical optoelectronic component parameters. (a) Extreme Value – mode = 22.42, scale = 2.36; (b) Normal – $\mu = 11.39$, $\sigma = 1.25$; (c) Extreme Value – mode = 31.22, scale = 1.00; (d) Normal – $\mu = 5.15$, $\sigma = 0.07$.

probability that a design random variable will take on a value between two limits by integrating the pdf between the two limits. If $f(x)$ is the pdf, then

$$\mathbf{P}\{v1 \le x \le v2\} = \int_{v1}^{v2} f(x)dx \qquad (11.3)$$

Closely related to the probability density function is the cumulative distribution function (Cdf). The cumulative distribution function is defined as the integral of the probability density function from the minimum allowed value to the value of interest; that is,

$$F(v) = \mathbf{P}\{x \le v\} = \int_{v\min}^{v} f(x)dx \qquad (11.4)$$

where $f(x)$ is the pdf and $F(v)$ is the Cdf. The value of the cumulative distribution function evaluated at v is equal to the probability of the random design variable taking on a value that is less than or equal to v. Therefore, the Cdf is even more useful than the pdf in that one can determine the probability that a random design variable will take on a value between two limits by simply subtracting the values

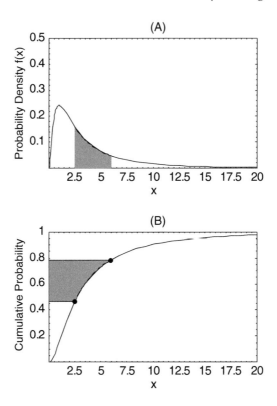

FIGURE 11.3 Relationship between the (a) probability density function (pdf) and (b) the cumulative distribution function (cdf).

of the Cdfs evaluated at the two limits, as shown in Figure 11.3b. This can be seen from

$$F(v2) - F(v1) = \int_{v\min}^{v2} f(x)dx - \int_{v\min}^{v1} f(x)dx = \int_{v1}^{v2} f(x)dx = \mathbf{P}\{v1 \le x \le v2\} \quad (11.5)$$

From the above expressions, it can be seen that if we associate a design variable with x and associate design specification limits (that is, lsl and usl) with $v1$ and $v2$, then the probability of the design variable taking on values between the design specification limits is equal to $F(\text{usl}) - F(\text{lsl})$. In the simple case where there is only a single output design variable subject to specification limits, this is equivalent to the definition of the yield of the process, provided that the number of parts run through the process is large. Therefore, if we can determine the Cdf function for the specified design variable of the finished part, then the yield can be predicted for any given set of design specifications, using Equation (11.5). Determining the Cdf is the role of Monte Carlo simulation.

An alternative, but equivalent, development can also be presented following Kalos and Whitlock [1]. Consider an integral, which is to be evaluated using Monte Carlo techniques, of the form

$$\int_{-\infty}^{+\infty} g(x)f(x)dx \tag{11.6}$$

where $f(x)$ has the form of a pdf. Then, from the theory of random variables [1, 24, 25], this integral can be interpreted as the expectation value of $g(x)$; that is,

$$\int_{-\infty}^{+\infty} g(x)f(x)dx = \langle g(x) \rangle \tag{11.7}$$

For a given pdf function $f(x)$, the expectation value of $g(x)$ can be estimated by a sampled sum, drawn at random from the probability distribution associated with $f(x)$, so that the integral may be approximated as

$$\int_{-\infty}^{+\infty} g(x)f(x)dx = \langle g(x) \rangle \cong \frac{1}{N}\sum_{x_i}^{N} g(x_i) \tag{11.8}$$

For the case of yield calculation, the integral of interest becomes

$$Y = \int_{lsl}^{usl} f(x)dx = \int_{x\min}^{x\max} R(x\,|\,lsl,usl)f(x)dx \cong \frac{1}{N}\sum_{x_i}^{N} R(x_i\,|\,lsl,usl) = \frac{S}{N} \tag{11.9}$$

where $R(x|lsl, usl)$ is the unit box function, defined as

$$R(x\,|\,lsl,usl) = \begin{cases} 1 & lsl \le x \le usl \\ 0 & \text{otherwise} \end{cases} \tag{11.10}$$

and S is simply equal to the number of x_i values occurring in the sum that lie within the interval $lsl \le x_i \le usl$. The advantage of this formulation is that it is readily extended to the multivariate case in which the yield Y is dependent on more than one output design variable (see reference [1] and Section 11.3.3).

11.1.3 PASSIVE ALIGNMENT, OPTOELECTRONIC ASSEMBLY, AND PACKAGING

This section considers the application of the Monte Carlo method to the modeling of optoelectronic assemblies. Generally speaking, there are two broad categories of alignment methods currently used in optoelectronic assembly; namely, active alignment and passive alignment. The primary interest here is in the application of Monte Carlo analysis to passive alignment, for the reasons given earlier.

However, it is of some interest to briefly compare the two approaches to establish the advantages and disadvantages of each as well as to provide a brief introduction to the two methods. Next, the use of binning will be considered as another application of Monte Carlo analysis, followed by a brief discussion on the trade-offs between performance and yield. As in all of these analyses, the ultimate interest is in predicting the average yield and cost associated with an optoelectronic assembly given a specific assembly design, component characteristics, and assembly process characteristics.

11.1.3.1 Passive vs. Active Alignment

In traditional optoelectronic assembly, especially for single-optical mode products, active alignment of optical components has been an essential process for achieving high yields; in particular for products that require very efficient optical coupling between a laser and an optical fiber. Quite generally, in any active alignment process, one or more of the "active" optoelectronic devices in the assembly is operated while the alignment process takes place. By monitoring a suitable optical parameter that is affected by the alignment, usually coupled with power in the fiber, the parameter can be optimized. Therefore, active alignment is essentially a feedback process, in which the alignment affects the monitored parameter, which in turn is used to modify the alignment, and so on. Normally this active alignment process is automated, with high-resolution micromanipulators controlling the alignment and a computer algorithm determining the feedback optimization. After the desired alignment is achieved, the component parts are then fixed in position, typically using some kind of welding, soldering, or adhesive method.

Because active alignment is a feedback process, the feedback algorithm can guarantee that the characteristics of the finished subassembly will lie within the design specification, and therefore very high yields to the design specification are achievable. However, within the design specification limits, considerable variation may still occur, depending on the nature of the computer algorithm, resolution of the micromanipulators, time response of the feedback loop, and so forth. Moreover, dealignment may occur during or after the fixing process, leading to postalignment offsets and variations from part to part. Thus, there is still a characteristic distribution or pdf of the output design variables associated with the active alignment process. Actively aligned assemblies that cannot be aligned (usually within some allowed time) are rejected as scrap, and so active alignment also functions as an in-process test and sorting operation.

With the advent of newer precision technologies such as silicon micromachining and intelligent vision systems, passive alignment methods have come to the forefront of optoelectronic assembly production. In a typical passive alignment assembly operation, precision mechanical alignment features or optical fiducials are fabricated directly onto the component parts, usually employing the same photolithographic techniques used by the silicon IC industry, which can achieve submicron accuracy and precision. Therefore, the design variables associated with these alignment features (and fiducials) can have very narrow distributions with submicron variability. These component parts are then aligned either by using direct contact mechanical registration [5] or by noncontact optical registration [13]. Submicron placement of the parts is accomplished using ultraprecision micromanipulators. Fixing of the aligned components is often still achieved by using some kind of welding, soldering, or adhesive method.

In both types of passive alignment processes, the optoelectronic devices are not operated, thus allowing some flexibility in the order in which components may be assembled. For passive alignment methods using mechanical registration, there is no process feedback whatsoever, and the variations realized in the final alignment are completely determined by the accuracy of the mating alignment features and the quality of the "mate." For passive alignment methods using optical registration, in general there is a feedback optimization during the optical registering step, and so the variations in the final alignment are determined by the optical system, computer algorithm, and micromanipulation factors similar to those mentioned previously for active alignment.

Because passive alignment methods using mechanical registration do not require complicated vision systems or ultraprecision micromanipulators, relatively simple equipment may be used to implement the process, as compared with active alignment or optically registered passive alignment. In the mechanical registration method, the accuracy and precision of placement are limited by the variations in the alignment features, the mating process, and the fixing process. In the optical registration method, the placement is limited by the variations in the alignment features, the alignment algorithm, the placement process, and the fixing process. To ascertain the cost effectiveness of each of these different approaches, Monte Carlo modeling can be used to evaluate achievable yields for each method, given reasonable estimates of the underlying part and process variations.

11.1.3.2 Binning

In a typical manufacturing process, the usual goal is to produce a product with a narrow distribution of performance characteristics, centered on the nominal design values specified for the product. To achieve this goal, variation of the intermediate design specification variables (see Figure 11.1) must be reduced, either by in-process testing and sorting or by process improvement that reduces the natural variation at each step. In-process testing and sorting reduces variation by truncating the natural distributions but increases cost because of the additional testing and sorting operations and increased scrap material. Process improvement, in contrast, narrows the natural distributions associated with the manufacturing processes, thereby reducing cost by eliminating the need for in-process testing and minimizing scrap. This latter approach is the goal of most quality improvement programs in manufacturing operations.

In some instances, another approach is also possible. In the case in which there are a number of distinct product categories, having final product design specifications that span a contiguous or nearly contiguous range of performance specifications, then "binning" may become a viable manufacturing approach. This is

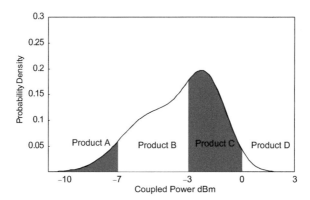

FIGURE 11.4 Relationship between specification limits and the pdf illustrating product binning.

illustrated in Figure 11.4. For example, some simple laser component product lines specify distinct power ranges, −10, −7, −3, and 0 dBm, with all other parameters essentially being equal. In this approach, nonfailure related in-process testing and sorting is eliminated, reducing cost, and process variation is allowed (possibly even encouraged), so that the final performance characteristics of the produced parts span the range of various product specifications, as indicated in Figure 11.4. A final test and sorting procedure is required to "bin" each part according to its performance. To the extent that the various product specifications abut or overlap, scrap is minimized, thereby also reducing cost. A significant disadvantage of this approach, however, is that the product "mix" is determined by the distribution characteristic of the overall process and may not be easily tailored to the product demand in each category. Therefore, it is of considerable interest, if this approach is being considered, to be able to model and predict what the final distribution of performance characteristics will be, and how changes in the design or process could be used to alter this distribution to affect the product mix.

11.1.3.3 Yield, Performance, and Cost Trade-Offs

Within the context of the simple model just developed, a simple relationship between cost, yield, and performance of an optical subassembly can be formulated. For a given assembly process step, the total material cost, labor costs, and fixed costs per unit can be designated as M_i, L_i, and F_i, respectively. If the process step has an overall yield of Y_i, then the step cost per compliant part is

$$Cost = \frac{M_i + L_i + F_i}{Y_i} = \frac{M_i + L_i + F_i}{\int_{lsl}^{usl} f(x)dx} = \frac{M_i + L_i + F_i}{[F(usl) - F(lsl)]} \qquad (11.11)$$

If x is a performance-related variable, then as the required performance level increases and becomes more stringent, the specification limits shift such that the yield decreases, thereby increasing cost. Often M_i, L_i, and F_i may also increase because of more expensive materials or increased labor being required or because of more expensive or additional equipment being needed for the higher-performing assembly. Because a Monte Carlo analysis can estimate yields, and in particular the Cdf in Equation (11.11), it is also possible to model cost as a function of performance requirements provided the additional materials, labor, and fixed overhead information are available.

11.2 SOFTWARE PACKAGES

To perform any useful Monte Carlo analysis, a suitable software program or environment must be selected that can provide the necessary random-number generators, mathematical functions, programming capability, and graphical output features necessary for the analysis undertaken. Fortunately, there are a number of software packages commercially available with the desired capabilities, and these are described below.

11.2.1 @RISK AND CRYSTAL BALL

@RISK and Crystal Ball are add-in programs for Microsoft Excel and Lotus 1-2-3 that are designed for performing risk-analysis simulations in a spreadsheet environment. Essentially, these programs are Monte Carlo engines that use the programs' spreadsheet capabilities to compute and display their results. In addition, the programs add a comprehensive collection of random-variable-generating functions for all of the important probability distributions likely to be required. They also provide an extended set of graphical displays and control functions specifically adapted for easy implementation of a Monte Carlo analysis. In the author's view, these programs provide the lowest investment in time and dollar expense to get started in Monte Carlo analysis.

Crystal Ball 4.0 is available from Decisioneering Inc. for use with Microsoft Excel or Lotus 1-2-3 with Microsoft Windows and for use with Microsoft Excel in an Apple Macintosh environment [29]. The program creates a new toolbar within the spreadsheet for defining random variable cells, output cells, graphical output, and controlling the simulation. Some simple correlation capability is also provided by the use of correlation coefficients, and both standard Monte Carlo as well as Latin Hypercube sampling is supported. A comprehensive distribution-fitting capability is also provided, which allows the user to determine a best-fit distribution form to a set of measured data. A typical display of a Monte Carlo simulation using Crystal Ball is shown in Figure 11.5.

@RISK is available from Palisade Corporation for use with Microsoft Excel or Lotus 1-2-3 in a Microsoft Windows environment [30]. This program also creates a new toolbar within the spreadsheet for controlling the simulation and graphical output. Input random variables are explicitly defined in the spreadsheet

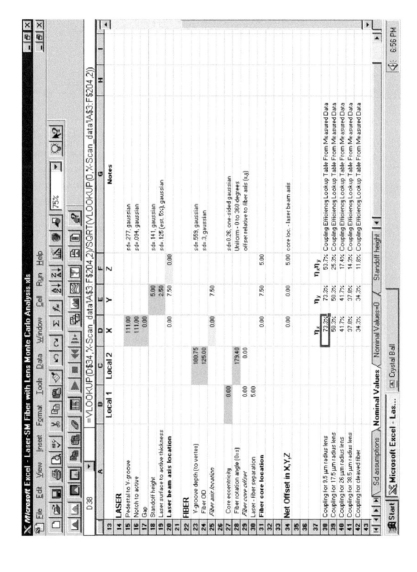

FIGURE 11.5 Example of a Monte Carlo implementation using Crystal Ball 4.0 in MS Excel.

cells, using probability distribution functions entered directly in the cells. Correlation capability is provided by the use of a correlation coefficients matrix, and both standard Monte Carlo as well as Latin Hypercube sampling is supported. An explicit cumulative distribution function is also provided for the analysis of output variables. A typical display of a Monte Carlo simulation using @RISK is shown in Figure 11.6.

11.2.2 Mathematica

Mathematica 3.0, available from Wolfram Research, Inc., is a very powerful, comprehensive, symbolic and numerical mathematical computation package for both Microsoft Windows and Apple Macintosh environments [31]. Unlike spreadsheet modeling, Mathematica provides an environment or shell in which the Monte Carlo model is written out in more conventional mathematical notation and simple structured programs. Extensive built-in graphical and statistical functions are available for analyzing simulation results. Standard add-in packages provide a comprehensive set of probability distributions for random variable generation. In general, somewhat more work is involved in setting up a Monte Carlo model within Mathematica, as compared with the add-in/spreadsheet approach described above, because of the general nature of the program environment. However, the user has far greater control over the details of the simulation and analysis. For example, more explicit correlation relationships between random variables can be defined, and it is easier to perform multiple simulations in which the input distribution characteristics are parametrically varied. A typical display of a Monte Carlo simulation using Mathematica is shown in Figure 11.7.

11.2.3 Spreadsheets

Even without the add-in programs described above, it is possible to perform simple Monte Carlo simulations in a standard spreadsheet like Microsoft Excel or QuattroPro, using the built-in statistical functions provided in the spreadsheet. However, usually the random-number-generation capability is limited to Uniform or Normal probability distributions. In addition, either complicated macros must be created for sample iterations or large columns of the simulated intermediate variables must be created, requiring very large memory requirements. In general, the use of just a spreadsheet for Monte Carlo simulations is limited from a practical standpoint to simple modeling problems involving only a few Normally or Uniformly distributed random variables.

11.2.4 Programming Languages

Conventional programming languages like Fortran, C, or Basic can also be used for Monte Carlo modeling of passive alignment problems. Indeed, before the development of PC-based applications, all Monte Carlo modeling was programmed in this way. There is a vast collection of statistical and graphics subroutine packages available for virtually any computer/operating system platform.

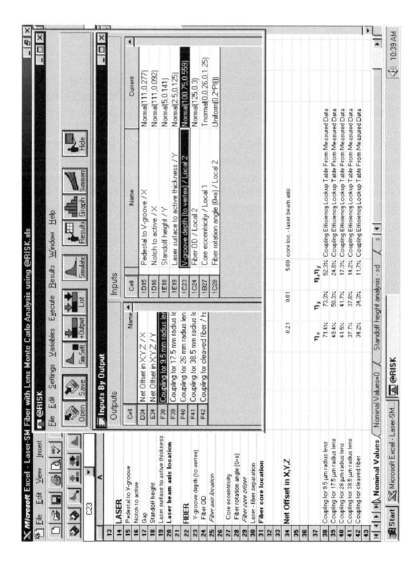

FIGURE 11.6 Example of a Monte Carlo implementation using @RISK in MS Excel.

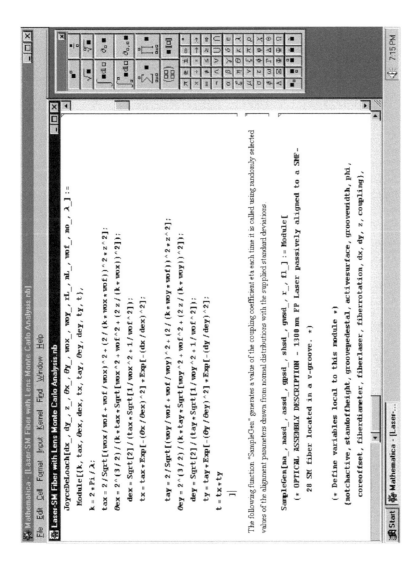

FIGURE 11.7 Example of a Monte Carlo implementation using Mathematica 3.0.

11.3 BASIC METHOD

We consider an individual process step, as shown in Figure 11.8. The basic method for defining and building a Monte Carlo analysis consists of five steps:

1. Identify the dependent output design variable or variables of interest for the problem and establish the acceptable limits (design specifications) for these variables. Typically, these will be performance characteristics of the electro-optical subassembly, such as coupled optical power, cross-talk, wavelength, and so on. Because these characteristics have well-defined in-process or customer specifications associated with them, any subassembly in which the characteristic falls outside of the specified range is rejected (or set aside for rework), resulting in a yield loss.

2. For each dependent output variable, identify the relevant input parameters and design variables on which it depends. Typically these will be design characteristics that are specified in the subassembly design but that are subject to variation either through assembly process variability or random deviations in the subassembly component parts as they are received from a supplier.

3. Create a deterministic model for the subassembly, in which the value of each dependent output variable is quantitatively related to the value of each of the relevant input variables. Typically this takes the form of a series of linked mathematical equations, however complex. In many cases this relationship may not be continuous, nor even analytic, but it may consist of a lookup table of values. In principle, it can also be or contain a definite integral, evaluated numerically.

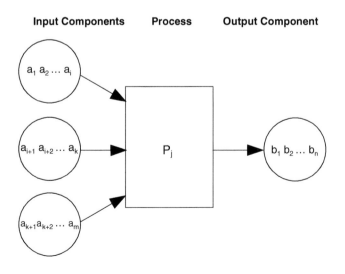

FIGURE 11.8 Individual process step in a manufacturing process flow diagram.

4. For each input variable subject to random behavior, identify an appropriate distribution type and estimate the required distribution parameters. Typical distributions are Gaussian or Uniform, with mean and standard deviation as the distribution parameters; however, other types of distributions can and do occur (see Figure 11.2). In many cases, these distribution types can be determined on the basis of sound physical reasoning or, better yet, by direct measurement. Some examples of the types of distributions that often occur, and methods of parameter estimation, are described below.

5. Implement the above four steps into an iterative sampling and test algorithm, as shown in Figure 11.9, to generate distributions of the output variables and evaluate yield.

11.3.1 DISTRIBUTIONS AND THE ESTIMATION OF PARAMETERS

The variability associated with assembling component parts can be divided into two basic categories: variability associated with the characteristics of the parts themselves, and variability associated with the process of putting them together. In the first case, the nature of the variability is determined by the sum total of the processes and subparts used to fabricate it. If the part is supplied from a vendor, these processes and subparts may be unknown, but in many cases the variations have already been determined by the vendor who produces it.

For example, consider a semiconductor diode laser used in a hybrid optical assembly. Some of the critical characteristics of this diode that ultimately affect the overall assembly performance are threshold current, slope efficiency, and the parallel and perpendicular farfield angles of the emitted beam, as well as any mechanical features fabricated into the device for passive alignment purposes. Each of these parameters will vary from chip to chip, with each parameter varying according to its own characteristic or natural distribution function. In Figure 11.2, we show some typical distributions of beam divergence, threshold current, and slope efficiency for several different devices. Depending on the type of device, how it is manufactured, and so forth, it can be seen that these distributions may deviate considerably from simple Uniform or Gaussian distributions. In some cases, especially for highly critical component parts such as laser diodes, strict performance specification limits are imposed on the devices by the vendor, resulting in truncated distributions. As a result of truncation, broad, initially normal, distributions may be converted into distinctly nonnormal distributions, sometimes approaching a Uniform distribution in shape. Significant errors can be introduced into a Monte Carlo simulation if the assumed input variable distributions are not accurately characterized.

The processes used to assembly component parts can introduce variability of their own. Consider the x, y, z placement of a single-mode optical fiber in a V-groove and a subsequent attachment process using solder or epoxy. In the absence of any mechanical alignment stops along z (groove-fiber axis), the fiber can be passively z aligned optically by a human operator or vision system. This will likely result in

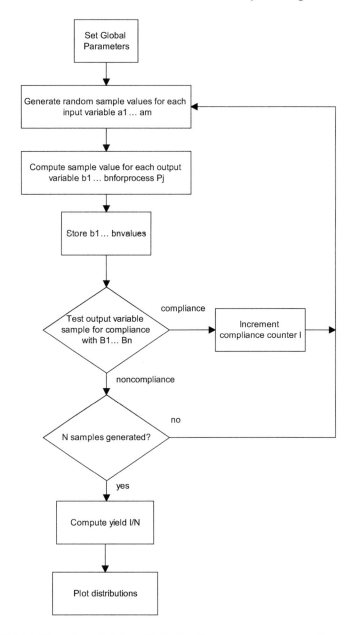

FIGURE 11.9 Flow chart of a Monte Carlo algorithm to compute passive alignment yields.

a z alignment variability with a Gaussian-like distribution with a half width determined by the limitations of either the optical resolution of the system used for alignment (typically on the order of the wavelength of visible light) or the characteristics of the mechanical system used to manipulate the fiber such as step size, backlash, slippage, and so on. Along the x direction (perpendicular to the groove-fiber axis), the variability is primarily determined by the variation in the fabricated parts before attachment. In the y direction (normal to the bonding surface), process-induced variability might be dependent on the amount of solder or epoxy remaining between the fiber and the V-groove walls, the shifting of the fiber position during solder solidification or epoxy curing, and any bending of the fiber because of induced stresses in the fiber during the attachment process. These latter kinds of variations can be distinctly non-Gaussian.

In general, nearly all component design variability ultimately comes from some kind of process-related variability, because inevitably some kind of fabrication process is used to create the features of the device that determine the component design characteristic. For example, variability in the optical characteristics of a laser diode such as divergence angle result from variations in the laser waveguide dimensions that are ultimately determined by variations in the epitaxial growth process, mesa etching process, or other processes used to create the waveguide. Similarly, variations in passive alignment features created on silicon submounts or on active devices are also, in the final analysis, determined by the material removal process (such as wet or dry etching) or material addition process (such as growth) that is used to create them. Even an alignment process itself, such as a mask alignment process, is actually a combination of addition and subtraction processes as the mask is jockeyed back and forth before exposure.

Several types of distributions naturally arise asymptotically, as predicted from the Central Limit Theorem [24, 25], when a large number of underlying random variables combine together in different kinds of processes [27], such as simple addition or subtraction processes, in which multiple random variables contribute to a sum constituting the final parameter tends to result in Normally distributed random variables, independent of the nature of the underlying contribution variables. The mask alignment process mentioned above is an example of this, where a large number of independent back-and-forth movements sum together to determine the final location. In the case where multiple random variables contribute to a product (i.e., are multiplied together), constituting the final parameter, the resulting variable pdf asymptotically tends toward a Lognormal distribution [32] as the number of variables increases. The same distribution results when a design variable is an exponential function of an underlying variable that is Normally distributed. Certain active device parameters as well as epitaxial growth and chemical etching processes can fall into this latter category. Finally, in the case in which algebraic sums or sums of exponentials of random variables constitute a final parameter, the resulting variable pdf asymptotically tends toward an extreme value distribution [33] as the number of variables increases.

11.3.2 Optical Coupling Models

One of the most important aspects of a Monte Carlo simulation of passive alignment is the modeling of the optical coupling between components, especially in single-mode assemblies, as this is usually a yield- and cost-limiting parameter. Because of the extremely tight tolerances required to achieve high optical coupling between single-mode optical components, even small variations in component or process design parameters can lead to large variations in the final optical power transferred through the assembly, leading to high yield losses and corresponding high manufacturing costs. Thus, it is of paramount importance to have a clear understanding of the origin and relative influence of the component and process variations leading to optical coupling loss. The key to this understanding is the optical coupling model.

Coupling models used in Monte Carlo analysis can take several forms: a closed-form equation that gives the desired coupling coefficient explicitly as a function of the design specification variables and parameters, an integral form dependent on the design specification variables and parameters that must be solved numerically, and a lookup table, usually derived from experimental data or an off-line numerical calculation.

A relatively simple, yet very effective optical coupling model is derived using the Gaussian Beam approximation, first developed by Kogelnik [34–38]. Using the Gaussian Beam approximation, closed form formula have been derived which can be used for computing the single mode coupling efficiency for fiber-fiber, laser-fiber, laser-lens-fiber, and laser-lensed-fiber as a function of tilt, axial, and lateral offset. Two of the most useful formulations for Monte Carlo simulation derived from Gaussian Beam analysis have been given by Joyce and DeLoach [39] and Sakai and Kimura [40]. The Joyce and DeLoach model gives the coupling coefficients for axial, lateral, and angular offset for two displaced Gaussian modes. The Sakai and Kimura model extends this analysis to include a hemispherical lens at the location of one Gaussian mode, taken to be a single-mode optical fiber. Other coupling models for both single-mode [41–43] and multimode [17, 44, 45] applications are also available.

In Figure 11.10a and Figure 11.10b, the coupling efficiency of a typical Fabry-Perot laser to a lensed single-mode fiber is calculated from the Sakai–Kimura model as a function of axial separation (z), and lateral offset (x), respectively, for three different values of the lens radius. In this example, the mode field radius of the fiber is 4.65 µm, and the x and y mode field radii of the laser are 1.1 and 0.95 µm, respectively [18]. In Figure 11.10b, the z position is set to the peak value, except for the cleaved fiber case, where $z = 0$. For high coupled optical power applications using conventional Fabry-Perot lasers, lensing the fiber tip is essential to achieve high coupling efficiency. However, high coupling efficiency is obtained at the expense of lower tolerance to lateral offset, as measured by the full width at half maximum (FWHM) of the curves in Figure 11.10b. Random variation of the lateral offset will therefore lead to greater variations of the coupling efficiency in small-radius lensed parts as compared with designs using a large-radius lens or flat, cleaved fibers. These variations in coupling efficiency will directly affect the performance specification yield, and the point at which this tradeoff becomes uneconomical can be answered easily with a Monte Carlo simulation for any given design.

Another type of semiconductor laser device of growing importance is the large spot size laser [46–48; also see Chapter 10] in which a tapered mode-converting waveguide section in the device results in an emitted Gaussian beam that is

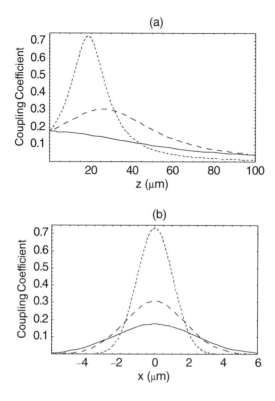

FIGURE 11.10 Optical coupling coefficient using the Sakai–Kimura model: (a) vs. z; (b) vs. x. Solid curve: $r = \infty$; large dashed curve: $r = 10$ μm; small dashed curve: $r = 25$ μm.

effectively mode-matched to a single-mode optical fiber. In Figure 11.11, we compare the axial and lateral coupling dependence of a large spot size laser coupled to a cleaved single-mode fiber through an index-matching gel with two of the cases presented in the previous figure. Again, the Sakai–Kimura model is used, but with a modification to account for the index-matching gel in the large spot case. As seen in Figure 11.11a and Figure 11.11b, the large spot device enables very high coupling efficiency while simultaneously achieving lateral tolerances that are very large, as compared with both the lensed and unlensed fiber or conventional laser case. Therefore, this device is preferred for applications where high coupling efficiency is required.

11.3.3 ERROR ESTIMATION

One question that naturally arises when performing a yield simulation is how to determine the accuracy of the result. Fortunately, there is a simple way to estimate the expected variance of a calculated yield value from a Monte Carlo simulation.

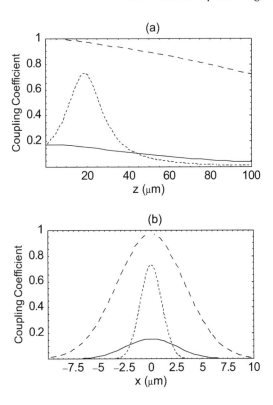

FIGURE 11.11 Optical coupling coefficient using the Sakai–Kimura model: (a) vs. z; (b) vs. x. Solid curve: $r = \infty$ with Fabry-Perot (FP) laser; large dashed curve: $r = 10$ μm with FP laser; small dashed curve: $r = \infty$ with large spot (LS) laser and index-matching gel.

Let x_i be the sampled values of an output random variable x in a simulation; for example, the optical coupling coefficient or optical power in the fiber. Letting $f(x)$ be the pdf for x, the actual yield is given by

$$Y = \int_{lsl}^{usl} f(x)dx = \int_{x\,min}^{x\,max} R(x\,|\,lsl, usl)f(x)dx \qquad (11.12)$$

where $R(x|lsl, usl)$ is the unit box function defined as

$$R(x\,|\,lsl, usl) = \begin{cases} 1 & lsl \le x \le usl \\ 0 & \text{otherwise} \end{cases} \qquad (11.13)$$

As discussed previously, the Monte Carlo estimator for the yield Y is

$$Y_N = \frac{1}{N}\sum_i^N R(x_i\,|\,lsl, usl) = \frac{1}{N}\sum_i^N R(x_i) \qquad (11.14)$$

Here the dependence on lsl and usl has been dropped in the last expression for simplicity but is still implied. As discussed by Kalos and Whitlock [1], the Monte Carlo estimator for the variance of Y is

$$Var(Y_N) \cong \frac{1}{N-1}\left[\frac{1}{N}\sum_i^N R^2(x_i) - \left(\frac{1}{N}\sum_i^N R(x_i)\right)^2\right] \qquad (11.15)$$

but from Equation (11.13), $R^2(x_i) = R(x_i)$ for any x_i, and using the above definition of Y_N in Equation (11.14), we have

$$Var(Y_N) \cong \frac{1}{N-1}\left[Y_N - Y_N^2\right] \quad \Rightarrow \quad \sigma(Y_N) \cong \left[\frac{Y_N - Y_N^2}{N-1}\right]^{1/2} \qquad (11.16)$$

where $\sigma(Y_N)$ is the estimated standard deviation of Y. This equation predicts that the expected standard deviation associated with a yield value calculated from N sampled values of x will decrease with increasing N approximately as $1/\sqrt{N}$. This is demonstrated in Figure 11.12, where 50 independent yield calculations were performed for optical coupling (to a particular specification) and the yield values obtained are plotted as a histogram. Three histograms were generated with $N = 100, 1000,$ and $10,000$ samples used, respectively, for each individual yield computation. Means and standard deviations for each histogram were computed and compared with Equation 11.16 in Table 11.1. As can be seen from the comparison, the agreement is excellent. Equation 11.16 is very convenient for estimating the uncertainty in a Monte Carlo yield simulation, as it only depends on the value of the predicted yield and the number of samples used in the calculation.

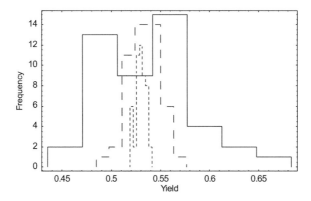

FIGURE 11.12 Histogram of 50 independent yield computations for different sampling sizes. Solid curve: $N = 100$; large dashed curve: $N = 1000$; small dashed curve: $N = 10,000$.

TABLE 11.1
Comparison of Standard Deviations Obtained from
Computation and Estimated from Equation (11.16)

Sample, N	Mean Y_N from Monte Carlo	$\sigma(Y_N)$ from Monte Carlo	$\sigma(Y_N)$ from Equation (11.16)
100	.531	.055	.050
1000	.533	.017	.016
10,000	.533	.005	.005

11.3.4 CORRELATION EFFECTS

One complication associated with Monte Carlo simulations in general is the problem of correlated random variables. In a real situation, like an optoelectronic manufacturing line, correlation is a common phenomenon. For example, it is often the case that, when one parameter on a manufactured part tends to be on the low side, some other parameter on the same part has a tendency to be on the high side, and so on. For a simulation to accurately reflect these phenomena, these correlation effects must be explicitly accounted for in the model. Often, the correlation effects are known or conjectured to be related to dependencies on underlying parameters, common to the variables exhibiting the correlated behavior, and in many cases they can be mathematically described, thus enabling such effects to be accurately taken into account during the simulation.

11.3.4.1 Origin of Correlation Effects

So far, in the discussion of design random variables, there has been an assumption of statistical independence. In many cases, this is a good assumption, especially when the variables are of a distinctly different type, originating from very different fabrication processes. For example, variations in the epilayer thicknesses of a laser that control the active area placement in the Y dimension are expected to be completely independent of the variations in the X dimension of a dry-etched notch for which the position is dependent on a photolithographic patterning process on the surface of the device. However, certain subgroups of design variables may exhibit correlated behavior, with either a positive or negative correlation. Correlated behavior can occur for a number of different reasons. In general, a given design variable will depend on a number of other underlying random variables. If a pair of design variables have one or more underlying variables in common, then a random change in this variable will cause a random, but correlated, change in the two dependent design variables. This is especially true when both design variables are strongly dependent on a single underlying variable, the variations in this variable are large, and the variations in the remaining underlying variables are small.

As a example, consider an etching process in which two distinct mechanical registration features are fabricated in the same etching step, such as two standoff pedestals, one determining the X position and the second determining the Z position for an optical component that is to be passively aligned. Here, any global variations in the lateral etch rate of the etching process will lead to variations in both standoff heights in the same way, leading to correlation between the two design variables. In fact, in this case, even if the standoffs are etched at different steps in the process sequence, variations in the lateral etch rate that persist between steps could also lead to correlation. Moreover, even local variations in lateral etch rate will give rise to correlation because adjacent pedestals will tend to etch at the same local rate, even if there are large variations at different locations on the wafer.

11.3.4.2 Introducing Correlation Effects into Monte Carlo Models

For the purposes of Monte Carlo modeling, it is important to take variable correlations into account, especially if the correlations are strong. Even in the simplest cases, correlation effects can lead to dramatic changes in the shape of the output variable distributions, and thus affect projected yields.

Correlation effects can be implemented in one of several different ways, depending on the nature of the correlation and the software used in the Monte Carlo model. There are three basic methods: (1) using correlation coefficients that "link" the random-number-generation functions provided in the software. This method is a standard feature in the the add-in programs Crystal Ball and @RISK and is most useful when the physical understanding of the correlation is unknown or very complicated but empirical correlation coefficients have been determined between design specification variables. Such may be the case as for parts obtained from a vendor. (2) Using a coordinate or variable transformation that converts a set of independent underlying variables into the correlated set. This is most useful for geometrically related design variables from which explicit equations can be derived relating the variables. (3) "Programming in" the desired correlation relationships directly into the model. This is most useful in cases in which the correlation relationships are known from either geometric or physical considerations.

11.3.4.3 Yield and Correlated Output Variables

In cases in which there are multiple output variables, each of which is yielded against an appropriate design specification, the method used to calculate overall yield is dependent upon whether the output variables are correlated or not. For output variables that are uncorrelated, the Cdf method may be used in the same manner as for a single-output variable to obtain an individual yield for each output variable. The overall yield is then given by the product of the individual yields. However, if the individual output variables are correlated, which they usually are, then a different method must be used.

This problem can be more adequately discussed in terms of probabilities. The probability that a given part will pass the yield criteria for a single-output design variable is given by $\mathbf{P}\{S\} = \mathrm{Cdf(usl)} - \mathrm{Cdf(lsl)}$, where S denotes success. In the case of two design variables, we have $P_1\{S\} = \mathrm{Cdf}_1(\mathrm{usl}_1) - \mathrm{Cdf}_1(\mathrm{lsl}_1)$ and $P_2\{S\} = \mathrm{Cdf}_2(\mathrm{usl}_2) - \mathrm{Cdf}_2(\mathrm{lsl}_2)$. If variables 1 and 2 are statistically independent, then the probability of success of both—that is, of passing both yield criteria—is given by the product of $P_1\{S\} \times P_2\{S\}$. However, if variables 1 and 2 are not statistically independent, then this simple multiplication of probabilities to obtain the overall probability of success (i.e., overall yield) is no longer valid [25]. Correlation between variables, as we have considered above, is one form of statistical dependence between variables.

As a final example to demonstrate the problem of yield calculation when correlated multiple output variables are required, consider the following two-output variable sample set (v_1, v_2), where S and F denote success and failure to meet the yield criteria, respectively:

$$(S, F)\ (S, F)\ (S, F)\ (F, S)\ (F, S)\ (S, F)\ (F, S)\ (S, F)\ (S, F)\ (S, F)$$

The probability of success for variable 1 is $7/10 = 0.70$. The probability of success for variable 2 is $3/10 = 0.3$. If variables 1 and 2 were statistically independent, then the calculated overall probability of success (i.e., the yield) would be given by the product $0.7 \times 0.3 = 0.21$, whereas the actual probability of simultaneous success is zero. The reason is that in this small sample the variables are perfectly correlated, such that success for variable 1 implies a failure for variable 2. In an analogous fashion, statistical dependence, or correlation, can lead to erroneous estimates of yield in Monte Carlo simulations of passive alignment.

The simple model above does, however, suggest one simple alternative method for computing the overall yield for multiple output variables, even when correlation effects are significant. As each set of output variables is generated, $(V_1, V_2, V_3, V_N) \ldots$ each may be tested against the yield criteria. Then, only if all variables meet the yield criteria, does the sample contribute to overall yield, and a success counter is incremented. After all samples have been tested, the overall yield is then given by the ratio of overall successes to the total number of samples.

11.3.5 ANALYTICAL SOLUTIONS

It is worthwhile pointing out that in simple cases involving independent variables and Gaussian or uniform distributions, it is sometimes possible to derive closed-form solutions for the yield by performing the multivariable integration of the input variable distribution functions analytically (22). However, this approach becomes unwieldy very rapidly, as the complexity of an assembly increases, and does not even yield closed-form solutions for many distributions of interest, such as truncated Gaussians among others.

It is instructive, however, to consider an analytical solution for the simple case of the coupling of a Gaussian beam into an optical fiber, where the lateral offsets in one dimension are Normally distributed. The coupling efficiency in this case is given by Joyce and DeLoach [39] as

$$\tau = \tau_a \exp[-(d/d_e)^2] \tag{11.17}$$

where d is the lateral offset, τ_a is the axial z-dependent coupling coefficient, and d_e is related to the lateral tolerance of the coupling. The gaussian pdf of offset values is given by

$$f(d) = \frac{1}{(2\pi)^{1/2}\sigma}\exp[-(d-\mu)^2/2\sigma^2] \qquad (11.18)$$

Then, using standard random variable transformation techniques [24], the pdf function for τ is found to be

$$f(\tau) = \frac{(d_e/\sigma)}{\tau(8\pi\ln\tau_a/\tau)^{1/2}}\left(\frac{\tau}{\tau_a}\right)^{d_e^2/2\sigma^2}\exp[-\mu^2/2\sigma^2]\exp\left[\frac{d_e\mu}{\sigma^2}(\ln\tau_a/\tau)^{1/2}\right] \qquad (11.19)$$

Making the variable substitutions $\beta = d_e/\sigma$ and $\gamma = \mu/\sigma$, $y = \tau/\tau_a$ and setting $\tau_a = 1$ gives the normalized form:

$$f(y) = \frac{\beta}{y[8\pi\ln 1/y]^{1/2}}\,y^{\beta^2/2}\exp[-\gamma^2/2]\exp[\beta\gamma^{1/2}(\ln 1/y)^{1/2}] \qquad (11.20)$$

This equation is the probability density function of coupling efficiency for a single-mode optical fiber and a symmetric (equal mode field radii in x and y) laser diode with Normally distributed lateral variations along a single direction in the gaussian beam approximation. This distribution is fundamental to the form of the coupling efficiency of Equation (11.17). The parameter β measures the relative tolerance of the lateral optical coupling system in units of the variation standard deviation σ. Likewise, γ measures the mean displacement in units of σ. The parameter y is the fraction of the maximum possible coupling achievable τ_a. Equation (11.20) is plotted in Figure 11.13 for several values of the parameters β and γ. What emerges from this plot is the fact that when the offset standard deviation (σ) is larger than approximately half the lateral tolerance parameter ($d_e/2$), the distribution of coupling values is spread rather uniformly over the entire range of values, with peaks near zero and the maximum value τ_a (Figure 11.13a). As σ becomes smaller relative to d_e, distribution of coupling efficiencies becomes more sharply peaked toward the maximum value, as demonstrated in Figure 11.13d. As the mean (μ) of the offsets changes from zero, as seen in Figure 11.13b and 11.13c, the distribution becomes skewed toward lower values of the coupling efficiency, as expected. One interesting feature, seen in Figure 11.13d, is when both β and γ are large (finite μ and vanishing σ). In this case, the pdf becomes peaked about the coupling efficiency value corresponding to μ, and the FWHM is closely related to 2σ.

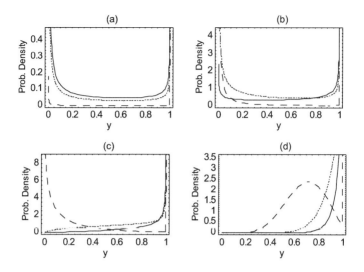

FIGURE 11.13 Plot of Equation (11.20) for different parameter values. (a) $\beta = 0.1$; (b) $\beta = 1.0$; (c) $\beta = 2.0$; (d) $\beta = 5.0$. Solid curve: $\gamma = 0$; small dashed curve: $\gamma = 1.0$; large dashed curve: $\gamma = 3.0$.

11.4 APPLICATIONS

In this section, results of Monte Carlo simulations are presented for some representative passive alignment cases of interest. These results will be presented in parametric form, using typical values of the parameters listed in Table 11.2.

11.4.1 SINGLE-MODE PASSIVE ALIGNMENT

11.4.1.1 Laser/Fiber

The alignment of a semiconductor laser to a single-mode fiber using the passive alignment scheme shown in Figure 11.14 is considered [18]. In this optical assembly, the laser is passively aligned to a precision-etched V-groove in a silicon submount. As shown in Figure 11.15, reactive ion etched side pedestals in the silicon mate with a precision-etched notch on the laser to effect the lateral alignment (along x) of the laser active region. The vertical alignment (along y) is achieved by precision-etched standoff pedestals in the silicon, which mate with the as-grown surface of the laser device. These alignments in x and y align the active region of the laser to the V-groove. The single-mode fiber, placed in the V-groove, is therefore passively aligned to the laser active region in x and y. The z alignment, along the z axis, is achieved by locating the fiber tip optically with a fiducial feature in the silicon near the V-groove edge.

The critical design variables for the alignment on the laser are the active to notch distance (l_1) and the active-to-surface distance (l_2) (see Figure 11.15). The variation in l_1 is determined by the random deviations of etch mask undercutting

TABLE 11.2
Parameter Values Used in Simulation of Laser/Fiber Coupling Results Shown in Figures 11.16–11.18

Component and Design Variable	Typical Probability Density Function	Typical Standard Deviation	References
Semiconductor laser			
Active-optical fiducial separation		0.5 μm	[12]
Etched notch-active separation	Gaussian	0.5 μm	[18]
Active depth		0.1–0.35 μm, depending on layer thickness	[12, 13, 18]
Threshold current		0.5–6 mA, depending on temperature and vendor	[8, 18]
Slope efficiency		0.01–0.03 watts/amp depending on temperature and vendor	[8, 18]
Silicon submount			
Solder bump height		0.5–2 μm	[9, 12]
V-groove to optical fiducial separation		0.5 μm	[13]
Vertical standoff height in Si	Gaussian	0.2 μm	[18]
Vertical standoff height in Au		0.7 μm	[10]
Pedestal to V-groove separation	Gaussian	0.5 μm	[18]
V-groove width	Gaussian	0.5–1 μm	[6, 8, 18]
Single mode optical fiber			
Fiber diameter	Gaussian	0.3 μm	[13, 18]
Core offset	One-sided Gaussian	0.3 μm	[12, 13, 18]
Lensed fiber radius			
Miscellaneous process related			
Laser facet-fiber tip separation	Gaussian	1.0–5.0 μm	[8, 18]
Fiber rotation angle	Uniform	$2\pi/\sqrt{12}$	[18]
Optical fiducial alignment	Gaussian	x 0.2 μm, θ 0.1°	[6]
Flip-chip optical fiducial alignment		0.5 μm	[12]
Self-aligned solder alignment		Lateral 0.3–1.5 μm, vertical 1.0 μm	[8–10, 13]
Cleave edge location		2.5–4 μm	[12]
Lapped wafer thickness		5 μm	

FIGURE 11.14 Passively aligned laser and tapered lensed fiber on silicon waferboard.

during the etching step of the notch and active mesa. In the silicon piece, the critical design variables are the V-groove-to-side pedestal distance (s_1), the standoff height (s_2), and the V-groove width (s_3), the latter controling the vertical position of the fiber axis through the geometry of the V-groove according to the expression

$$h_1 = \frac{f_1 - s_3 \cos(\theta/2)}{2\sin(\theta/2)} \tag{11.21}$$

where f_1 is the diameter of the fiber and θ is the full angle subtended by the V-groove, which is 70.53°. The remaining fiber-related design variables are the eccentricity (or offset) of the fiber core (relative to the geometric axis), f_2, and the rotational orientation of the fiber (f_3). The variables f_2 and f_3 correlate the variations along x and y, which in the absence of eccentricity would experience independent variability. A summary of the nominal values for these parameters and the distributions used for the model are shown in Table 11.3.

Shown in Figure 11.16 is the calculated pdf for the optical coupling coefficient for four different values of the fiber lens radius. The solid, dotted, small-dashed, and large-dashed curves correspond to $r = \infty$, $r = 5$, $r = 10$, and $r = 25$ µm, respectively. For the lenses having finite radius, the z position was set to correspond to the coupling peak value (see Figure 11.10a), and for the cleaved fiber case, $z = 10$ µm. For the cleaved fiber case, a narrow distribution is achieved, but with a relatively low mean value. As the radius is decreased, the mean and maximum value of the distribution increases, but the distribution width increases rapidly, resulting in coupling efficiencies varying over a wide range. For targeting a single product with low coupled power requirements, clearly the cleave fiber case would be the best choice. The $r = 5$ µm case might be a good choice if

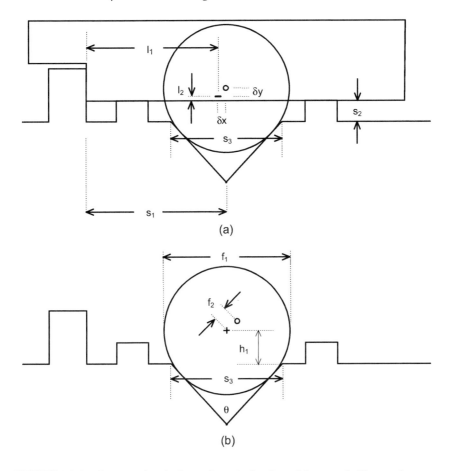

FIGURE 11.15 Cross-sectional view of passively aligned laser and silicon submount, showing design variables.

binning is an option, as the distribution is fairly uniform over the entire coupling range. Qualitatively, these results agree fairly well with the predictions of Equation 11.20 and Figure 11.13, with β increasing as r increases, but τ_a decreasing as r increases. The distributions are skewed to lower values for larger β (greater r) with a sharp cutoff, but they are very broad and uniform for small β. The sharp endpoints, especially near zero, seen in Figure 11.13, are washed out because of the two-dimensional and correlated nature of the displacements.

Shown in Figure 11.17 is the calculated yield for this subassembly as a function of the groove–side pedestal standard deviation. The curve dashing indicators are the same as those used in the previous figure for the same four lens radii. The yield is calculated for coupled power into the fiber at 85°C at the maximum drive current of 60 mA. The design specifications used for each figure are indicated. As predicted from Figure 11.16, only the $r = \infty$ case has significant yield at the lowest performance specification (see Figure 11.17c). In Figure 11.17b,

TABLE 11.3

Parameter Values Used in Simulation of Laser/Fiber Coupling Results Shown in Figure 11.19

Component and Design Variable	Nominal Value	Assumed Distribution	Distribution Parameters (Standard Deviation)
FP semiconductor laser			
Etched notch	500 μm	Gaussian	0.5 μm
Active depth	4 μm	Gaussian	0.1 μm
Threshold current at 25°C	13.3 mA,	Gaussian	2.3 mA
at 85°C	31.6 mA		2.7 mA
Slope efficiency	0.18 W/A	Gaussian	0.02 W/A
	0.10 W/A		0.01 W/A
Mode field radius	1.10 μm		
// and ⊥	0.95 μm		
Vacuum Wavelength	1.3 μm		
Maximum drive current	60 mA		
Silicon submount			
Vertical standoffs in Si	6 μm	Gaussian	0.2 μm
Groove-pedestal separation	500 μm	Gaussian	0.5 μm
V-groove width	138.7 μm	Gaussian	0.5 μm
V-groove angle	70.53°		
SM optical fiber			
Fiber diameter	125 μm	Gaussian	0.3 μm
Core offset	0 μm	1-sided Gaussian	0.26 μm
Mode field radius	4.65 μm		
Core index	1.4535		
Lensed fiber radius	5, 10, 25, ∞ μm		
Lens index	1.4535		
Process related			
Fiber rotation angle (0–2π)	π	Uniform	π
Fiber endface placement		Gaussian	1.0 μm

the narrow distribution of the $r = 25$ μm case enables it to dominate in yield for the intermediate performance specification. For the highest performance requirement, only the $r = 5$ μm case exceeds 50% yield. Two interesting features emerge from these plots. First, although for the highest performance devices, the yield tends to decrease with increasing standard deviation of the groove–pedestal separation s_1 (Figure 11.17a), for the lowest performance parts, the opposite is true (Figure 11.17c). For the finite radius cases, this is because as sd (s_1) increases, the tail of the distribution pushes farther into the specification range. The second point is that in spite of the prediction that the $r = 5$ case would cover all specification ranges roughly the same, suitable for binning, it turns out only to give significant

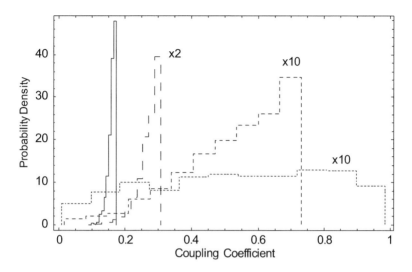

FIGURE 11.16 Calculated probability density functions using parameters in Table 11.3 for different lens radii and optimized z positions. Solid curve: $r = \infty$ and $z = 10\ \mu m$; dotted curve: $r = 5\ \mu m$ and $z = 10.6\ \mu m$; Solid curve: $r = 10\ \mu m$ and $z = 18.7\ \mu m$; Solid curve: $r = 25\ \mu m$ and $z = 26.1\ \mu m$.

yield for the 0- to 3-dBm specification. This is because of the logarithmic nature of the power specifications and the changes in shape to the distributions when laser power variations are added to the model.

Figure 11.18 shows the dependence of coupled power yield as a function of the distance between the laser facet and the fiber endface for the same conditions as the previous plots. For these plots, the standard deviation of s_1 was set to 0.5 μm. Here the effects of defocusing can be seen on yield, especially for the low performance specifications shown in Figure 11.18b to Figure 11.18d. For example, in Figure 11.18b, it can be seen that although the $r = 10\ \mu m$ case (medium dashed curve) has a very low yield to the {−3 dBm, 0 dBm} specification at the peak optical coupling position of $z = 18.7\ \mu m$, yields in excess of 80% can be achieved by moving the fiber either direction away from the peak power position. Similar effects are seen in the other cases. Even the flat-cleaved fiber case can be improved by moving the fiber from the previous position of 10 μm to one farther away, around 35 μm, as seen from the solid curve in Figure 11.18c. It should be pointed out that these kinds of Monte Carlo-generated plots give more information for achieving the optimum yield design for given set of specifications than other methods because all the random variation information is built in.

As a final examination of this example, the parameters were set as given in Table 11.3, and yields were calculated for each lensed radius case as a function of the specification limits in an effort to identify the highest yield specification range for each lensed radius design. In Figure 11.19, the upper specification limit was defined to be usl = lsl + 3 dBm, and the yields were plotted as a function of

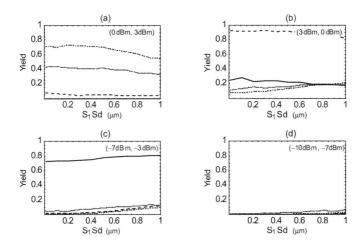

FIGURE 11.17 Calculated yield vs. V-groove/side-pedestal standard deviation using parameters in Table 11.3 for (a) {lsl = 0 dBm, usl = 3 dBm}; (b) {−3 dBm, 0 dBm}; (c) {−7 dBm, −3 dBm}; (d) {−10 dBm, −7 dBm}. Solid curve: $r = \infty$ and $z = 10$ μm; dotted curve: $r = 5$ μm and $z = 10.6$ μm; Solid curve: $r = 10$ μm and $z = 18.7$ μm; Solid curve: $r = 25$ μm and $z = 26.1$ μm. $N = 2000$.

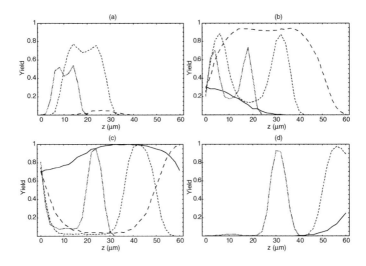

FIGURE 11.18 Calculated yield vs. z position of optical fiber tip using parameters in Table 11.3 for (a) {lsl = 0 dBm, usl = 3 dBm} (b) {−3 dBm, 0 dBm}; (c) {−7 dBm, −3 dBm}; (d) {−10 dBm, −7 dBm}. Solid curve: $r = \infty$; dotted curve: $r = 5$ μm; solid curve: $r = 10$ μm; solid curve: $r = 25$ μm.

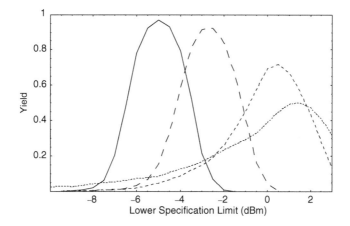

FIGURE 11.19 Calculated yield vs. lower specification limit (lsl) using parameters in Table 11.3. Upper specification limit (usl) is defined as usl = lsl + 3 dBm. Solid curve: $r = \infty$ and $z = 10$ μm; dotted curve: $r = 5$ μm and $z = 10.6$ μm; Solid curve: $r = 10$ μm and $z = 18.7$ μm; Solid curve: $r = 25$ μm and $z = 26.1$ μm. $N = 1000$.

the lower design specification limit. For each case, the yield-optimized product specification can be found. For example, for $r = 25$ μm (with $z = 26.1$ μm), the highest yielding product specification is predicted to be {5 dBm, −2 dBm}.

11.4.1.2 Fiber/Detector

In the next example we examine a passively aligned PIN photodetector coupled to a beveled single-mode fiber in a V-groove, as shown in cross section in Figure 11.20. The bevel on the endface of the fiber is to prevent reflections back into the fiber. The emitted beam from the fiber is reflected off the back (111) facet of the termi-nated V-groove, which is metallized. It is desirable to examine the yield of this design as a function of the z position of the fiber, the bevel angle, the diameter of the PIN active region, and the standard deviation of the passive alignment feature locating the position of the PIN chip. The yield is calculated to a responsivity specification. Gaussian beam analysis is used to propagate the refracted beam from the beveled fiber endface to the active region of the detector. At the detector the coupling efficiency of the beam to the active region geometry is given by

$$\eta = \frac{D^2}{\omega^2}\exp\left[-2\frac{\delta^2}{\omega^2}\right]\int_0^1 uI_0[2uD\delta/\omega^2]\exp[-u^2D^2/2\omega^2]du \qquad (11.22)$$

where D is the active diameter, δ is the displacement of the beam axis from the center of the active region, $I_0[2uD\delta/\omega^2]$ is a modified Bessel function of the first kind, and ω is the Gaussian beam-mode field radius at the active region.

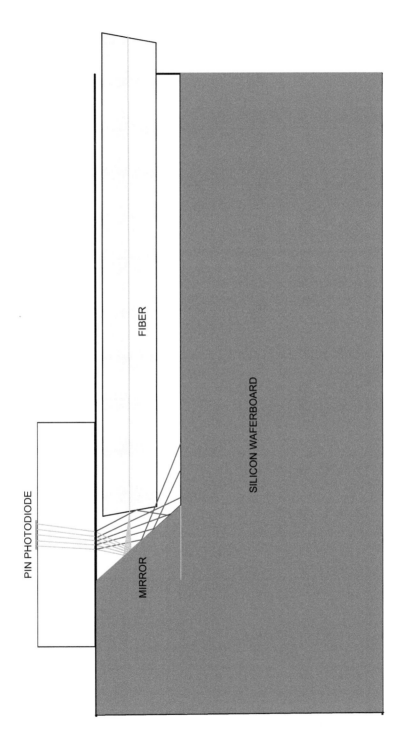

FIGURE 11.20 Passively aligned PIN photodetector with a beveled single-mode fiber using silicon waferboard.

TABLE 11.4
Parameter Values Used in Simulation Detector/Fiber Coupling Results Shown in Figures 11.21 and 11.22

Component and Design Variable	Nominal Value	Assumed Distribution	Distribution Parameters (Standard deviation)
InGaAs PIN photodetector			
Active center-cleaved edge separation	x 190.5, z 190.5	Gaussian	5 μm
Chip thickness	100 μm	Gaussian	5 μm
Active diameter	50 μm		
InP index of refraction	3.21		
Intrinsic responsivity	0.96 A/W		
Detection wavelength	1.3 μm		
Silicon submount			
Groove-x Pedestal Separation	190.5 μm	Gaussian	0.5 μm
Mirror-z pedestal separation	155.8 μm	Gaussian	0.5 μm
Mirror undercut	0 μm	Gaussian	0.25 μm
V-groove width	250.0 μm	Gaussian	0.5 μm
V-groove angle	70.53°		
Mirror angle	35.27°		
SM optical fiber			
Fiber diameter	125 μm	Gaussian	0.3 μm
Mode field radius	4.65 μm		
Core index	1.4535		
Bevel angle	8°		
Process related			
Fiber rotation angle (0–2π)	π	Uniform	π
Fiber endface placement	45 minimum	Gaussian	1.0

In Figure 11.21 we show the pdf of the PIN responsivity for four different values of the bevel angle and z position, with the remainder of the parameters having the values shown in Table 11.4. In Figure 11.21a, the bevel angle is set to zero, z is set at the minimum position of 45 μm, the active diameter is 50 μm, and the alignment sd is set to 5 μm, corresponding to a passively aligned cleaved edge. Figure 11.21b is the same as Figure 11.21a, except with $z = 100$ μm. Figure 11.21c is the same as Figure 11.21a, except with the bevel angel = 8°. Figure 11.21d is the same as Figure 11.21a, but with bevel angle = 8° and $z = 100$ μm. From these figures it can be seen that the effect of increasing the beam diameter, by either pulling back the fiber or deflecting the beam with the beveled endface, is to broaden the responsivity distribution pdf. Because of the assumed random orientation of the fiber in the assembly process, the effect of the bevel angle is especially pronounced, as the beam is diverted in a cone pattern around the active region center.

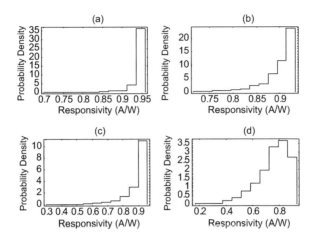

FIGURE 11.21 Calculated probability density functions using parameters in Table 11.4 for different bevel angles and z positions. (a) bevel angle = 0, z = 45 μm; (b) bevel angle = 0, z = 100 μm; (c) bevel angle = 8, z = 45 μm; (d) bevel angle = 8, z = 100 μm.

Figure 11.22 shows the dependence of yield on the design variables of interest. Increasing the bevel angle with z = 100 lowers the yield dramatically; however, about an 80% yield can be achieved at 8° with z near the minimum value. Increasing the active diameter to 75 μm allows nearly 100% yield to be realized with an 8° bevel even for z = 100 μm (Figure 11.22c). From Figure 11.22d, it is

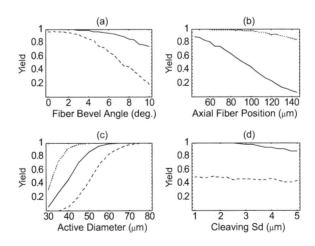

FIGURE 11.22 Calculated yield using parameters in Table 11.4. (a) vs. bevel angle. Solid: z = 50 μm; dashed: z = 100 μm. (b) vs. fiber position z. Solid: bevel angle = 8°; dashed: bevel angle = 0°. (c) vs. active diameter. Solid: bevel angle = 8° and z = 50 μm; dashed: bevel angle = 8° and z = 100 μm; dotted: bevel angle = 0° and z = 50 μm. (d) vs. cleaved edge sd. Solid: z = 50 μm; dashed: z = 100 μm. N = 1000.

clear that the passive alignment standard deviation has little effect on the yield, at least over the range considered.

11.4.2 MULTIMODE PASSIVE ALIGNMENT

Many low-cost applications require the use of multimode fiber and LEDs or vertical-cavity surface-emitting lasers (VCSELs) as the light source [49–51]. In contrast to single-mode systems where tolerances on the order of 1 µm are required, multimode systems typically work with optical spot sizes on the order of the diameter of the optical fiber, typically 62.5 µm. This considerably relaxes the mechanical tolerances required to keep the optical system in alignment, as compared with single-mode assemblies. However, cost and high-volume constraints on multimode products require the use of intrinsically higher-variability manufacturing processes, and so many of the yield-limiting assembly steps found in the single-mode case are mirrored in the multimode case. This is especially true in the coupling of the LED or vertical-cavity surface-emitting laser source to the fiber. However, unlike single-mode problems in which an expression for the coupling coefficient can be obtained in closed form, the multimode case requires the use of more complicated algorithms to compute the optical coupling efficiency for each geometrical configuration [17, 19, 44, 45, 52]. One common method used is a ray tracing approach [44], although other approaches are available.

Typically, in the ray tracing approach, an optical coupling efficiency is calculated by tracing a large number of rays emitted from the LED active region, usually assumed to be Lambertian, to the fiber endface. There, each ray is examined with respect to location and angle, to determine whether it falls within the fiber core, at or below the fiber acceptance angle. With appropriate normalization, counting the number of accepted rays into the core gives the coupling coefficient. By repeating this calculation for different values of a parameter (e.g., lens–fiber distance, x offset, etc.) a multidimensional optical coupling efficiency table can be generated. Once a tabular coupling efficiency table is generated, then the random nature of the assembly processes is modeled using worst-case analysis or Monte Carlo methods in the usual way.

The problem with this approach is that it is very inefficient. Essentially, the ray tracing calculation of each coupling coefficient is a Monte Carlo yield calculation. This can be seen as follows: Rays are sampled from a Lambertian distribution and traced through the system to generate a transformed distribution at the fiber endface. Upper and lower specification limits are then applied to the distributions of ray location and angle to obtain the "yield" of rays accepted into the fiber core. Moreover, the random variables describing the rays are rigorously statistically independent of the random variables describing the manufacturing variations of the optical assembly components. Therefore, a substantially more efficient approach is to incorporate the random variation of the optical assembly components simultaneously into the Monte Carlo computation of the coupling efficiency. What is then obtained is a table of coupling efficiencies averaged over the assembly process variations. The statistical independence between the ray and the assembly parameters guarantees the success of the procedure. Optimization is then achieved by identifying the highest value of coupling efficiency from the table. What is lost in this procedure is the ability to obtain the assembly yield to a specific set of coupling efficiency specification limits. To obtain that information, the distribution of coupling efficiencies must be computed, which can only be obtained by the previously described method.

11.5 SUMMARY AND CONCLUSIONS

It has been demonstrated in this chapter that Monte Carlo analysis is a powerful and effective method for yield prediction and analysis of optoelectronic assemblies and products. Although emphasis has been placed on the optical coupling aspect of single-mode fibers to sources and detectors, the methods presented here are equally applicable to other yield-limiting factors in optoelectronic assembly and testing. By the use of parametric modeling and suitable graphical display, optimization can be carried out in a straightforward manner using yield as the primary criteria as opposed to individual performance criteria. This provides for a more "global" optimization process, which automatically takes into account the requirements for manufacturability and business success of the final product.

11.5.1 LIMITATIONS OF MONTE CARLO ANALYSIS

The primary limitations of Monte Carlo analysis generally fall into three categories: the limitations on computer memory and processor speed, the requirements for complete and accurate input variable distribution data, and the requirements for analytic or closed-form relationships between the output and input variables. Although available computing power is increasing at a very rapid rate, parameterized Monte Carlo models of even moderately complex assemblies can consume considerable memory and computation time to achieve accurate results. Variations on the basic method have been introduced to improve "convergence," which require a more sophisticated and individualized approach to each problem. Often a compromise between computation time and final accuracy is required.

The requirement for detailed input information is somewhat more severe. The Monte Carlo method, similar to many other computational methods requiring adequate input data, is subject to the familiar "garbage in–garbage out" phenomena. In practice, the major difficulty is that good statistical data over a wide number of parameters is desired, which can only be obtained from preproduction or production processes. Often many of these processes are not well characterized. Monte Carlo analysis is most valuable as a yield prediction tool early in the product development cycle, to optimize design parameters and identify inadequate process capability. However, in these early stages, extensive statistical preproduction data are usually not available. Hence, many input variable distributions must be estimated from limited sample size runs. Judgement and care must be exercised to not over- or underestimate the process capabilities.

Finally, a fairly thorough knowledge of the cause-and-effect relationships between input and output variables is required. For example, in the cases presented here, accurate and complete coupling models are available in closed form. However, in some cases, such as etched fiber endfaces [53, 54], which form diffracting mode transforming lenses, closed-form coupling models have not yet been reported. Also, for multimode problems or cases with unique geometries, coupling formulae require the use of complicated summations or integral solutions that are more complicated to implement and time consuming to evaluate. For these cases, approximate methods can be used, but they must be verified on a case-by-case basis.

11.5.2 EXTENSION TO OTHER PROBLEMS FOR FUTURE RESEARCH

11.5.2.1 Cost Modeling

As outlined in Section 11.1.2.3, Monte Carlo analysis can be incorporated into product cost models, where cost becomes the final optimization parameter. Although yield and cost are closely related, labor and material factors can significantly alter the relative importance of different yield points. Although ambitious, linking Monte Carlo Analysis at limiting yield points via a comprehensive cost model would allow final product cost optimization.

11.5.2.2 Production Flow

Because yield ultimately affects throughput, it is also possible that Monte Carlo analysis could play a role in optimization of production flow balancing and capacity planning.

11.5.2.3 Linking to Optimization Programs

The approach in this chapter has been to generate graphical output and visually identify optimum values of the varied parameters. However, once a fully parametric Monte Carlo model has been developed, it is relatively straightforward to configure the model as a subprogram to an optimization algorithm program and to allow optimization to be performed simultaneously to achieve global optimization.

11.5.2.4 DFB Lasers for ITU Grid Applications

As dense wavelength division multiplexing (DWDM) applications continue to be deployed, the need for laser modules with wavelengths aligned to the International Telecommunication Union (ITU) grid is expanding rapidly. It is of considerable interest for laser manufacturers to be able to optimize the production yield of distributed feedback (DFB) lasers to these wavelength requirements. Monte Carlo analysis can clearly play an important role for this problem.

11.5.2.5 DFB Laser Characteristics

In the production of DFB lasers there are several yield-limiting process steps that can be analyzed using Monte Carlo analysis to optimize yield for application specifications. Examples include the coupling coefficient (κ), facet reflectivity, Bragg grating period (Λ), and so forth.

11.5.2.6 External Cavity Lasers

At present, there is renewed interest in external cavity lasers, in which a Bragg grating plays the role of the wavelength selective element. In devices of this sort it is expected that mechanical/packaging variations will play an important role in the yield to application specifications.

11.5.2.7 Monolithic Array Devices

In devices of this sort, there are expected to be strong correlation effects between devices on the array, providing an interesting Monte Carlo analysis problem. Some parameters of interest to model are optical coupling, crosstalk, wavelength, and optical power/drive current characteristics. Both vertical-cavity surface-emitting lasers and in-plane laser devices fall into this category.

REFERENCES

1. Kalos, M. H. and Whitlock, P. A., "Monte Carlo Methods," Wiley, New York, NY, 1986.
2. Sobol, I. M., "A Primer for the Monte Carlo Method," CRC Press, Boca Raton, FL, 1994.
3. Rubinstein, R. Y., "Simulation and the Monte Carlo Method," McGraw-Hill, New York, NY, 1982.
4. Metropolis, N. and Ulam, S., "The Monte Carlo Method," J. Am. Stat. Assoc., 44, N247, 335–341, 1949.
5. Armiento, C., et al., "Passive Coupling of an InGaAsP/InP Laser Array and Single Mode Fibers Using Silicon Waferboard," Electron. Lett., 27, 1109–1111, 1991.
6. Goto, A., et al., "Hybrid WDM Transmitter/Receiver Module Using Alignment Free Assembly Techniques," Electronic Components and Technology Conference, San Jose, CA, pp. 620–625, 1997.
7. Tanaka, N., Arai, N., Takahara, H., and Ando, Y., "3.5 Gbit/sec × 4 ch Optical Interconnection Module for ATM Switching System," Electronic Components and Technology Conference, San Jose, CA, pp. 210–216, 1997.
8. Joo, G., et al., "Fabrication of Optical Tx/Rx Subscriber Modules Incorporation Passive Alignment Technique," Electronic Components and Technology Conference, Orlando, FL, pp. 37–41, 1996.
9. Itoh, M., et al., "Use of AuSn Solder Bumps in Three-dimensional Passive Aligned Packaging of LD/PD Arrays on Si Optical Benches," Electronic Components and Technology Conference, Orlando, FL, pp. 1–7, 1996.

10. Tan, Q. and Lee, Y. C., "Soldering Technology for Optoelectronic Packaging," Electronic Components and Technology Conference, Orlando, FL, pp. 26–36, 1996.

11. Collins, J. V., et al., "The Packaging of Large Spot-Size Optoelectronic Devices," Electronic Components and Technology Conference, Orlando, FL, pp. 640–644, 1996.

12. Ambrosy, A., Richter, H., Hehmann, J., and Ferling, D., "Silicon Motherboards for Multichannel Optical Modules," Electronic Components and Technology Conference, Las Vegas, NV, pp. 570–576, 1995.

13. Kurata, K., et al., "A Surface Mount Type Single-Mode Laser Module Using Passive Alignment," Electronic Components and Technology Conference, Las Vegas, NV, pp. 759–765, 1995.

14. Lee, S., et al., "Optical Device Module Packages for Subscriber Incorporating Passive Alignment Techniques," Electronic Components and Technology Conference, Las Vegas, NV, pp. 841–844, 1995.

15. Dautartas, M. F., et al., "Optical Performance of Low-Cost Self-Aligned MCM-D Based Optical Data Links," Electronic Components and Technology Conference, Las Vegas, NV, pp. 1254–1262, 1995.

16. Wale, M. J., "Self Aligned, Flip Chip Assembly of Photonic Devices with Electrical and Optical Connections," Electronic Components and Technology Conference, Las Vegas, NV, pp. 34–40, 1990.

17. Wang, S. C., et al., "Coupling Efficiency of an Alignment-Tolerant, Single Fiber, Bi-Directional Link," Electronic Components and Technology Conference, San Jose, CA, pp. 30–36, 1997.

18. Wilson, R. B. and Boudreau, R. A., "Single-Mode Laser/Fiber Coupling Yields Using Silicon V-Groove Passive Alignment," AMP J. Technol., 4, 41–49, 1995.

19. Sutherland, J., George, G., and Krusius, J. P., "Optical Coupling and Alignment Tolerances in Optoelectronic Array Interface Assemblies," Electronic Components and Technology Conference, Las Vegas, NV, pp. 577–583, 1995.

20. Zaleta, D, et al., "Alignment Tolerancing of Free-Space MCM-to-MCM Optical Interconnects," Electronic Components and Technology Conference, Las Vegas, NV, pp. 1286–1293, 1995.

21. McGroarty, J., et al., "Statistics of Solder Joint Alignment for Optoelectronic Components," IEEE Trans. Comp., Hybrids, Manuf. Technol., 16, 527–529, 1993.

22. Ando, Y., "Statistical Analysis of Insertion-Loss Improvement for Optical Connectors Using the Orientation Method for Fiber-Core Offset," IEEE Photonics Tech. Lett., 3, 939–941, 1991.

23. AT&T, "Statistical Quality Control Handbook," Delmar Printing Company, NC, 1985.

24. Freund, J. E., "Mathematical Statistics," 2nd ed., Prentice-Hall, Englewood Cliffs, NJ, 1971.

25. Feller, W. "An Introduction to Probability Theory and Its Applications Vol I," Wiley, New York, NY, 1950.

26. Chambers, J. M., Cleveland, W. S., Kleiner, B., and Tukey, P. A., "Graphical Methods for Data Analysis," Duxbury Press, Boston, MA, 1983.

27. King, J. R., "Probability Charts for Decision Making," Industrial Press, New York, 1971.

28. Smith, W. M., "Worst Case Circuit Analysis—An Overview," Proceeding of the Annual Reliability and Maintainability Symposium, Las Vegas, NV, pp. 326–334, 1996.

29. Crystal Ball is available from Decisioneering Inc., 1515 Arapahoe Street, Suite 1311, Denver Colorado 80202, 800-289-2550. Contact information is available from their Web site at www.decisioneering.com.

30. @Risk is available from Palisade Corporation, 31 Decker Road, Newfield, NY 14867, 800-432-7475. Contact information is available from their Web site at www.palisade.com.

31. Mathematica is available from Wolfram Research Inc., 100 Trade Center Drive, Champaign, Illinois 61820-7237, 217-398-0700. Contact information is available from their Web site at www.wolfram.com.

32. Aitchison, J. and Brown, J. A. C., "The Lognormal Distribution," Cambridge University Press, New York, NY, 1973.

33. Castillo, E., "Extreme Value Theory in Engineering," Academic Press, London, 1988.

34. Kogelnik, H., "Coupling and Conversion Coefficients for Optical Modes," Microwave Research Institute Symposia Series 14, Polytechnic Press, New York, 1964, pp. 333–347.

35. Kogelnik, H. "On the Propagation of Gaussian Beams of Light through Lenslike Media Including Those with Loss or Gain Variation," Appl. Optics, 4, 1562–1569, 1965.

36. Kogelnik, H. and Li, T., "Laser Beams and Resonators," Appl. Optics, 5, 1550–1567, 1966.

37. Davis, C. R., "Lasers and Electro-Optics—Fundamentals and Engineering," Cambridge University Press, Cambridge, 1996.

38. Marcuse, D., "Light Transmission Optics," 2nd ed., Van Nostrand Reinhold, New York, 1982.

39. Joyce, W. B. and DeLoach, F. C., "Alignment of Gaussian Beams," Appl. Optics, 23, 4187–4196, 1984.

40. Sakai, J. and Kimura, T., "Design of a Miniature Lens for Semiconductor Laser to Single-Mode Fiber Coupling," IEEE J. Quantum Elec., QE-16, 1059–1066, 1980.

41. Saruwatari, M. and Nawata, K., "Semiconductor Laser to Single-Mode Fiber Coupler," Appl. Optics, 18, 1847–1856, 1979.

42. Sumida, M. and Takemoto, K., "Lens Coupling of Laser Diodes to Single-Mode Fibers," J. Lightwave Technol., LT-2, 305–311, 1984.

43. Hillerich, B., "Efficiency and Alignment Tolerances of LED to Single-Mode Fibre Coupling—Theory and Experiment," Opt. Quantum Electron., 19, 209–222, 1987.

44. Deimel, P. P., "Calculations for Integral Lenses on Surface-Emitting Diodes," Appl. Optics, 24, 343–348, 1985.

45. Wagner, R. E. and Tomlinson, W. J., "Coupling Efficiency of Optics in Single-mode Fiber Components," Appl. Optics, 21, 2671–2688, (1982).

46. Lealman, I. F., et al., "1.56μm InGaAsP/InP Tapered Active Layer Multiquantum Well Laser with Improved Coupling to Cleaved Singlemode Fibre," Elect. Lett., 30, 857–858, 1994.

47. Lealman, I. F., et al., "Low Threshold Current 1.6μm InGaAsP/InP Tapered Active Layer Multiquantum Well Laser with Improved Coupling to Cleaved Singlemode Fibre," Elect. Lett., 30, 973–974, 1994.

48. Lealman, I. F., et al., "InGaAsP/InP Tapered Active Layer Multiquantum Well Laser with 1.8 dB Coupling Loss to Cleaved Singlemode Fibre," Elect. Lett., 30, 1685–1686, 1994.

Index